Terahertz Technology

Terahertz (THz) technology is an active area of research, but only in recent years has the application of THz waves (T waves) in food and agricultural industries been explored. *Terahertz Technology: Principles and Applications in the Agri-Food Industry* describes the operating principles of THz technology and discusses applications and advantages of the THz regime of the electromagnetic spectrum for use in the agri-food industry. The agri-food industry is focusing on the development of non-destructive quality evaluation techniques that can provide accurate analysis quickly and are environmentally friendly. Among such techniques is THz technology that provides a novel noninvasive approach to quality assessment and safety assurance of agri-food products. The low energy of T waves is best suited for the analysis of sensitive biomaterials and does not cause photoionization. Therefore, THz imaging is complementary to X-ray imaging. Although accessing the THz spectrum is tedious by conventional devices, the combination of optics and electronics principles has opened a dimension of research in this field.

This book provides an overview of THz spectroscopy and imaging, system components, types of THz systems, and applications and advantages of THz for applications in the agri-food industry. It describes the basic working mechanism, operating principle, operation modes, and system components of THz spectroscopy and imaging. Various advancements in THz technology related to agricultural and food applications are discussed that could serve as a guidebook for all those working and interested in non-destructive food assessment techniques.

Key Features:

- Explores broader applications of the THz regime in the agri-food sector
- Describes system components, different forms of THz systems, and the working principle of T waves for spectroscopic and imaging techniques
- Provides insights on future research needs for industrial implementation of THz technology
- Complements the knowledge of other existing non-destructive spectroscopy and imaging techniques for food analysis

Although books on biomedical applications of THz have been published, no book is available that deals with applications in the agri-food industry. Hence, *Terahertz Technology* is beneficial for undergraduate and graduate students and those food industry professionals involved in research related to non-destructive quality assessment and imaging techniques.

Terahertz Technology
Principles and Applications in the Agri-Food Industry

T. Anukiruthika and Digvir S. Jayas

CRC Press
Taylor & Francis Group
Boca Raton London New York

CRC Press is an imprint of the
Taylor & Francis Group, an **informa** business

First edition published 2024
by CRC Press
6000 Broken Sound Parkway NW, Suite 300, Boca Raton, FL 33487-2742

and by CRC Press
4 Park Square, Milton Park, Abingdon, Oxon, OX14 4RN

CRC Press is an imprint of Taylor & Francis Group, LLC

© 2024 T. Anukiruthika and Digvir S. Jayas

ISBN: 978-1-032-04051-6 (hbk)
ISBN: 978-1-032-05317-2 (pbk)
ISBN: 978-1-003-19701-0 (ebk)

DOI: 10.1201/9781003197010

Typeset in Adobe Garamond
by KnowledgeWorks Global Ltd.

Contents

Preface

Food processing plays an inevitable role in today's modern world through the automation of unit operations. Consumers are becoming more conscious of the quality and safety of processed foods. The quality of the final product is considered with utmost priority by major players in the supply chain and distribution including producers, processors, and consumers. The existing conventional quality assessment procedures are destructive, laborious, and time consuming. To overcome these limitations, advanced analytical techniques were introduced. Some of the recently developed analytical techniques include enzyme-linked immunosorbent assay and chromatographic techniques like thin-layer chromatography, column chromatography, high-performance liquid chromatography, paper chromatography, gel permeation chromatography, ion exchange chromatography, gas chromatography-mass spectrometry, and affinity chromatography. Although these analytical methods are rapid in the quality assessment, these are sample destructive and require long sample preparation time, sophisticated lab facilities, solvent preparations, and laborious analytical procedures. These attributes limit their applications and cannot be applied for real-time industrial measurements. Considering the industrial need, spectroscopy and imaging technology have emerged as promising techniques for onsite quality assessment of raw and processed foods.

The past two decades have witnessed the emergence of various spectroscopy and imaging techniques in many fields including the agri-food sector. The scientific exploration of the electromagnetic spectrum has resulted in many advancements in diverse industrial areas such as communication, sensing, and security monitoring. Some of the common spectroscopy and imaging techniques for agri-food applications are Fourier transform infrared spectroscopy, thermal imaging, near-infrared spectroscopy, fluorescence spectroscopy, nuclear magnetic resonance spectroscopy, Raman spectroscopy, hyperspectral imaging, and X-ray imaging. In line with these techniques, THz spectroscopy and imaging are newly emerged novel technologies. The distinct characteristic features of THz having shared properties of microwave and infrared regions have led to the emergence of THz spectroscopy and imaging technologies. The low photonic energy and nonionizing nature of T waves have made this a very appropriate analytical technique for biomaterial characterization. The THz technology is one of the simple and feasible methods providing valuable information at the molecular level. Despite the evident progress of THz science in other industrial sectors, its application in the agri-food industry is in the nascent stage. Most of the research on THz spectroscopy and imaging applications in the agri-food industry are at the lab scale. The present book is the first of its kind to explore the vast potential capabilities of THz spectroscopy and imaging applications in the agri-food industry. This book provides comprehensive knowledge on the novel but less explored T waves for agri-food applications with up-to-date information on research progress.

This book has been organized to provide a detailed discussion of principles and applications of THz technology in the agri-food industry. The first chapter broadly discusses the applicability of the electromagnetic spectrum for spectroscopy and imaging with an introduction to T waves, presenting a brief discussion on an overview of THz sources, detectors, and other system components along with recent research advancements. The second chapter describes the use of THz spectra emphasizing various types of spectroscopic systems. The third chapter is focused on the principles of THz imaging and different imaging methods. The fourth chapter highlights the need for understanding the acquisition of spectral data and signal processing methods. The remaining chapters from 5 to 8 are focused on THz agri-food applications dealing with molecular characterization as an inspection and identification tool, agricultural applications, and quality monitoring and control. The last chapter (Chapter 9) provides a summarized discussion of the emerging trends, practical implementation, prospects, opportunities, and challenges of THz science in the agri-food sector. We hope the readers will find this book to be a valuable resource for future research and development of applications of T waves for noninvasive contact-free detection and quality assessment of agri-food materials.

T. Anukiruthika

Digvir S. Jayas

Acknowledgments

The authors sincerely thank the University of Manitoba for providing a great research infrastructure and facility for studying approaches to minimizing food spoilage and monitoring its quality along the supply chain from production to consumption. This infrastructure has been made possible through funding from the federal and provincial governments as well as producer groups and the food storing, handling, and processing industries. We gratefully acknowledge this funding support. We express our sincere thanks to our colleagues and friends for their valuable support during the writing of this book. We would like to express our heartfelt gratitude to our family members for their continued moral support throughout this project. Dr. Jayas, who conceptualized the content of this book and worked collaboratively with T. Anukiruthika in its completion, extends his sincere thanks to his wife, Mrs. Manju Jayas, children (Dr. Rajat Jayas, Dr. Ravi Jayas, and Dr. Rahul Jayas), and grandchildren (Priya Jayas, Isabella Jayas, Rohan Jayas, Gabriel Jayas, and Leon Jayas) for their understanding of work commitment and for allowing him time away from them in completing this project. The first author, Ms. Anukiruthika, thanks the almighty God for giving the strength to complete the project. She would like to express her gratitude to her mother, S. Revathi, and father, U. Thangarasu, for their love, affection, care, and moral support. She also would like to extend her sincere thanks to Dr. Jayas for his valuable guidance, supervision, and moral support during the progress and completion of this book. She also would like to express her sincere gratitude to Dr. Jeyan A. Moses, Assistant Professor and In-Charge, Computational Modeling and Nanoscale Processing Unit, National Institute of Food Technology, Entrepreneurship and Management – Thanjavur, Ministry of Food Processing Industries, Tamil Nadu, India, for his valuable advice and moral support.

Our sincere thanks to Laura Piedrahita (project coordinator), Stephen Zollo (acquiring editor), and Paul Boyd (production editor), at CRC Press/Taylor & Francis Group, as well as Kavitha Sathish, project manager at KnowledgeWorks Global Ltd., for agreeing to publish our work and guiding us through the process of publication.

About the Authors

T. Anukiruthika is currently pursuing her Ph.D. from the Department of Biosystems Engineering, University of Manitoba, Canada. Her doctoral dissertation is on the determination and prediction of stored grain insect movement in one-dimensional grain columns under different storage conditions. She completed her undergraduate and postgraduate programs in Food Process Engineering from the National Institute of Food Technology, Entrepreneurship, and Management (NIFTEM) – Thanjavur, India. During her graduation studies, she worked in diverse emerging areas of Food Science and Technology, Food Process Engineering, and Biosystems Engineering. This proficiency helped her to apply her knowledge in a broad disciplinary area that is evident from her publications in several high-impact journals dealing with engineering, technology, and food sciences. She has coauthored two books, four invited book chapters, 12 conference proceedings, and 16 peer-reviewed publications with two manuscripts under review. She has made conference presentations and has received many awards from several organizations including the Jawaharlal Nehru Memorial Fund, the Shree Vijayalakshmi Charitable Trust, and the Indian National Academy of Engineering (INAE) for M.Tech. research work on "Design and Fabrication of Customized Foods using 3D Printing". She was also honored with the Certificate of Appreciation under Individual Awards Category – Eat Healthy theme for M.Tech. research work by the Eat Right Research Awards and Grant, Eat Right India, Food Safety and Standards Authority of India (FSSAI), Government of India. She was one of the recipients of the "Price Graduate Scholarships for Women in Engineering" for the academic year 2021–2022. She was also awarded with the "University of Manitoba Graduate Fellowship" during her Ph.D. She has worked as a trainee on the processing of shrimp, fruits, and vegetables, the aseptic processing of fruits and vegetables, harvesting and rearing of shrimp, and processing of dairy products. She has also worked on the by-products utilization of waste streams of the food processing line with the aim of reducing the processing loss and to enhance implementation of sustainable approaches in food industry.

Dr. Digvir S. Jayas is currently serving as the President and Vice-Chancellor of the University of Lethbridge. He is also a Distinguished Professor Emeritus of Biosystems Engineering at the University of Manitoba. He was educated at the G.B. Pant University of Agriculture and Technology in Pantnagar, India; the University of Manitoba, Canada; and the University of Saskatchewan, Canada. On September 30, 2022, he completed an 11.75-year term as Vice-President (Research and International) at the University of Manitoba. Before assuming the position of Vice-President (Research and International), he was Vice-President (Research) for two years, and Associate Vice-President (Research) for eight years. Prior to this, he was Associate Dean (Research) in the Faculty of Agricultural and Food Sciences, Head of the Department of Biosystems Engineering, and Interim Director of the Richardson Centre for Functional Foods and Nutraceuticals. For a year, he served as Interim President of the Natural Sciences and Engineering Research Council of Canada (NSERC), and for 4.5 months, he served as Interim Director (CEO) of TRIUMF-Canada's particle accelerator center. He is a Registered Professional Engineer and a Registered Professional Agrologist.

Dr. Jayas is a former Tier I (Senior) Canada Research Chair in Stored-Grain Ecosystems. He conducts research related to the drying, handling, and storing of grains and oilseeds and digital image processing for grading and processing operations in the agri-food industry. He has collaborated with researchers in several countries and has had significant impact on the development of efficient grain storage, handling, and drying systems in Canada, China, India, Ukraine, and the United States. He has authored and co-authored over 1000 technical articles in scientific journals, conference proceedings, and books dealing with issues of the storing, drying, handling, and quality monitoring of grains and foods.

Dr. Jayas has received awards in recognition of his research and professional contributions from the Agriculture Institute of Canada, Applied Zoologists Research Association (India), American Society of Agricultural and Biological Engineers (ASABE), Canadian Institute of Food Science and Technology, Canadian Academy of Engineering, Canadian Society for Bioengineering, Engineers Canada (formerly Canadian Council of Professional Engineers), Engineers Geoscientists Manitoba (formerly Association of Professional Engineers and Geoscientists of Manitoba), Engineering Institute of Canada, Indian Society of Agricultural Engineers, Manitoba Institute of Agrologists, National Academy of Agricultural Sciences (India), National Academy of Sciences (India), and Sigma Xi. He was the recipient of the 2017 Sukup Global Food Security Award from ASABE and the 2008 Brockhouse Canada Prize from NSERC. In 2009, he was inducted as a Fellow of the Royal Society of Canada, and in 2018, he was appointed as an Officer of the Order of Canada for "his advancements to agricultural practices worldwide and for his promotion of academic and scientific research in Canada". In 2019, he received the RSC Sir John William Dawson Medal for "important contributions of knowledge in multiple domains". In 2022, Dr. Jayas was inducted into the Manitoba Agricultural Hall of Fame and the Canadian Agricultural Hall of Fame in recognition of his engineering contributions to grain preservation.

Dr. Jayas has served/serves on the boards or committees of many organizations such as ArcticNet, Churchill Marine Observatory (CMO), Centre for Innovative Sensing of Structures (SIMTReC), Genome Prairie, GlycoNet, Manitoba Centre for Health Policy, North Forge Technology Exchange, NRC Council, NSERC Council, Research Manitoba, and TRIUMF. He has served as the President of the Agriculture Institute of Canada, the Canadian Institute of Food Science and Technology, the Canadian Society for Bioengineering, Engineers Canada, Engineers Geoscientists Manitoba, and the Manitoba Institute of Agrologists. He chaired the NSERC Council, the board of directors of TRIUMF, the board of RESOLVE, a prairie research network on family violence, and the Smartpark (the University of Manitoba's Research and Technology Park) Advisory Committee.

Chapter 1

THz Technology: A Non-Destructive Approach

1.1 Introduction

Consumers are becoming more conscious about the safety and quality of the foods that are meant for consumption. Food safety and quality assurance are the two major critical processes that ensure the safe delivery of final products to the consumers from farm to fork (Pandiselvam et al. 2020). The changing consumer attitude is making the food industries enforce strict regulations and stringent practices for monitoring the quality of processed foods (Hoffmann et al. 2019). For instance, detection of pesticide residues, antibiotics, and adulterants; monitoring of freshness of the produce, identification of insect infestation and microbial spoilage; tracing of cross-contamination with metal or other hard pieces are the commonly encountered issues that are strictly monitored and addressed during the processing of foods (Jha 2015). Thus, high quality is one of the most important criteria that acts as a driving force promoting the growth of the agri-food industry. Generally, the quality of agri-food products can be evaluated based on extrinsic and intrinsic factors (Arefi et al. 2015). Different quality assessment techniques followed in the agri-food industries are compression test, penetration test, sensory analysis, shear and tensile testing, enzyme-linked immunosorbent assay, and chromatographic techniques like thin layer chromatography, column chromatography, high-performance liquid chromatography, paper chromatography, gel permeation chromatography, ion exchange chromatography, gas chromatography-mass spectrometry, and affinity chromatography. Most of these conventional analytical methods available for quality evaluation in the agri-food industry are based on the destructive approach that requires sophisticated lab facilities, solvent preparations, laborious analytical procedures, and time-consuming process steps (Sanchez et al. 2020). Hence, the

DOI: 10.1201/9781003197010-1

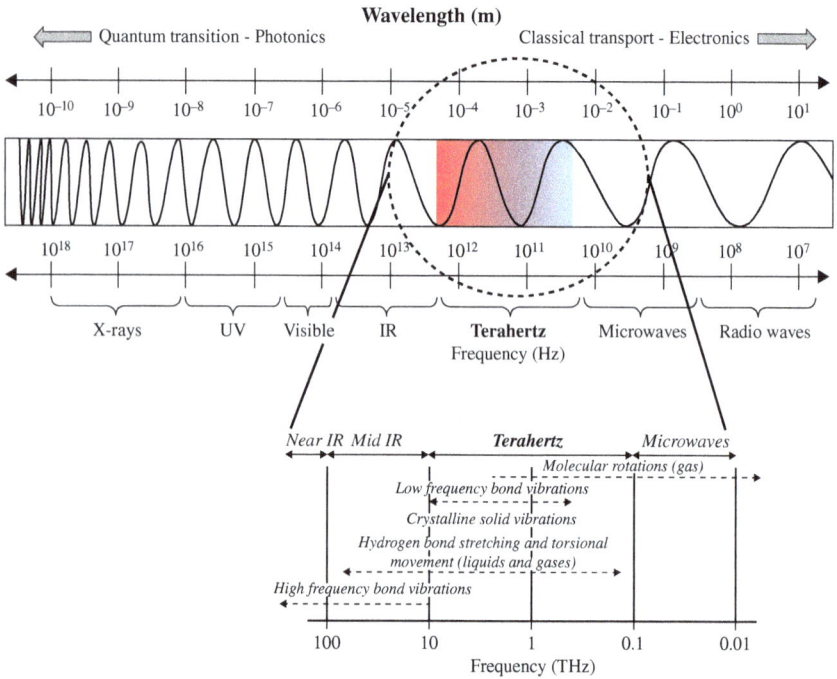

Figure 1.1 THz regime and molecular transitions in the electromagnetic spectrum.

qualitative and quantitative analytical techniques based on the non-destructive approach are the need of the hour for the real-time industrial measurements.

The major advantages of the non-destructive testing methods are their simplicity, accuracy, speed, performance efficiency, and most importantly, the feasibility of integration with the process line for online monitoring (Yin et al. 2020; Zhou et al. 2020). Recently, spectroscopic and imaging techniques have gained interest as an emerging approach for detection and quantification of components in the raw and processed food products (Vadivambal and Jayas 2016). One such non-destructive approach using far-infrared rays is terahertz (THz) spectroscopy and imaging (Figure 1.1). Due to the fact that THz waves share the properties of microwave and infrared waves, they are gaining recent research potential as a non-destructive quality evaluation tool (Li et al. 2020a). In this regard, there exists a vast scope for applicability of THz in process control as well as in communication making them more appropriate for industrial applications. The acquisition of the THz spectrum of the biomaterials aids in the characterization of these materials (Ghann and Uddin 2017).

Although THz waves have vast research potential, still there exists a challenge in accessing this particular portion of the electromagnetic spectrum. Due to which the THz region of the electromagnetic spectrum is often referred to as the "terahertz gap" (Figure 1.2) as this region is hardly accessed by electronic and optical methods (Mantsch and Naumann 2010). Fortunately, the emergence of novel

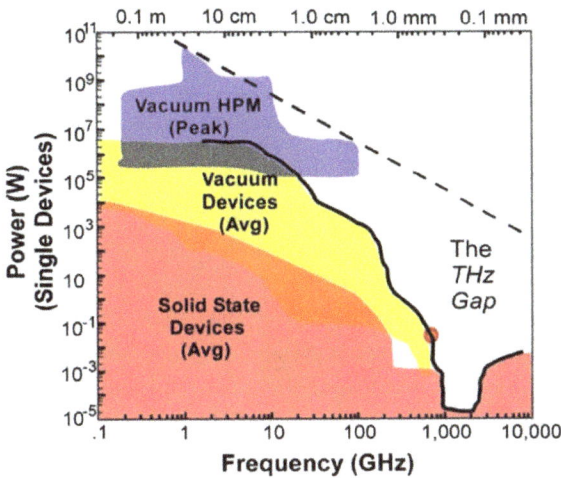

Figure 1.2 Plot of power versus frequency of solid state and vacuum electronic devices (HPM – high power microwave. Result shows plot with average power device. The circle data point denotes a regenerative traveling wave tube oscillator). (Source: Booske 2008.)

THz sources and detectors eliminates the difficulties associated with accessing this region. During the past decades, research on THz science had increased significantly because of the spectroscopy and imaging applications in various fields such as biotechnology, solid-state physics, drugs and pharmaceuticals, telecommunication systems, and security screening. However, the emergence of THz waves as a non-destructive quality evaluation tool in the agri-food industry is still in its infancy. This book provides a comprehensive review of the applications of the novel and less explored THz waves and the state-of-the-art of imaging technique for food safety and quality control. In addition, the scientific insights on principle, working components, and types of THz spectroscopy and imaging systems are summarized. Further, this book attempts to streamline the vast list of applications with up-to-date information on the efficient utilization of THz waves in the agriculture and food industry as a non-destructive spectroscopic and imaging technique. Thus, this book attempts to fill the research gap existing in spectroscopy and imaging using THz waves emphasizing the prospective merits of adapting THz waves for quality assessment in the agri-food industries.

1.2 Applicability of Electromagnetic Spectrum

The term electromagnetic spectrum covers the entire range of existence of light starting from radio waves to gamma rays. Most of spectra are invisible to the naked eye; however, they possess different unique electrical and magnetic properties.

Various spectra with different frequencies and wavelengths are distributed over the entire range of the electromagnetic spectra making them exert unique fingerprint characteristics. The electromagnetic radiation of 1 THz is characterized by 1 picosecond (ps) and 300 μm wavelengths with photonic energy of 4.1 millielectron volt (meV), wave number of 33 cm^{-1}, and temperature of 47.6 K (Zhang and Xu 2010a). With these attributes, the THz waves have the inherent capability to transmit signals for generating images and to translate information. THz waves are also referred to as T waves that cover the tera region (10^{12}) of the electromagnetic spectrum. The THz waves are not the newly invented region of the electromagnetic spectrum as we encounter an abundance of THz sources in our daily life ranging from cosmic radiation to the blackbody emission of objects at room temperature (Li et al. 2012). Most of the existing THz sources are incoherent and are not fully utilized. Due to which a major realm of the THz spectrum is not utilized as it is not compatible and supportive for the emission and transmission of the signals to retrieve information of the physical objects. Hence, this portion of the THz spectra is commonly known as the THz gap (Sirtori 2021). Over the past decades, THz technology has undergone multiple transitions with the development of backward wave oscillator (BWO), gas laser, GaN-based Gunn diode, quantum cascade laser (QCL), and pulsed THz sources. Various unique features of the THz radiation over the other electromagnetic spectra are transient, coherent, transmission, broadband, low energy, fingerprint spectrum, and water absorption property (Qin et al. 2013). Therefore, THz remains as an active area of research that is widespread with applications in biomedicine, pharmaceuticals, microelectronics, forensic science, and so on. Although the behavior and electromagnetic properties of the THz waves are governed by Maxwell equations, the unique positioning of the THz waves in specific locations makes them difficult to handle than other regions (Lee 2009). Based on the literature, the scientific research on electromagnetic waves falls under two categories: electromagnetics and optics. The working principle, method of generation, and operating tools are quite different for electromagnetics and optics. THz waves are not well aligned to fall into either of these categories because the approximations used for electrical waves and optical waves may not remain the same when falling into the THz regime. Hence, THz technology is considered as an emerging science that has been explored for identifying its principle, working mechanism, and tools for the complete utilization of its applications.

Some of the current challenges with the THz waves are the incompatible waveguides as the metallic waveguides, optical fibers, and dielectric waveguides suffer from greater dispersion loss (Nallappan et al. 2020). Other than mechanisms of blackbody radiation and synchrotron radiation, most of the emitted light is a result of the transition among the different energy states. Due to its low photonic energy, the thermal relaxation hinders the distinction between the energy states where the existing difference equals 1 THz photonic energy. The electrical waves are generated because of the drifting charge carriers; however, the conventional radio-frequency sources are not capable of generating high-frequency radiation greater than a hundred GHz. This is due to the mismatch of motion of carriers with the speed of the THz oscillation in devices (Zhang and Xu 2010a). Despite these challenges, THz waves possess some of the unique advantageous features over the other regimes of the electromagnetic spectrum that creates the need for research studies focusing on utilization of THz for diverse applications. The low photon energy of THz waves makes these suitable for biological materials rather than X-rays as THz waves do not cause photoionization. Hence, THz waves are considered safe not only in terms of application to samples but also to worker's safety (Ergun and Sonmez 2015). Unlike microwaves, THz waves cannot penetrate the human body as penetration is limited to the surface of the skin only (~0.3 mm at 0.4–0.6 THz) (You et al. 2018). Compared to infrared waves and visible spectrum, THz waves have relatively longer wavelengths. Due to which it is clear that THz waves are least affected by nonmolecular scattering (also termed as Mie scattering) (Zhang and Xu 2010a). In addition, the THz waves are translucent to the dielectric materials such as plastic, fabrics, wood, and paper. Therefore, THz technology has gained great scope for non-destructive quality evaluations. The coherent THz signals are detected through mapping of the transient dielectric field under the time domain considering the phase and amplitude. This enables applications of dispersion and absorption THz time-domain spectroscopy. Most of the molecules exert strong absorption and dispersion pattern as a result of vibrational and rotational transitions due to the dipole moment (Cui et al. 2015). The resulted transitions are more unique and specific to each molecule that allows for the spectroscopic applications of fingerprinting under the THz range. All these properties make the THz waves a suitable candidate for non-destructive quality evaluation of agri-food products.

1.3 Emerging Non-Destructive Analytical Methods

Various emerging non-destructive analytical techniques employed in agricultural and food processing industries include ultrasonic measurements, force impulse, and acoustic impulse response, Fourier transform infrared spectroscopy (FTIRS), thermal imaging, near-infrared spectroscopy (NIRS), fluorescence spectroscopy (FS), nuclear magnetic resonance spectroscopy (NMRS), Raman spectroscopy (RS), hyperspectral imaging (HSI), and X-ray imaging (Chen et al. 2013; Wang et al. 2017b). In line with these non-destructive approaches, THz technology is an emerging analytical technique with distinct electromagnetic properties. The space between the microwave (MW) and infrared (IR) region in the electromagnetic spectrum is referred to as the far-infrared region or THz gap (Figure 1.1). The wavelength of the THz waves is about 3.3–333.6 cm^{-1} with a frequency of 0.1–10 THz (Gong et al. 2020). As a single standalone technology, THz spectroscopy could provide useful information by combining the benefits of MW and IR spectroscopy that permit the use of interface techniques. The integration of THz spectroscopy with imaging helps in molecular studies utilizing the photonic energy in the THz region. Due to its nonionizing properties, THz waves can be used for characterizing the inner regions of a solid object (Wang et al. 2020a). The generated high-resolution images make THz technology to be the best substitute for X-rays for analyzing the internal quality of solid materials. Different biomolecular components exhibit varied absorption and dispersion behaviors under the THz region. This enables the use of the THz waves for molecular characterization. THz allows observing the low-frequency bond vibrations, torsional and rotational modes of the molecules. Water (being the major component of any biological material) is a strong absorber of the THz waves (Nie et al. 2017). The absorption behavior of polar molecules in the THz range is considered to be a nuisance that hinders the results of the analytical measurements. On the other hand, moisture-free nonpolar, nonmetallic materials are transparent to the THz waves that allows for enhanced accuracy and reliability in the measurements (Mathanker et al. 2013).

THz technology is one of the simple and feasible methods that provides valuable information at the molecular level. Among the spectral techniques, HSI is the most simple and rapid analytical technique. However, HSI has a limitation of its use in in-line processing because of its huge data processing procedures (Mahesh et al. 2015). On the other hand, NIRS and RS possess high process efficiency; these methods are expensive with long mapping times. While X-ray imaging is a simple cost-effective analytical technique, it has been limited for food applications due to its ionizing property. In this line, MRI requires large space for instrumentation, high cost, and long acquisition time with laborious data analysis procedure. Due to their coherent nature, THz waves are more sensitive and have a dynamic working range over the FTIRS (Afsah-Hejri et al. 2019). Techniques such as NIRS, FTIRS, and NMRS can only deliver spectral information and do not provide spatial information. On the other hand, HSI provides both spectral and spatial attributes of the objects which aid in the determination of the surface morphology and textural features of the food products (Hussain et al. 2019). The major advantage of THz waves over other quantification techniques is the low characteristic energy of the hydrogen bond that reduces the possible damage of the samples when exposed to THz radiations. In terms of the performance, techniques like FS and RS may affect the quantification efficiency while NIRS requires a known standard concentration of analytes as reference material for testing (Cortes et al. 2017; Nekvapil et al. 2018). Thus, the selectivity of the analytical approach greatly depends on the robustness, accuracy, reliability, and sensitivity of the individual technique. Most of the biological materials possess a dominant hydrogen bond that lies in THz range. This allows direct detection of spectral resonance and molecular motions (Patil et al. 2022). In addition, the low energy of THz waves corresponds to a few meV (4 meV for 1 THz) that allows for the characterization of the materials without disturbing the state and the nature of the material under study (Son 2013). Thus, THz waves provide a promising solution in the quality evaluation of sensitive biological materials such as agricultural products and foods.

1.4 THz Radiation and Material Characterization

The THz radiation is made of submillimeter waves having extremely high frequencies with 1 mm wavelength that proceeds to even shorter. Hence the term "submillimeter radiation" is most common in research related to astrophysics (Patil et al. 2022). As mentioned earlier, the THz spectroscopy works in the range between the gigahertz (GHz) and T waves. The two types of generation mechanisms of T waves are pure rotational and combined rotational and vibrational transitions. The research on THz was started in the early 1960s with the generation of the first astronomical image using THz waves (Sethy et al. 2015). Later, the applications of THz for the space research were reported (Chantry 1971). The incoherent light was applied for characterizing the liquids at the molecular level in four-wave mixing experiments based on THz detuning oscillations (Dugan et al. 1988). The spectroscopic characterization of the water vapors using time-domain spectroscopy was reported in 1989 (Exter et al. 1989). This application has been considered as the most significant in the progress of THz research. Gradually, the THz spectroscopy was applied for measuring the extent of dispersion and absorption characteristics of chemical compounds. With the advancements of THz research, the THz waves were successfully used for imaging applications. This progress in the history of THz has led to numerous applications in testing the quality of physical entities like detection of foreign bodies, an inspection of packages, and mapping of chemical components to name a few (Mathanker et al. 2013). Like microwaves and infrared radiations, THz radiations are nonionizing and could travel in the field of vision. THz waves are capable of penetrating through a wide range of nonconductive materials like ceramics, paper, wood, fabrics, masonry, cardboard, and plastics. However, the depth of penetration is relatively small compared to microwaves (Mantsch and Naumann 2010). The THz radiation cannot permeate through liquids and metals; and shows a narrow permeation through fog and cloud (Pawar et al. 2013). The intramolecular and intermolecular motions under the spectral responses in the THz frequency range are beneficial in analyzing the chemical structure of single molecule and its arrangement. Since THz radiation has low photonic energy, it significantly reduces the chances of damage to the sample. Further, the coherent nature of THz waves remains to be advantageous over other spectroscopic techniques such as FTIRS (Puc et al. 2018). In addition, THz spectroscopy is highly sensitive and dynamic than FTIRS. The traditional THz frequency domain spectroscopy has been augmented with the development of powerful time-domain THz spectroscopy. Unlike the infrared and optical spectroscopy that measures the light intensity at particular frequencies, the time-domain THz spectroscopy could compute both the amplitude and phase of THz pulse enabling the refractive index and the coefficient of absorption measurements.

With the emergence of the applications of femtosecond lasers for generating coherent THz waves, researchers are quite interested in the development of high-power laser systems with broader bandwidth. The rising interest on the exploration of

THz applications started with the realization of improved performance of the THz time-domain spectroscopy in the far-infrared range other than FTIRS. This allows analyzing the temperature-dependent vibrational low-frequency spectra below 3 THz that gives additional details on vibrational dynamics and the solid structures (Davies and Linfield 2007). The study of the low-frequency vibrations of the molecules helps in the determination of the functionalities of proteins, amino acids, and nucleic acids based on the atomic arrangements and the conformational changes. In addition, the hydrogen-bonded networks are also considered as significant in this context as they are highly susceptible and easily broken allowing the structural changes than covalent bonds. Various techniques like surface field generation, electro-optic rectification, and ultrafast switching photoconductive emitters are explored for improving the bandwidth and power of THz spectral systems (Ferguson and Zhang 2002). Among these, photoconductive emitter is one of the efficient techniques in the conversion of visible/near-infrared radiation to THz beam and is mostly applied in spectroscopic and imaging systems. The process involves the generation of the electron-hole pairs (referred to as charge carriers) in semiconductor crystal with acceleration of the photoexcited charge carriers by the applied electric field. The physical dissociation of the charge carriers forms a space-charge field oriented against the biasing field. The applied external electrical field is processed resulting in fast-changing temporal electric field that generates a transient current. The induced current then assists in the generation of the THz pulses. It was reported that the use of the electric field screening model of theoretical simulations yields sub-100 fs electrical pulses (Davies and Linfield 2007). However, the experimental results suggested that 200 fs electrical pulses and 350 fs free space THz pulses were the shortest pulses obtained from gallium-arsenide emitters with a characterized frequency of 4 THz. A brief description of the system components, instrumentation, and research progress toward design of components is provided in the following section. On realizing the potential applications of THz, recently several industrial firms are showing interest in offering THz sensing systems that have diverse significant applications in noninvasive testing, remote sensing, quality inspection, security screening, structural analysis, and biomedical imaging (Bogue 2018). Such aspects of non-food and food applications of THz systems are briefly presented in the subsequent sections of this chapter (Sections 1.7.1 and 1.7.2).

1.5 Overview of THz Sources and Detectors

The classification of THz sources based on the spectral region with frequency range of 0.1–10 THz are as follows: solid state oscillators and gas and QCL. These system components work in the frequency range located at low and high frequency ends of THz domain, respectively (Gallerano and Biedron 2004). Other sources of THz radiation evident with technological advancements are laser-driven emitters and free electron-based sources. This section describes various research progress on the THz system components outlining different kinds of THz sources and detectors.

1.5.1 Solid-State Oscillators

The solid-state devices can be of either two- or three-terminal modules (Haddad et al. 2003). The two-terminal modules generate more power and are used as oscillators for THz mixers and derivers for frequency multipliers. On the other hand, three-terminal modules are used as amplifiers for broadband low noise applications. The electronic solid state instruments including oscillators and amplifiers are frequency limited because of carrier's transit time through semiconductor junctions (Grubin et al. 2013). This would result in high-frequency deployment. Devices like resonant tunneling diodes (RTD) (Figure 1.3) and transferred electron devices (TED) or Gunn are the two-terminal devices while tunnel injection transit time diode (TUNNETT) (Figure 1.4) and impact avalanche transit time diode (IMPATT) (Figure 1.5) are the transit time devices. All these solid-state components exhibit negative differential resistance for amplification and oscillation. Due to submicrometric gate length, the three-terminal device like field effect transistors (FET) (Figure 1.6) are applied for high-frequency operations. The heterojunction bipolar transistor (HBT) (Figure 1.7) is another three-terminal device having promising applications as low noise and broadband transistors. The high-frequency solid-state devices like Gunn, TUNNET, and IMPATT are compact and rugged devices used for continuous wave (CW) operations at room temperature with a narrow linewidth of 10^{-6}. For applications with frequencies in the range of 0.2–1 THz, frequency multipliers with two or more diodes are used. This would result in the average power level of 0.1–1 mW (Gallerano and Biedron 2004). The solid-state devices have a growing application on THz science due to their compactness. The common solid-state devices for THz generation like QCLs, RTDs, HBTs, and high-electron mobility transistor (HEMT) are limited with output power in the range only few µW. Active research progress in the 21st century, made researchers to explore the advantage of wide-bandgap (WBG) material-based IMPATT oscillators as frequency radiators (Biswas et al. 2018). The two WBG semiconductor

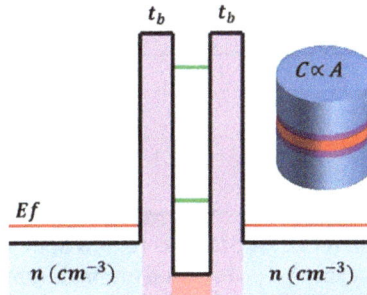

Figure 1.3 Schematic diagram of a double barrier resonant tunneling diode conduction band structure (n – doping density, t_b – quantum barrier thickness, C – capacitance, A – tunneling area, and E_f – Fermi level). (Source: Jacobs et al. 2015.)

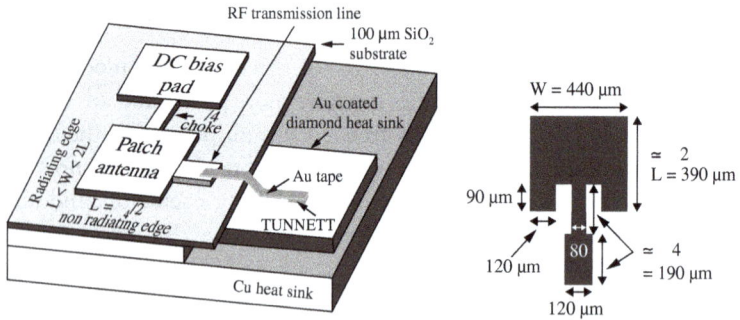

Figure 1.4 Schematic diagram of patch antenna coupled TUNNETT oscillator. (Source: Balasekaran et al. 2010.)

materials Wurtzite-GaN (Wz-GaN) and 4H-SiC attracted research interest due to their capability of high power generation from IMPATT sources in mW range (Acharyya and Banerjee 2014). Advancements in nanotechnology and material research paved a way for new generation of THz devices. Acharyya (2019) reported on the potential power of combining the capabilities of parallelly arranged IMPATT for enhanced power generation at THz frequency. In this study, a three-terminal planar IMPATT structure based on GNR on oxide (SiO$_2$) was proposed. The developed GNR IMPATT oscillator resulted in frequency range of 1–10 THz.

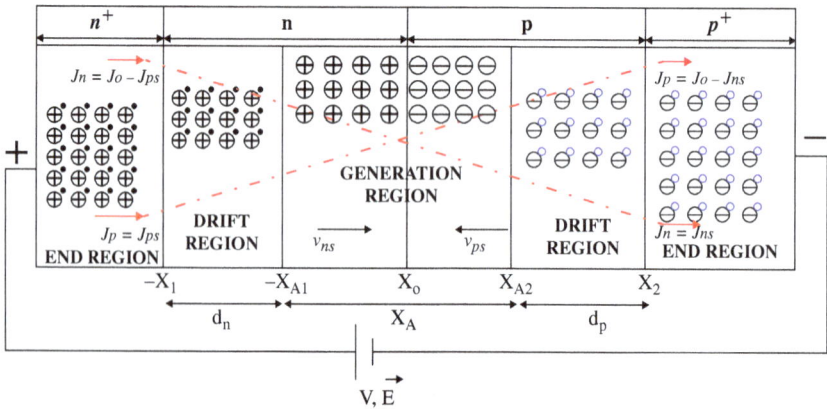

Figure 1.5 Mathematical model represents region of double drift impact ionization and avalanche transit time (X_0 – location of maximum electric field; W – total width of depletion; p and n are holes and electrons concentration, respectively with + sign denotes high doping; d_p and d_n are holes and electrons drift regions, respectively; J_p and J_n are holes and electrons current densities, respectively; J_0 – current density; v_{ps} and v_{ns} are hole and electron saturation velocities, respectively; and V – applied voltage). (Source: Ghivela and Sengupta 2021.)

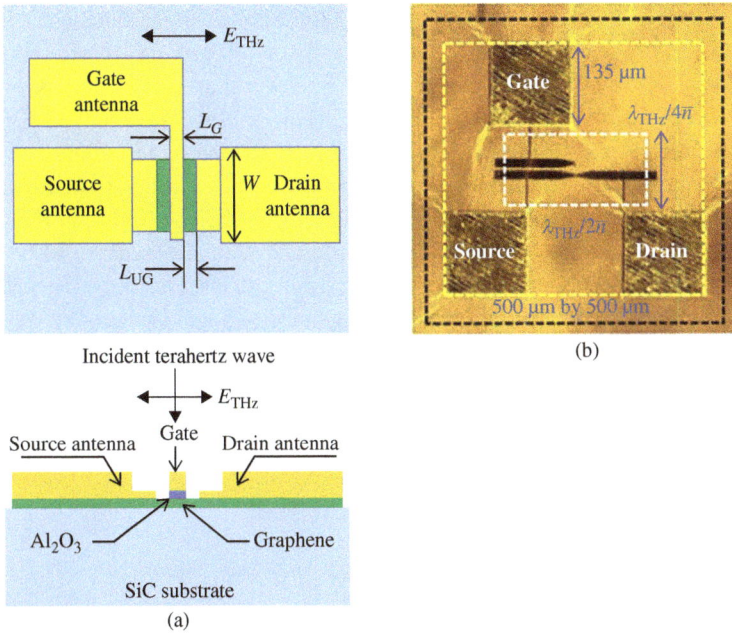

Figure 1.6 Bilayer graphene-based field-effect transistor: (a) schematic diagram of antenna coupled field-effect transistor and (b) optical micrograph of transistor. (Source: Qin et al. 2017b.)

1.5.2 Optically Pumped Gas Lasers

The far infrared (FIR) gas laser is the oldest coherent THz source applied for spectral imaging applications. This THz source utilizes optically pumped lasers often a CO_2 laser for excitation of gas molecules at millibar pressure range (Chevalier et al. 2019). The radiated CO_2 pumps laser into gas-filled cavity that

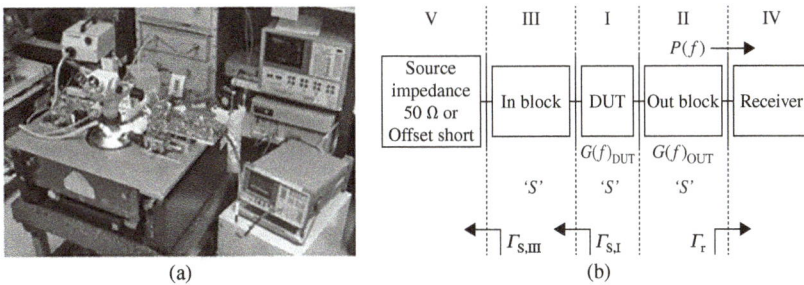

Figure 1.7 Diagram of heterojunction bipolar transistor (SiGe:C HBT): (a) noise setup equipment to characterize noise performances of SiGe:C HBT and (b) noise setup conceptual diagram. (Source: Ramirez-Garcia et al. 2019.)

emits radiation in THz frequency range. Methanol in its gaseous state is most widely used for powerful emission line typically of 100 mW at 118 μm (Gallerano and Biedron 2004). Companies like M Squared Lasers (United Kingdom), MeteroLaser Inc., (USA), B & W TEK Inc., (USA), Mesa Photonics LLC (USA), Brimrose Technology (USA), MIRO Analytical (Switzerland), Photon Systems (USA), Coherent Inc., (USA), IRsweep (Switzerland), Zurich Instruments AG (Switzerland), Chromacity (United Kingdom), Mirico (United Kingdom), Edinburgh Instruments (United Kingdom), and Vox Biomedical LLC (USA) offer commercial gas lasers for spectroscopy applications. Despite of their limited power, gas lasers are line tunable in the range of 0.3–5 THz frequency. Although the FIR lasers seem to have moderate development, they are the ideal sources for specific applications as in case of plasma diagnostics and heterodyne spectroscopy. A miniatured gas laser of dimensions 75 cm × 30 cm × 10 cm weighing 20 kg has been reported to generate a power of 30 mW at 2.5 THz (Saeedkia and Safavi-Naeini 2008). The gas lasers are often mixed with tunable microwave sources to develop tunable THz sources.

1.5.3 THz Semiconductor Lasers

The THz semiconductor lasers yield stimulated emission based on population inversion between discrete energy levels. A single crystal of p doped Ge sandwiched between the mirrors was the first developed THz semiconductor laser. It can be used for low temperature operation of 20 K with large applied magnetic field of 1 T (Hovenier et al. 2000). This semiconductor laser can be tuned to THz frequency in the range of 1–4 THz. The QCL is a newly developed coherent THz semiconductor source (Figure 1.8). The QCL is a unipolar laser where transition of

Figure 1.8 View of dual quantum cascade laser (QCL) spectrometer (the pulsed QCLs are placed in blue housings, MPC – multi-pass sample cell, RC – reference cell, and Det – nitrogen cooled detector). (Source: Curl et al. 2010.)

carriers occurs among the discrete energy levels within either conduction or valence band. The electrons are injected in an electrical bias into the periodic structure of superlattice. The excitation through multiple wells by resonant tunneling results in inter-sub-band transitions (Gallerano and Biedron 2004). This would result in cascade process as the name of THz source implies. Different variations in the design of THz QCLs are chirped superlattice, bound to continuum configuration, and resonant phonon active region (Vitiello et al. 2006). The waveguiding can be done based on metal-metal waveguides and semi-insulating surface plasmon wave-guides. Although the radiation efficiency is weak and out-coupled waves are scattered in space, the metal-metal waveguides provide better temperature performance than semi-insulating surface plasmon waveguides (Kohen et al. 2005). Some of the challenges that exist in developing QCLs are increased operation temperature, frequency tuning range, output power, and wall-plug efficiency; and reduced spectrum linewidth, minimum operation frequency; and enhanced quality of radiation pattern (Saeedkia and Safavi-Naeini 2008).

1.5.4 Laser-Driven THz Emitters

The laser-driven emitters are the most used THz sources. The THz pulsed radiation sources that use laser-driven emitters which work on the scaling down frequency from the optical domain. The two common techniques in practice for the THz generation are as follows: first is the use of short femtosecond Ti:Sapphire laser pulses (Figure 1.9) and second is the application of sub-picosecond pulses to the

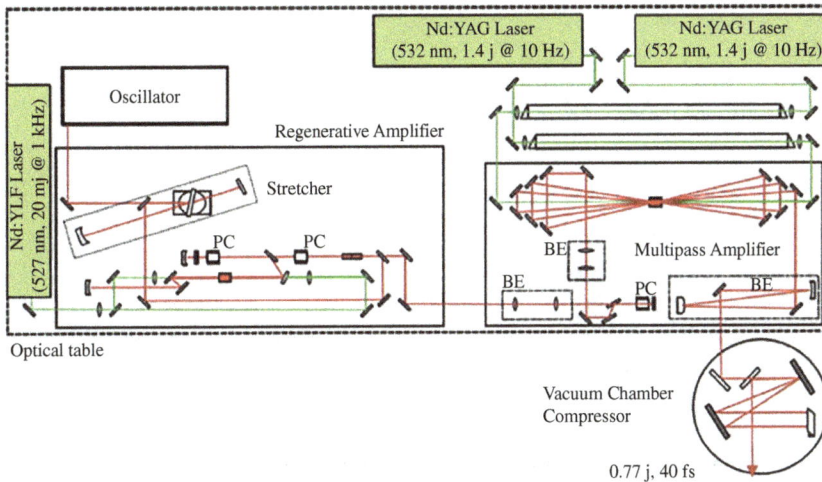

Figure 1.9 **Schematic diagram of the Ti:Sapphire laser system (20 terawatt) (PC – pocket cells and BC – beam expander; the compressed beam has 770 mJ of energy, 35–40 fs pulse duration with 10 Hz repetition rate). (Source: Nam et al. 2015.)**

crystal like ZnTe with great second-order susceptibility. The first method involves the illumination of the electrode gap by Ti:Sapphire laser on the semiconductor generating carriers that accelerates by the applied bias field of 100 V. The resultant transient current is coupled to the radio frequency antenna using stripline that radiate the THz frequency at broadband corresponding to the Fourier transform of the time profile laser pulse. The second method involves the application of the picosecond laser pulses to the crystal (ZnTe) (Figure 1.10). This results in photomixing from nonlinear response of crystal producing time-varied polarization that in turn leads to emission of THz waves. The quick crystal response and the lack of stripline or conductor eventually assist in the operations at higher frequencies. In addition, the laser-driven solid state emitters include CW photomixers. Here, the off-set frequency locked continuous laser source is focused on a photoconductor under bias. Later the induced photocurrent is modulated at the frequency difference and coupled to antenna emitting continuous THz radiation at narrow band (Brown

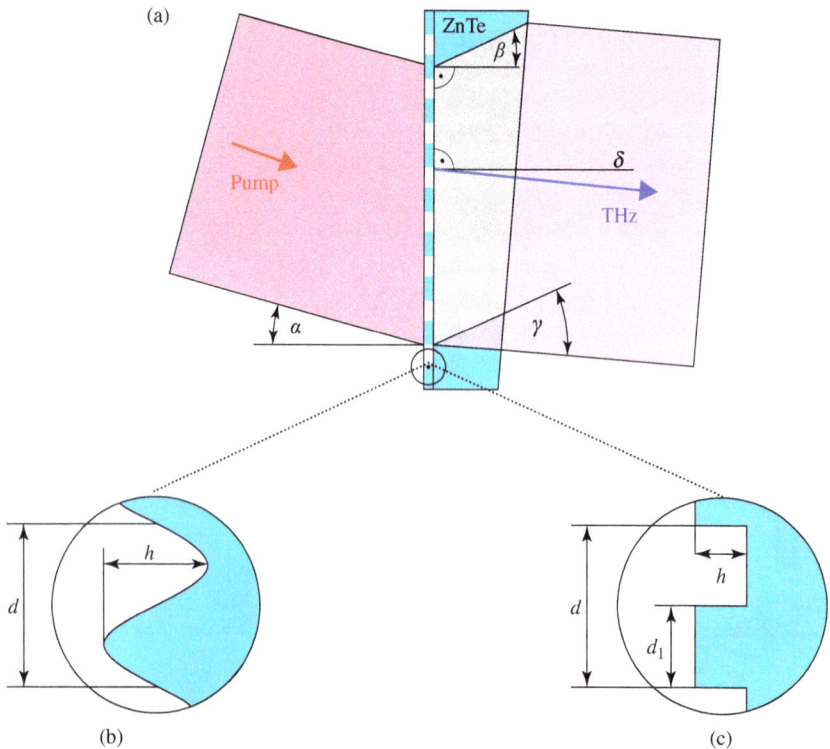

Figure 1.10   Schematic sketch of ZnTe based contact grating THz source: (a) ZnTe based contact grating setup, (b) sinusoidal profile, and (c) grating profile (d – grating period, h – trench depth, α – angle of incidence, β – diffraction angle, and γ – angle of generated THz beam that is equal to γ−β). (Source: Ollmann et al. 2014.)

et al. 1995). The optical frequency of the drive laser is shifted to help tuning the frequency at broader range. Based on the pulse parameters of the laser, the solid-state laser-driven emitters would cover 0.2–2 THz or even above (Gallerano and Biedron 2004). The resulting average power levels lies in the range of nanowatts to 100 μW with pulse energies of femtojoule to nanojoule range.

1.5.5 Free Electron-Based Sources

The operation of electron beam-based devices is based on high energy electron beam interaction with strong magnetic field inside waveguides or resonant cavities (Biedron et al. 2007). This results in energy transfer between the beam of electrons and electromagnetic wave. The devices based on electron sources are free electron lasers, gyrotrons (Figure 1.11), and BWOs. These devices generate high power signals at THz frequency range. A free electron laser is a gyrotron operating at high

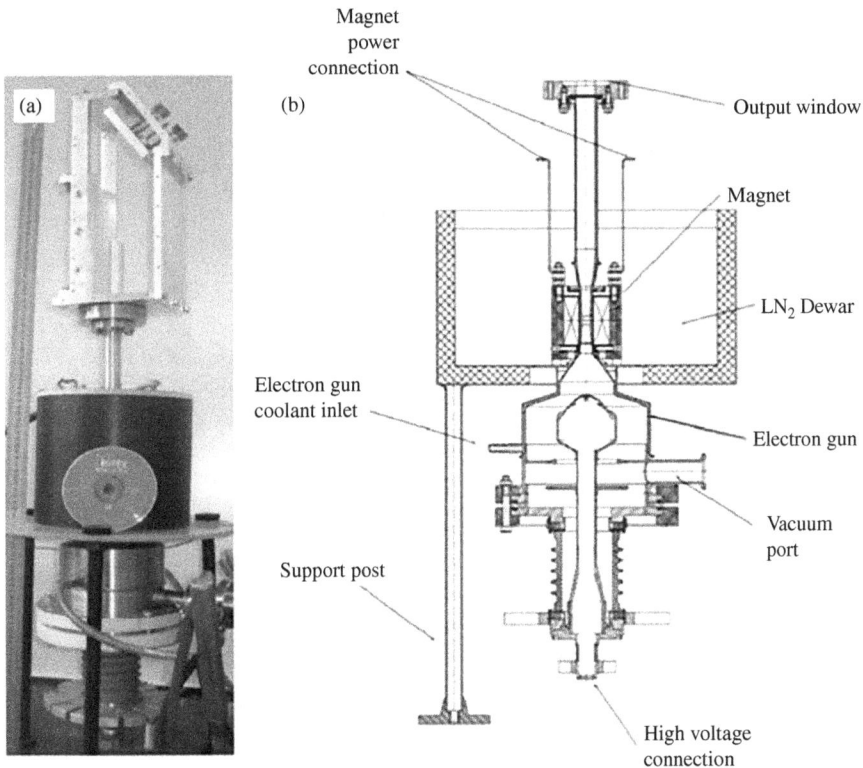

Figure 1.11 **Sub-THz/THz gyrotron system: (a) the 670 GHz gyrotron with the 28 T pulsed solenoid, mirrors, and Vlasov converter (1 m ruler and compact disc are shown for scale) and (b) schematic diagram of the gyrotron. (Source: Glyavin et al. 2012.)**

and wide tunable frequency range. This would result in high energy acceleration from deriving the electron beam using optical cavity. The electron-based lasers are operated virtually over the entire range of electromagnetic realm from microwaves through ultraviolet and extended to the X-rays (Saeedkia and Safavi-Naeini 2008). In case of BWOs, the phase velocity of the electromagnetic waves is opposite to that of the group velocity. The BWOs are tuned electrically over a bandwidth of higher than 50% of their operating frequencies resulting in generation of power up to 50 mW at 300 GHz and could go below to a few mW at 1 THz (Schmidt et al. 2002). The complete system is large and heavy that requires high biased voltage and water-cooling facility. Technologies like microfabrication and micro-assembly have emerged as a promising approach used for size reduction of the electron beam sources more suitable for THz applications (Srivastava 2015). Some studies have reported on analyzing the feasibility of the developed tabletop free electron laser systems (Liu et al. 2017b; Zhang et al. 2022a).

1.5.6 Frequency Multipliers

As the name implies, the frequency of the employed drive source is multiplied in a nonlinear instrument for the generation of higher order harmonic frequencies (Saeedkia and Safavi-Naeini 2008). Considering the advantage of GaAs substrate less technology, the Planar-Schottky diodes are most used for reducing the loss of substrate in frequency multipliers (Figure 1.12). The drive sources can be either solid-state devices like IMPATT and Gunn oscillators or BWOs with high power of 0.05–0.15 THz. The monolithic microwave integrated circuit technology used for fabrication of microwave synthesizers along with high power amplifiers could generate high output power above 0.1 THz (Samoska et al. 2000). The heterostructure barrier varactor diodes showed promising results as high efficiency frequency multipliers. More efficient THz frequency multipliers could be achieved using series of frequency doublers and triplers (Siegel 2002). The maximum achievable frequency

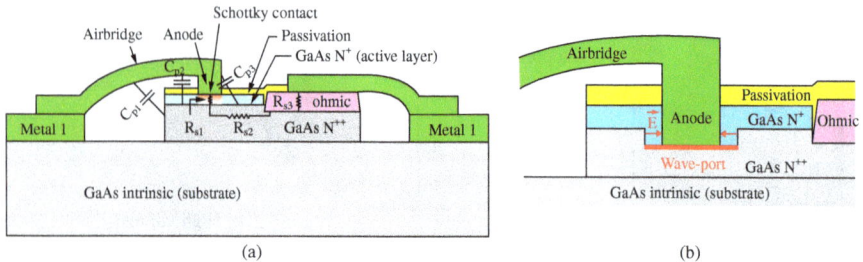

(a)   (b)

Figure 1.12   Schematic diagram of a planar Schottky diode: (a) sketch of planar Schottky diode and (b) wave port at Schottky contact (C_p – parasitic capacitance, R_s – series resistance, R_{s3} – series resistance due to ohmic contact at DC). (Source: Maestrini et al. 2010.)

from solid-state frequency multipliers is limited to 2 THz. However, it could be possible to generate THz signals above 2.5 THz using hybrid systems consisting of chain of frequency multipliers integrated with BWO.

1.6 Instrumentation, Methodology, and Techniques

The THz regime is often referred to as the THz gap because of the lack of compatible sources and detectors to study material characteristics under this region. In contrast, a certain range of sources and detectors are available for both the electronic end (low-frequency side) and the optical end (high-frequency side) of the THz gap. Both generation and detection of THz waves are quite challenging (Mantsch and Naumann 2010). In the early 1990s, the THz research was led by an optical generation of THz waves through continuous or pulsed wave lasers. This generates ultrafast photocurrents in a semiconductor or photoconductive switch as a result of the acceleration of the electric field carriers. With recent progress in THz research, the complexity involved in the generation of THz waves is significantly reduced. Nowadays, the THz waves are most commonly generated by nonlinear optical methods such as difference frequency generation, optical rectification, and backward optical parametric oscillation methods (Kitaeva 2008). Some of the effective materials that are currently used for the generation of THz radiations are GaP, GaAs, CdTe, ZnTe, and $LiNbO_3$. Another interesting method for the THz generation is ambient air plasma generation. This method is gaining recent research interest, especially in security screening applications. This method involves the use of a high-intensity pulsed laser to generate the air plasma that in turn emits the THz waves. The distant laser beams are used for controlling the source of THz radiation remotely (Cook and Hochstrasser 2000). In context with the THz detectors, deuterated triglycine sulfate crystals (DTGS) or conventional bolometers are commonly used. Recently, the low temperature grown GaAs received greater attention to be used as detector material due to their high resistivity, high breakdown voltage, carrier mobility, and fast capture time of carriers (Singh et al. 2021). The pulsed THz techniques sense the THz radiation through nonlinear optical processes. Here, the photoconductive antennae are used as both emitters as well as detectors, and their performance is determined by their trapping time and carrier mobility (Blanchard et al. 2010). The THz radiations are focused using lenses made from Teflon and silicon. Sometimes, the gold-plated mirrors are used to collimate and focus the THz beams. The resolution of the captured image can be defined by the diameter of the THz beam with a particular wavelength. The higher the frequency components in the THz system, the greater the resolution of the obtained image. The semiconductor-based QCL is a novel source of generation of THz radiation. It generates THz waves through electron relaxation between bands of quantum wells (Zeng et al. 2020). Advancements in photonics paved the way for the development of solid-state electronic devices like photodiodes that generate

THz waves through the photo-mixing technique. The powerful broadband THz sources based on the free-electron laser are also available. This type of THz source, available at the Jefferson Laboratory, Virginia, could generate high power THz beams of up to 10 W or even greater that is ideal for fundamental studies to characterize molecular structures (Williams 2002). With these available resources, scientists are working toward the development of low-cost THz system for imaging applications. One such attempt is the coupling of dielectric image line with the THz disk resonator in whispering gallery mode. This approach provides a means of low-cost THz instrumentation that can be used for off-site applications. Most of the techniques employed for the generation of THz radiations are quite complex requiring multicomponent systems such as external pump lasers, cryogenic cooling and vacuum technology. Some of the systems used for commercial THz imaging and sensing employ IMPATT diodes, photoconductive antennas, BWOs, and QCLs. Each of which has its own merits and demerits in view of capacity, size, cost, cooling requirements, power consumption, power output, stability, tunability, and frequency range (Bogue 2018). The research progress in near-infrared and ultrafast pulsed lasers helped in guiding coherent generation and detection of broadband sub-picosecond THz pulses of 0.05–10 THz (Smye et al. 2001). The eventual progress led to research on THz spectroscopy and imaging techniques. Thus, THz spectroscopy and imaging have a remarkable application in the study of biomaterials. Certainly, further improvements in the emission and detection of THz waves would open a new window for the non-destructive quality assessment of materials at molecular level. Some of the key research advancements and progress in the evolution of THz sources and detectors have been briefly presented in this section. However, different types of basic system components of the THz spectroscopy and imaging systems are discussed in the subsequent chapters (Chapters 2 and 3) of the book.

1.6.1 Progress in THz Sources

Research progress in THz technology remains active and global efforts are being made to develop novel sources with improved THz features. This involves converging of diverse techniques of nanotechnology, spintronics, metamaterials, and plasmonics. The scientists of the University of Michigan, University of California, and the Dublin City University have confirmed the room temperature generation of wavelength tunable high power THz source by combining the plasmonic photo mixer with digitally distributed feedback diode lasers (Yang et al. 2015). The THz system based on this approach comprises of ultrafast photoconductor integrated with the spiral antenna substrate (ErAs:InGaAs). The stimulation of surface plasmon waves generates a large fraction of photocarriers within the electrode proximity. This phenomenon occurs when the dual mode-distributed feedback laser (D-DFB) beam incidents on the photomixer anode. A large amount of electrons would drift toward the anode by increasing the incidence of the laser beam at the contact electrodes. The induced photocurrent is applied to the spiral antenna that generates THz radiation within the spectral peaks of laser of broad frequency in the range of 0.15–3 THz. In a single cycle of continuous-wave radiation, an optical pump with an average power of 100 mW could produce 0.44 THz, 1.20 THz, and 2.85 THz corresponding to 1.3 mW, 106 μW, and 12 μW radiation powers, respectively (Yang et al. 2015). The future work of this study aimed at the investigation of the inclusion of elements of plasmonic photo mixing within bimodal lasers as a single chip THz source.

Ongoing research works are being focused on the applicability of metamaterials like split-ring resonators for THz emissions. The collaborative research of Karlsruhe Institute of Technology, Iowa State University, and Ames Laboratory successfully demonstrated the broadband THz generation through optical excitation of metamaterials (Figure 1.13) (Luo et al. 2014). The employed metamaterial resembled a U-shaped structure consisting of 40 nm gold split-ring resonators. A tunable near-infrared radiation of 1,100–2,600 nm was generated by pumping an optical parametric amplifier through directing the output from a Ti:Sapphire amplifier with 35 fs pulse duration, 800 nm center wavelength, and 1 kHz repetition rate. The induced generation beam was used to drive the metamaterial. Later it was realized that the pumping of a metamaterial array resulted in the emission of string THz

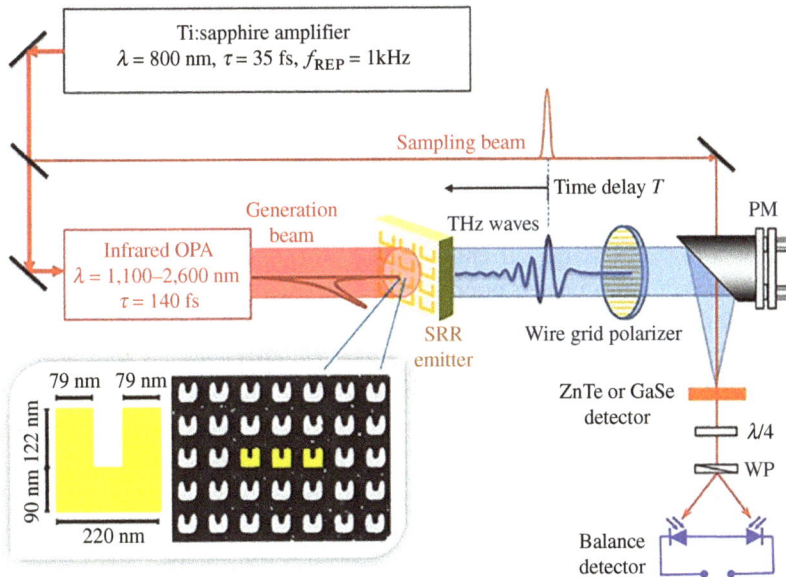

Figure 1.13 Experimental setup used for broadband THz generation from meta-materials (WP – Wollaston prism and PM – parabolic mirror, the magnified image at the bottom left is SEM of split ring resonators array along with yellow split ring resonator and its geometrical design overlay). (Source: Luo et al. 2014.)

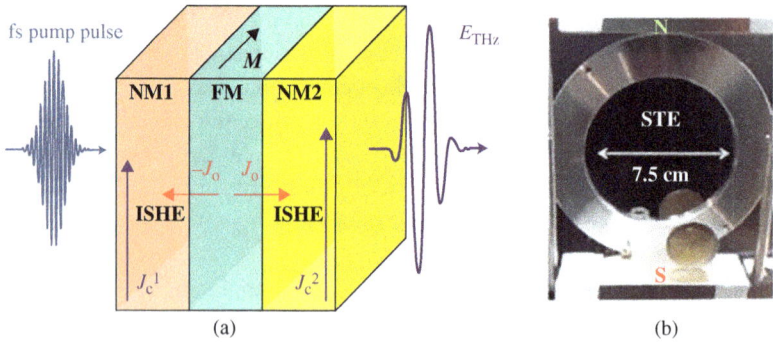

Figure 1.14 Schematic diagram of high electric field spintronic THz emitter: (a) sketch of operating principle and (b) photograph of spintronic THz emitter (FM – ferromagnetic layer, M – in-plane magnetization, NM – nonmagnetic layer, and ISHE – inverse spin Hall effect. In (b), two magnetic bars N and S provide a magnetic field of ≥ 10 mT across the entire emitter area. Scale reference – 2 €coin). (Source: Seifert et al. 2017.)

broadband wave of 0.1–4 THz frequency. Another group of research collaboration team from Johannes Gutenberg University and the Fritz Haber Institute had come up with the new concept of generation of THz waves from spintronic devices (Figure 1.14) (Seifert et al. 2017). Compared to traditional electronic devices, spintronics possesses electron spins in addition to the charge that allows for further degrees of freedom. This represents a simple approach to generate spin-polarized current that allows the induced current to pass through the ferromagnetic layer. A 6-nm thick W/CoFeB/Pt trilayer on a glass substrate was excited by laser pulses of 5.5 mJ and 800 nm wavelength for 40 fs. The applied pulses drove the spin currents into adjacent nonmagnetic layers. The spin currents were transformed into orthogonal in-plane charge currents because of inverse spin Hall effect. The design of the W and Pt layers possess opposite spin angles causing constructive superposition of the charge currents (Seifert et al. 2017). Eventually, it results in the emission of broadband THz pulses into the optical far-field characterized with 5 nJ energy and 230 fs thereby exhibiting a continuous spectrum at 10% maximum amplitude from 1 to 30 THz.

1.6.2 Progress in THz Detectors

Compared to the THz generation their detection is less complex and feasible with existing instruments such as bolometers (Figure 1.15), optical acoustic Golay cells, pyroelectric detectors, Schottky-barrier diodes, photoconductive antenna, and modified FETs (Bogue 2018). As mentioned earlier, photoconductive antenna could act both as generators and detectors of THz radiation and are most commonly used in commercial products. The firing of ultrashort laser pulses over a thin film of highly

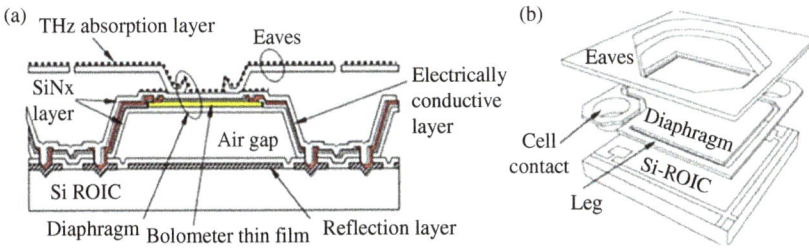

Figure 1.15 THz focal plane array representing the pixel structure of uncooled bolometer: (a) cross-sectional view and (b) oblique view. (Source: Oda 2010.)

resistive semiconductor material containing III–V of compounds like GaBiAs, GaAs, and InGaAs combined with the electrical pads in contact generates THz radiation. The laser pulses create pairs of holes and electrons from the substrate that are accelerated by the applied voltage. This would result in the mobility of the electric field delivering pulses of THz waves. The whole sequence of the process comprises of generation of THz radiation and the detection involves the reverse generation. The rebounding pulse when reaches the detector, the generated electron-hole pairs get accelerated onto an electrode and resulted electrical current is measured. The scientists from the Universities of Manchester and Oxford along with the Australian National University had designed and developed a photoconductive antenna-based InP nanowire by growing through the vapor phase epitaxy (Peng et al. 2016). The developed detectors provide THz time-domain spectrum of 0.1–2.0 THz bandwidth. Similar to photoconductive antenna, the RTDs are capable enough to perform the dual role of emission and detection of THz. The RTDs are based on mechanical quantum tunneling. Since tunneling through thin layers of substrates does not take much time, the RTDs are a rapid process capable of ultra-high-speed operations. A THz sensor based on RTDs was reported by the scientists of Osaka University along with a waveguide photonic crystal cavity as a resonator. In this design, the module was grown on a semi-insulating InP substrate and a GaInAs/AlAs double barrier quantum well that constitutes a tunneling region (Figure 1.16) (Diebold et al. 2016; Okamoto et al. 2017). Ultimately, these arrangements yield a compact sensing system. During the assessment, a dielectric tape of different thicknesses was fastened to the photonic crystal cavity and the resulting variation in the resonator's refractive index was determined. The advantage of measurement of refractive index sensing was an order higher than the magnitude of metallic metamaterial resonators.

Research works are still in progress to explore the applications of nanotechnology and incorporation of nanomaterials like nanowires, carbon nanotubes, graphene films, metal nanoparticles, and semiconductor nanoparticles in FET-based detectors. Based on this approach, researchers from the Chinese Academy of Engineering Physics and Sun Yat-Sen University developed an InN nanowire with incorporated FET device through a transfer printing approach (Figure 1.17)

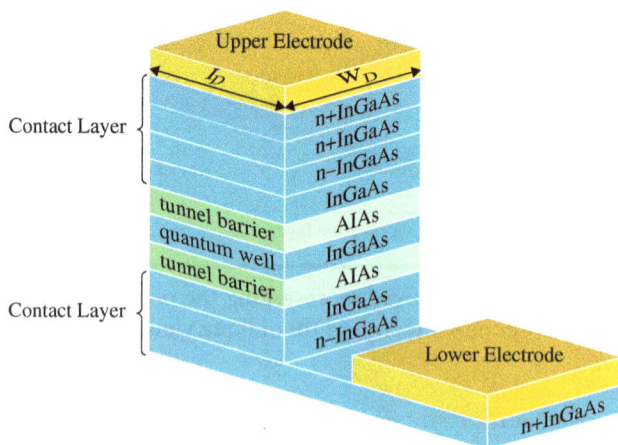

Figure 1.16 Typical resonant tunneling diode heterostructure including contact layers and electrodes. (Source: Diebold et al. 2016.)

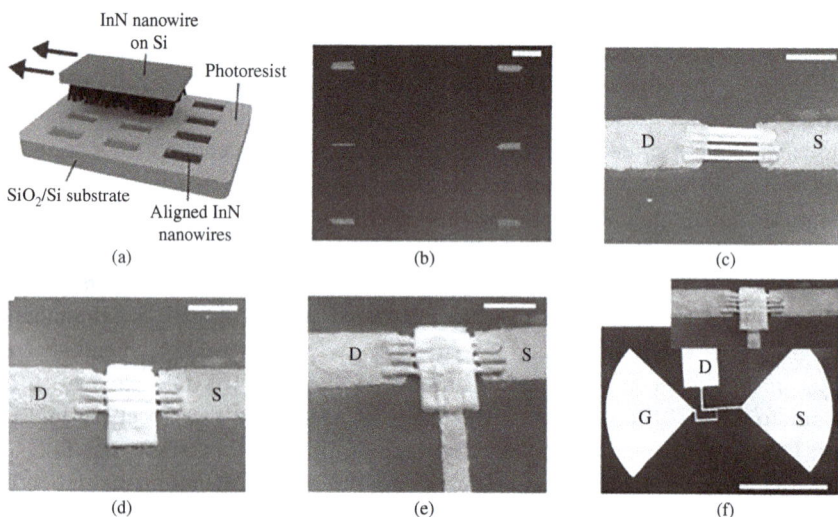

Figure 1.17 THz detectors based on array of orderly aligned InN nanowires: (a) schematic process of transfer printing, (b) SEM micrographs of InN nanowires (5 μm scale bar), (c) SEM micrographs of the aligned nanowires (2 μm scale bar), (d) SEM micrographs of the nanowire device with gate, (e) SEM micrographs of the nanowire FET (2 μm scale bar), and (f) SEM image of the whole device (100 μm scale bar) (D – drain, S – source, and G – gate). (Source: Chen et al. 2015a.)

(Chen et al. 2015a). An array of FETs was developed by photographic etching combined with focused ion beam electrode deposition. Results showed that detectors in series connection led to the generation of stronger output photovoltage. A greater response of 1.1 V/W was evident from interconnected module on illumination with a signal of 290 GHz. Another research group from the University of Electronic Science and Technology (China) established modified microbolometer-based THz imaging system (Wang et al. 2016). It consisted of a thin film of nanostructured Ti acting as absorber in integration with the micro-bridge module of a vanadium oxide (VO_x) microbolometer. This resulted in a continuous THz imaging system employing a 2.52-THz far-IR CO_2 laser source and micro-bridge/VO_x microbolometer focal plane array. The developed structure was employed to produce images of circular metallic washer covered by fabric and paper clip placed in an envelope.

1.7 Novel Trends and Applications of THz Waves

With advantages of the combined optics and electronic principle, THz waves gained recent research interest as a noninvasive testing aid. A plethora of rising industrial applications of THz technology including quality testing and monitoring of industrial processes and products have been identified. The applications of THz waves are being explored with the emergence of novel sources and detectors in THz spectroscopy and imaging systems. Although some of the applications of THz technology such as polymer testing, detection of foreign bodies, and characterization of molecules are quite well known, the applications of which in the agri-food sector are at their nascent stage (Wietzke et al. 2010). On realizing the merits of THz waves over other electromagnetic radiations, researchers are now increasingly focused on exploring the applicability and implementation of THz technology in the agricultural and food industries. This includes quality parameter extraction, packaging, polymer testing, determination of fiber orientation, measurement of water content, food inspection, hydrology monitoring of crops and plants, security systems, and detection of liquid explosives. In order to get familiar with the current applications of THz, this section gives a brief introduction to the THz applications in the non-food and food sectors.

1.7.1 Non-Food Applications

THz technology offers a superior method of quality testing of materials, especially as inspection and monitoring tools. Over time we have witnessed the excessive use of polymers and plastic materials due to their higher flexibility and ease of processing than metals and glasses. The rising applications of polymers require stronger demand for quality assurance from industrial and consumer points. The common quality testing techniques employed for polymer testing are based on specimen examination and often based on destructive methods (Jansen et al. 2012). This poses serious issues in terms of material loss and obtained results would deviate to represent the quality of the whole batch since the physiochemical attributes of plastics are greatly subjected to the method of processing. Therefore, total inspection of fabricated materials is often recommended because a failing part of the whole component eventually causes higher loss than production costs. Considering these aspects, THz technology is seemed to be a highly efficient technique in the detection of the quality of polymers (Yakovlev et al. 2015). Most polymers are highly translucent to THz radiation that are capable to detect the presence of any additives or undesired metallic inclusions. The additives present in the polymeric matrix exert a higher dielectric contrast over pure polymers under the THz regime than with X-rays. This allows detecting the quality of the polymers. These merits of THz waves pave a way for the development of in-line systems for polymer compounding process that offers long-term stable performance, flexibility, and robust measurements (Fosodeder et al. 2020).

The next broad category of application of THz waves is in security screening and inspection. Right from the beginning of the exploration of THz waves, its unique features dictate the merits of THz waves in revealing metallic and non-metallic objects even beneath the fabrics. This immensely raised the interest of scientists and industrial firms to focus research on the investigation of THz in security imaging and scanning systems (Tzydynzhapov et al. 2020). The THz security systems have a great research scope and commercial interest in the establishment of standalone and mobile detection systems at checkpoints, especially in border crossing areas. Another interesting application of THz is the identification of explosives based on THz fingerprint (Palka 2013). This is because the liquid explosives pose a challenge in their detection with the existing methods. It was reported that the liquid present inside the closed containers can be categorized based on THz time-domain spectroscopy as either inflammable or harmless (Jepsen et al. 2008). These results provide a promising solution for the development of handheld security scanning devices for the identification of suspicious liquids at hospitals, harbor ports, airports, and other public premises. Hence more research and development are required in this direction for the efficient utilization of THz systems for security screening purposes.

Plastics are used significantly as construction materials, especially as pipes and tube fittings. This application requires the use of joining methods like welding to fix it with mainstream lines. Often the efficiency of the welding process is judged based on the mechanical integrity; however, the inclusion of air bubbles and delamination will eventually cause loss of weld strength and structural integrity (Wietzke et al. 2009a). The transportation of liquids and gases in high-density polyethylene pipes is one such example where the failure of welding joints may associate with the high cost and health risks. Thus, it is most important to ensure the welding quality of the joints before its dispatch for end-use. Hence, welding integrity of the joints is crucial for ensuring the mechanical strength of the nonstructural applications. The evaluation of the plastic weld quality of the materials remains to be a great challenge from the past. Few of the novel non-destructive testing methods based on ultrasound and X-rays are not sufficient enough to detect the delamination and inclusions that insist on the adequacy of the quality evaluation in careful monitoring of the process parameters of welding to attain low failure rates. In this context, non-destructive imaging using THz waves provides a promising scope in analyzing the textural surface quality of weld joints (Wietzke et al. 2007b). Although this method of quality evaluation is in progress, there is a great scope of industrial applications in near future. Taking into the advantage of the transmission of the THz fingerprint spectra, it is possible to find the differences that exist between the properly welded region and the delaminated areas. It has been reported that the transmitted THz intensity was integrated at frequency range from 0.2 to 0.3 THz in revealing the contrast that exists between the welded and non-welded portions (Wietzke et al. 2009a). These THz imaging systems are quite useful in the detection of the contamination of weld joints with sand and metal that possess a great scope for mobile unit applications that can be directly used in industrial manufacturing sites.

As stated earlier, plastics have a great industrial significance that has applications in nearly all products ranging from small commodities to high-tech products. In this context, measurement of the water content of plastics is a necessity to determine their end-applications (Farhat et al. 2020). The measurement of the water resistance of the hygroscopic materials such as polymers and their composites are required in the determination of the useful life of the materials when exposed to intended atmospheric conditions. This is mainly because of the physical and mechanical attributes of plastic materials that are significantly affected by the percentage of water present in it. Hence, the uptake of water especially in wood-plastic composites becomes a significant parameter to be studied in evaluating its quality (Nakanishi and Takahashi 2018). The THz technology finds its applications in assessing the sorption of water into the polymer and its composites due to the strong dispersion and high absorption coefficient of water under the THz frequency range. Based on this aspect, a model was developed for assessing the dielectric behavior of the materials using THz spectroscopy system. In this study, the modeling of the dielectric behavior of the material considers the free and bound water, effective medium theory, and the density differences based on swelling characteristics of the materials (Jordens et al. 2010). There exists a good concurrence between the measured and simulated outputs with a slight nonlinear behavior of the refractive index mainly because of the increase in effective density. Thus, the THz sensitivity serves to be a noninvasive probe for the determination of the water content of the polymeric composites (Figure 1.18). This could also be explored for applications in the agri-food materials as most of these are hygroscopic.

Another interesting application of THz in the plastic industry is the assessment of fiber orientations. Fiber reinforcement of materials is practiced to enhance their specific mechanical attributes like high tensile strength and low specific gravity. These kinds of reinforced materials found their applications in the civil engineering

Figure 1.18 **THz image showing water distribution in a wood plastic composite plate. (Source: Jordens et al. 2010.)**

and automobile sector where THz testing can be used for determining the strength and quality of fiber reinforcement of plastics (Xu et al. 2020). The absorption spectra of the composite materials vary based on material composition. THz imaging was used for evaluating the extent of the fire-damaged fiber in carbon composites (Karpowicz et al. 2006). Since carbon fiber composites are characterized by strong absorption characteristics, the measurements using THz waves are limited with reflection geometries. On the contrary, the fiberglass composites are reasonably transparent to THz waves which can be assessed in transmission mode. A non-destructive quality testing of glass composites for the detection of fiber delamination, voids, thermal and mechanical defects using THz was reported (Stoik et al. 2008). The mechanical properties of the glass material are greatly influenced by fiber content and their orientation hence understanding of these parameters is significant regarding safety aspects of composite materials. Fiber orientation into the host material results in the birefringent behavior of composites under THz waves. Studies were reported on the potential of THz time-domain spectroscopy (THz-TDS) for analyzing fiber-reinforced polymers and THz imaging to detect the temporal point of the THz pulse that distinguish the dominance of ordinary and extraordinary THz beam (Valk et al. 2005; Rutz et al. 2006, 2007). Thus, THz waves remain a promising tool in analyzing the quality of materials of economic significance.

1.7.2 Food Applications

Food safety and quality assurance remain to be the top priorities among consumers and food manufacturers. Despite advancements of the science and technology, there exist multiple food threats in food processing lines such as exceeded limits of additives, preservatives, presence of traces of pesticides, and foreign bodies. It is a must to ensure the quality of the processed agri-food products before reaching the consumers. Currently, various emerging non-destructive food quality evaluation methods based on X-ray imaging, thermal imaging, infrared imaging, ultrasonic imaging, fluorescence imaging, and electrostatic techniques are being exploited for the discrimination and identification of foreign bodies in food products (Vadivambal and Jayas 2016; Wang et al. 2020a). For total quality assessment of food products, no single technique can be employed for identifying multiple food safety threats as each of these techniques has its merits and demerits. In simple terms, no single non-destructive testing method could fulfill all the quality demands of the agri-food industry in analyzing complex food safety threats and foreign bodies such as paper, plastics, peel, stone, bone, glass, and so on. For instance, the application of machine vision is limited to characterizing the materials having similar physical attributes. Hence, the machine vision can be applied for distinguishing the samples with alike quality characteristics such as color, size, shape, and volume (Vithu and Moses 2016). Similarly, X-rays are not capable of detecting foreign objects with lower densities than food samples (Li et al. 2015b). Often the applications of X-rays are limited in the agri-food industry due to safety concerns as it causes ionization that is harmful to the operators as well as could damage the food samples. Similarly, applications of infrared imaging are also restricted in the food and agricultural sector due to its low penetration depth (Almeida et al. 2006). Considering these limitations, the exploration of novel technologies is mandatory to meet the urgent needs of the food industry to ensure safety and quality monitoring. In this view, THz technology is the novel emerging non-destructive quality evaluation technique for agri-food applications. In contrast to X-rays, THz waves do not cause photoionization and operate in regime of low energy that is comparatively safe for both samples and for working personnel (Zhang and Xu 2010a). Since the THz regime comes between the microwave and mid-infrared realms of the electromagnetic spectrum (Figure 1.1), the applications of THz are widened as it inherits the advantages of both microwaves and infrared waves. Due to its characteristic features of water absorption, coherency, transmission, low energy, and fingerprint spectra properties, THz waves are more suitable for quality assessment of agri-food products (Qin et al. 2013). Broadly, the applications of THz in the agri-food industry are categorized into four major classes: spectroscopy, imaging, sensing, and communication that are generally applied in quality control, inspection, and monitoring in agri-food products. In these aspects, THz technology is quite useful in molecular characterization of macro and micro food constituents, detection of foreign bodies and toxic compounds, screening of antibiotics and adulterants, analysis of microbial contamination, discrimination of transgenic seeds, hydrology and water status in plants, inspection of oil seeds and crops, food packaging, operation, and process control to name a few (Afsah-Hejri et al. 2020).

It has been proven that THz waves are highly sensitive to soft organic materials than X-rays where the high-resolution THz image was used to distinguish the defective red ginseng (Kim et al. 2012). Results confirmed that defects were identified using THz waves but not using X-rays. Notably THz spectral band is characterized by a longer wavelength than near-infrared rays that causes less scattering and imparts higher resolution in the THz imaging. As an emerging technique, THz waves are conveniently used for transmitting and receiving signals for displaying the vibration of the chemical biomolecules. Hence, the chemical constituents of the materials can be achieved by analyzing the captured THz fingerprint spectra which is nearly impossible with other conventional imaging methods. Recent developments of THz sources and detectors had transformed THz spectroscopy and imaging to the next level of analytics as a rapidly evolving non-destructive assessment method that has a great scope in the agri-food sector. Research advancements have been made in the identification of foreign bodies in foods using THz waves, such as plastic and insects in chocolates (Ok et al. 2019), stones and granite pieces in powdered instant noodles (Lee et al. 2012), metal traces in sausages (Wang et al. 2019a), and somatic cell counts in milk (Naito et al. 2011). Jiang et al. (2019) explored reflection THz-TDS imaging for the identification of foreign bodies in whole wheat grains and wheat flour under varied sample depths. Similarly, the optical attributes of fish and cereals with insect foreign objects were studied based on complex refractive index mapping of the samples (Shin et al. 2018a). In another study, THz imaging was used for discriminating the shells from kernels of walnut based on absorption spectra at different concentrations (Wang et al. 2020a). The undesired materials (shells) were distinguished by employing principal component analysis that classified the THz spectra followed by THz imaging to distinguish the kernels from shells. Thus, this work proved the successful application of THz technology to identify the low-density foreign objects mixed with food samples thereby proving to be a promising tool in food quality assessment. Similar studies must be carried out to explore THz waves for the detection of endogenous contamination in foods. THz-TDS imaging was used for identifying the metallic and nonmetallic contaminants from chocolate samples (Jordens and Koch 2008). Based on the THz pulse delay, the acquired image clearly reveals the contamination of chocolate bar with glass splinter. In this regard, considering the shape of the waveform of the obtained THz signals, it is possible to detect undesired inclusions such as stones, plastic, fabric, polymer, and glass into the desired food products such as nuts, cereals, and confectionery items through smart data processing algorithms and spectra classification (Wietzke et al. 2009a). With these merits, THz imaging system has a great scope for the direct integration with main production line. Currently, the published research works were being carried out under laboratory conditions based on raster scanning method and research works are in progress for the development of THz sensor arrays. However, the measurement speed is an important factor to consider in context with the industrial applications of THz waves in agricultural and food sector. Future investigations on the design of THz spectroscopy and imaging systems like quasi-optic machine parts for multi-focus operations are required for implementing THz systems on large scale (He et al.

2016; Jia et al. 2019). For which a thorough knowledge of the operating principle and system components of THz are required. In this regard, this book focuses on summarizing the various applications of THz technology, principle, and operation of spectroscopy and imaging systems based on THz waves with a special emphasis on current research and recent trends on THz research in agri-food applications.

1.8 Recent Research and Advancements

THz technology remains the emerging research area that explores the nature of different types of materials under low excitation and ultrafast dynamics. Research progress on the QCLs and metamaterials has expedited the progress of the THz technology (Almond et al. 2020). In addition, the development of system components and devices for photo mixing, optical ultrafast switching, and sub-THz electronics broadens the application of THz waves in diverse industrial sectors such as process monitoring, operation control, security screening, food safety, and quality assessment (Zhang et al. 2021a). The three major classes of THz science are the application of THz waves, THz photonics, and THz electronics (Tonouchi 2005). A typical THz system comprises photo switches and femtosecond laser for the generation of THz waves. The generated THz waves are used to characterize the fingerprint spectra of the materials under the time domain. This helps in quick and fast determination of the complex parameters such as the dielectric constant of different test materials. The major advantage of the THz waves is that it can be considered as an optical beam enabling the imaging of materials of our interest. Considering the practical implementation, the laboratory THz systems are quite large and costly thereby limiting the applications of THz systems. To overcome this limitation, research works are being carried out in developing simple, convenient, compact, flexible, mobile, stable, and on-site monitoring THz systems for large-scale industrial applications. In view of the characterization of materials, the strong absorptive nature of the water restricts the THz spectral applications of biomaterials and chemical solutions under moist atmospheric conditions. However, the development of THz generators and detectors aids in overcoming this limitation. So, these kinds of materials can be characterized using a THz attenuated total reflection spectroscopy system (Arikawa et al. 2008). Based on the modes of operation, THz spectroscopy can be operated in reflection and transmission modes. The high-frequency THz components can be attenuated in the transmission mode while the reflection mode has low attenuation (Mathanker et al. 2013). Compared to the transmission optics, the alignment of the reflection optics in THz system is difficult. In the early 1990s, most of the research works on THz spectroscopy were focused on time-domain spectroscopy. Only in recent times, other forms of THz spectroscopy like time-resolved spectroscopy, gas THz spectroscopy, and attenuated total reflection spectroscopy have emerged. Similarly, the THz imaging systems are grouped based on forms of THz waves as pulsed or continuous THz imaging. Further, the pulsed forms of THz imaging systems are categorized as far-field THz imaging and near-field THz imaging. The operation principle of each form of THz spectroscopy and imaging systems are described in the subsequent chapters of this book (Chapters 2 and 3). Another variant is laser THz emission microscope (LTEM) induced by active excitation and specific emission of THz depending on the nature of the materials (Klarskov et al. 2017). Thus, THz systems possess multiple and interesting novel applications that must be explored. For which the basic understanding of optronics is quite adequate that unwinds the research feasibility of novel metamaterials and nanomaterials in enhancing the potential application of THz in industrial sectors.

In recent times, the THz spectroscopy and imaging systems are emerging as an efficient quantification tool due to their convenience, rapidness, flexibility, label-free sensing, and noncontact applications in the agri-food industries. The THz spectra are unique fingerprints that reveal information about the characteristic nature of the food components. Water forms a major part of foods that exists in either free or bound forms. Water has strong THz absorbing power that aids in the determination of the nature of water that exists in foods. The hydration levels of water affect the vibrational modes of proteins and peptides (Gong et al. 2020). This helps in the determination of the molecular structure and chemical state of the biomaterials. However, the high absorptivity of water limits THz waves for food applications as it interferes in the characterization of specific components such as amino acids and fatty acids. The application of THz technology for inhomogeneous food materials would significantly affect the accuracy of the spectral results. Sample thickness, shape, and particle size are the critical factors that greatly influence the absorption coefficient and refractive index. It was reported that samples of thickness <1 mm are best suited for THz spectroscopy while samples of thickness >1 mm result in opaqueness at frequencies of above 0.7 THz. Hence, it is recommended to analyze the samples at a relatively lower THz frequency. Sample pretreatments such as drying and freezing are also helpful in avoiding the interaction of liquid water by THz radiation (Qin et al. 2013).

In addition, the studies on material design and components of the THz system have yet to be optimized. This would help in the reduction of the overall cost of implementation of THz systems. Advancements in material design, metamaterials, nano-based materials are gaining research interest for effective transmission of THz signals with reduced noise levels (Qin et al. 2016; Han et al. 2020). The identification of the best classification models for the characterization of diverse food samples is adequate for the practical implementation of the THz systems. So, the combination of THz with high-end systems of novel THz sources and detectors would result in fast, accurate, and reliable detection for in-line monitoring. Improvements in data acquisition, scanning rate, and signal processing methods are needed to achieve precise results, to improve limit of detection (LOD) and accuracy of the prediction. Systems based on temperature and frequency-dependent sensing THz techniques based on guided-mode resonance (GMR) are useful in the detection of biomolecules. Currently, THz waves are used for the detection of microbial contamination as an alternative to conventional culturing methods, biosensors, and DNA-based detection techniques (Yang et al. 2017). More research must be directed in this area for developing a convenient and cost-effective, lab-on-chip device for microbial detection as a food safety tool.

1.9 Summary

The THz regime of the electromagnetic spectrum has the advantages of microwaves and infrared waves. Recently, more research is being done to explore the potential of the THz gap in quality monitoring, inspection, communication, sensing, detection, quantification, and characterization of biomaterials. This chapter describes the scope of THz waves and their unique capability to act as a fingerprint in determining the molecular and chemical structure of materials. Further, the nonionizing characteristics of THz waves made this technology to be used as an alternative to X-ray imaging suitable for agri-food applications. Non-destructive approaches based on THz technology especially in the detection of foreign bodies, toxic compounds, adulterants, and antibiotics; microbial examination; hydrology monitoring; seed inspection; process and operation control; and food packaging are quite evident. Thus, this chapter summarizes the scope and active research progress on THz systems that highlight the technological feasibility for the development of highly efficient, noninvasive, label-free, convenient, and cost-effective non-destructive detection approaches for agri-food products in near future.

Chapter 2

Terahertz Spectroscopy

2.1 Introduction

Recent years have witnessed a tremendous growth in terahertz (THz) research with significant advances in the assessment and diagnostics of biomaterials (Dey and Roy 2021). Typically, a THz system includes a high-power source generation device along with detectors that open a novel range of applications in the characterization of materials (Zhang et al. 2021b). A plethora of remarkable developments in THz systems broaden their applications to diverse industrial sectors such as biomedicine (Gong et al. 2020), material science and construction (Krugener et al. 2020), fabrics and textiles (Molloy and Naftaly 2014), automobiles, and aviation industries (Ellrich et al. 2020) to name a few. These diverse applications of THz have opened a novel range of opportunities and feasibilities for the next generation imaging and sensing systems for quality monitoring, security screening, and inspection (Patil et al. 2022). A range of well-established applications of THz waves (T waves) include high-temperature superconductor and semiconductor characterization, molecular imaging, label-free genetic analysis, and biological and chemical sensing (Zhang and Xu 2010a). These applications of T waves are well explored in the biomedical field that have thrust the THz research from relative obscurity into a limelight for agri-food applications. The THz regime is a part of electromagnetic spectrum that covers the frequency in the range of 0.3–30 THz, i.e., 10^{12} cycles per second and is located between the infrared and microwaves (Jansen et al. 2008; Alsharif et al. 2020). THz science has attained greatest research interest in the past decade; however, its high atmospheric absorption had limited the research interest in THz science. Despite of limitation and resistance of the THz radiation to adapt with the conventional techniques of its neighboring bands, the major applications of THz were researched by astronomers and chemists in early days of research on THz science (Ghann and Uddin 2017). The research on T waves started with

DOI: 10.1201/9781003197010-2

characterizing the vibrational and rotational resonances of simple molecular compounds under THz spectral regime. The recent revolution of the advanced materials research has brought up the THz research to the next level and has significantly increased the profile of the THz spectroscopy and imaging systems. Some of the noteworthy advancements in the field of THz are characterization of multiparticle interactions through THz spectroscopy, development of quantum cascade THz lasers, ultrafast near-infrared and pulsed visible lasers, and THz detection base-pair differences of DNA in femtomolar (fM) concentrations (Yang et al. 2016b).

The electromagnetic regime of the submillimeter wave is portrayed as a transition region of infrared and millimeter waves of electromagnetic spectrum. The submillimeter waves often termed as T waves share the electromagnetic properties of both infrared and microwaves (Li et al. 2020a). Thus, T waves have no notable characteristics on their own. This fact is true until today because combined optics and electronic nature of T waves still complicates the understanding of THz science. Despite of the emergence and discovery of novel unique features of T waves, the applications of THz for industrial and commercial purposes are still obscure. This clearly put forth the need for a distinct array of the techniques and THz systems with a specific community of the scientists to unwind the research opportunities in this emerging area of science. The early stage of development of characterizing systems based on T waves was started with the integration of microwave and optical techniques that gradually bridged up the gap that existed between the microwaves and optical regions of electromagnetic spectrum (Mittleman 2013). This resulted in the propagation of energy in forms of low-ordered mode or single-mode guided waves. On the other hand, the integration of optical and infrared techniques resulted in the beams characterized with multiple modes. Henceforth, better understanding of the THz regime and its neighboring bands would significantly provide useful information on the scope and potential applications of T waves in detection, sensing, and characterization of the materials at molecular level. Considering this aspect, this chapter aimed at describing the usefulness of THz spectra emphasizing the spectroscopic system components, generation, and detection. In addition, the limitations that exist with THz spectroscopy systems and their measurements are detailed along with the progressing research advancements to overcome these limitations.

2.2 Physics of THz Waves

Different forms of the molecular compounds, condensed matter, and gaseous vapors possess varied physical attributes in resonance with THz radiation; therefore, THz region shows a promising scope for spectroscopic and imaging applications. Various chemical compounds can be studied in detail, i.e., the chemical molecular information can be revealed easily through THz spectroscopic techniques (Gong et al. 2020). Further, the early motives for the THz spectroscopy were the development of high-sensitivity instrumentation for the astronomical studies and environmental sensing. With advanced THz research in recent years, the security and public safety sectors also get benefited through the artificial noses that are sensitive to THz which are used for detection of explosives and illicit drugs (Patil et al. 2022). A novel detection technique for the authentication of the integrated circuits was proposed using THz system (Figure 2.1) (Ahi et al. 2018). The developed THz system was used for quality monitoring in identification of packaged integrated circuits with misshaped die frames and bonded wires, contaminated and sanded surfaces, and reverse engineered and blacktopping layers. In addition, the applications of THz in bioscience specifically the THz spectroscopy of the biomolecules like DNA remains as the most interesting research yet to explore. Thus, THz science has a broader application in diverse fields of study. Hence, it is most important to understand the basics of T waves to explore their spectroscopic applications in material characterization. Therefore, this section attempts to provide a brief discussion about the basic principles of physics of interaction of matter with the T waves.

The T waves cover a spectral range of 300 GHz to 20 THz that occupies the spectral realms between the microwaves and infrared regions. One THz corresponds to wavelength of 0.3 mm, possesses a temperature of about 50 K, energy of about 0.004 eV, and a wavenumber of 33.3 cm^{-1} (Zhang and Xu 2010a). The working range of the commonly used bounds is described with the 0.1–10 THz frequency and 5–500 K temperature. This broad THz regime includes both limits $h\nu/kT \ll 1$ and $h\nu/kT \gg 1$ that corresponds to a diverse physical phenomenon. Similarly, the THz regime covers the wavelength ranging from 3 mm to 30 μm. Thus, the size considerations provide a low-order mode in the longer wavelength of THz domain as like microwave devices at one side and a high-order mode at its shorter wavelength as like laser-based devices on the other side (Zeitler 2016). In broad terms, the interactions of THz radiation are categorized based on Q of the resonances as: interactions with low pressure gases, interactions with liquids and solids, and interactions with gases at near atmospheric pressure (Mittleman 2013). The first class of THz interaction with liquids and solids has a very low Q that are often characterized by continuum interactions while the THz interactions with low pressure gases have a Q value as high as in the order of ~10^6. On the other hand, the THz gaseous interactions at near atmospheric pressure have a Q of the order of ~10^2. Most biophysical applications of THz spectra come under the second category of interactions with low pressure gases that has a higher Q. The gaseous THz interactions would result in the rotation of molecules that are about 10^3–10^6 times more intense than the interactions occurring under microwave regime. After a certain extent of mass-dependent spectral peak in the THz region, the strength of interaction decreases exponentially approaching the infrared region. This sharp peak of THz realm is responsible for most of unique attributes that are closely associated with the characterization and sensing applications using THz. On the other hand, the interaction of THz with liquids and solids has no molecular rotation

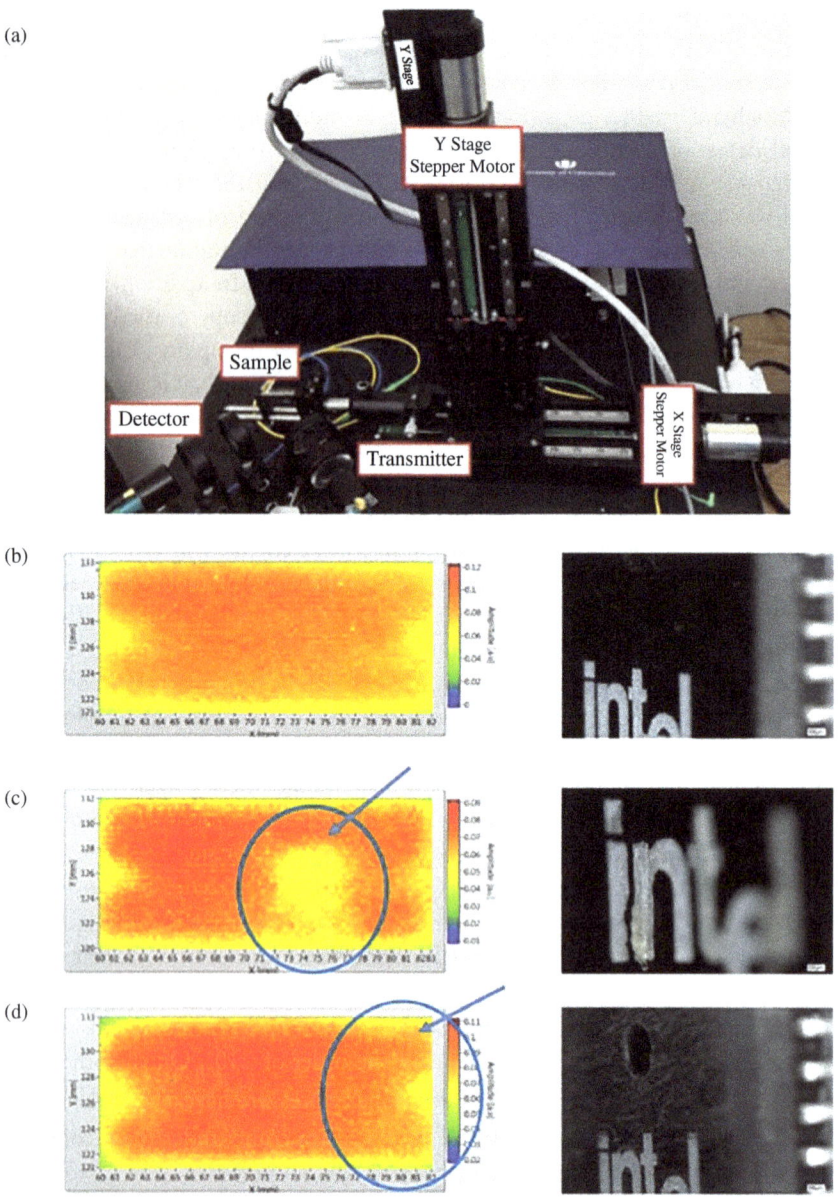

Figure 2.1 THz time-domain spectral imaging system for the authentication of packaged integrated circuits: (a) operation in reflection mode and (b), (c), and (d) are obtained THz images on left and optical images on the right of an authentic circuit, a counterfeit recycled integrated circuit, and integrated circuit sanded on one side, respectively (contaminated spots are obvious in THz images). (Source: Ahi et al. 2018.)

(Mittleman 2013). Sometimes the larger molecules may exert a collective motion resulting in energy level spacings at THz frequencies. The resulting resonances are much broader as the liquids and solids interact strongly with the THz radiation that leads to a continuum spectrum under THz region. This characteristic feature has a significant advantage on the applications of THz in characterizing molecules using fingerprint spectra.

2.3 Terahertz Spectroscopy

The most exciting THz application is the spectral assessment of the materials at the molecular level. As stated earlier, the THz frequency ranges from 0.1 to 10 THz that reflects the convergence of the electronics and photonics of the electromagnetic spectrum. The region of the THz radiations generally falls at high frequency end of the electromagnetic spectrum and are commonly referred to as submillimeter waves (Patil et al. 2022). The overlapping of the photonic energy of THz radiation with the energy of hydrogen bonds of biomolecules aids in the possible interactions among the organic solids and liquids. Further, it is possible to investigate the photonic vibrations and the intermolecular motions like bend and torsional motions of the organic molecules at the THz frequency range (Zeitler 2016). The T waves are characterized with greater penetration through a broad range of the insulating materials like polymers and plastics, paper, wood, cardboard, and ceramic materials; nonionizing characteristics; high reflectivity against metallic objects; and large absorptive power by water (Wang et al. 2018a). All these unique properties transform and widen the applications of THz radiation in diverse industrial sectors. The complex interaction of T waves with the biological materials results either in absorption, reflection, penetration, and scattering effects that make T waves to be used as diagnostic systems ranging from process monitoring and sensing to quality assessment and safety assurance (Garbacz 2016). The mechanisms of the generation of the T waves are based on the rotational as well as combined rotational-vibrational molecular transitions. The range of the THz frequency lies between GHz and T waves. The THz spectroscopy yields spectral data that requires adequate processing for extracting the physical, chemical, and structural features of the materials. The intramolecular and intermolecular motions generated by the spectral response of the materials to T waves reveal the information about the chemical state and the structural arrangements of the molecules (Xie et al. 2014). In recent years, the processing of acquired THz spectral data is made easier through the integration with chemometric methods. The implementation of chemometric methods with spectroscopy greatly reduces the amount of process variables facilitating analytical examination of the samples. The spectral data processing using the chemometric methods speeds up analysis time and improves the process efficiency (Gong et al. 2020). The application of chemometrics in THz technology helps in removing the undesirable and irrelevant optical parameters. The chemometric methods are usually applied during data preprocessing and in the multivariate analysis (Wang et al. 2017b). The common chemometric methods that are combined with THz spectroscopy include support vector machine (SVM), principal component analysis (PCA), least-squares support vector machine (LS-SVM), partial least squares (PLS), linear discriminate analysis (LDA), principal component regression (PCR), back propagation neural network (BPNN), and artificial neural network (ANN) to name a few (Hussain et al. 2019; Gong et al. 2020). This section describes the basic working mechanism, critical components, and available THz source technologies.

2.3.1 Instrumentation and Working Principle

The THz technology allows to capture the spectral data and the analysis of these spectral data provides information about the structural features of biomaterials. The THz instrumentation consists of three major parts: THz radiation source, accessory components, and detector (Ren et al. 2019a). Advancements in the technology pave a way for the upcoming modern generation THz sources that can produce THz radiation in a broader frequency range. Different types of THz sources used for producing THz radiation are based on the electronics and optics principles. The typical light sources, commonly used in THz spectroscopy, are thermal-based, electronic-based (vacuum and solid-state electronic), and laser-based (pulsed, continuous, optical rectification, mechanical excitation, and transient currents) (Patil et al. 2022). Some of which including the free-electron and solid-state electronic sources are combined with the frequency multipliers for enhancing the dynamic frequency range of the THz radiation. Currently, the generation of THz radiation based on optical methods using plasma is gaining a great scope in the security screening and quality inspection (Ren et al. 2019b). Thus, the selection of the source is based on the process requirements and the end-use applications.

Other accessory components include mirrors, lenses, and polarizers. These are the critical components of the THz spectroscopy that determine the detection efficiency and accuracy (Lewis 2017). As like in any other spectroscopic technique, mirrors are used for converging and focusing the THz beam from the source. Conventionally metals like silver, aluminum, and gold are used as mirrors (Feng and Otani 2021). Other than the metals, semiconductors like doped as well as undoped GaAs, hybrid polypropylene, and silicon tunable mirrors can also be used as mirrors (Lewis 2014). The next significant component that forms a major part of THz instrumentation is the lenses. Lenses are used for achieving the highest possible resolution for precise results during the analysis. The advantage of the greater transmission of visible light of the polymers like polyethylene makes plastics to be conventionally used for making lenses. Commonly used materials to make lenses are polymethylpentene (TPX) and polyethylene (PE). Further, lenses help in proper optical alignment and increase the sensitivity of detection. With the advancements in the manufacturing practices, lenses are being produced by compacting metallic powders at micro scale using tabletop hydraulic press, and other materials used are 3D printed diffractive lenses, Fresnel zone plates, variable focal length lenses, plasmonic resonances, and THz lenses made out of paper (Ren et al. 2019a). For adjusting the focal length, the cavity of the lenses is filled with the medical white oil that acts as a lubricant.

Polarizer is a filter that permits only particular polarized light to pass through while blocking the undesired beams. Most used polarizers are wire grid type and liquid crystal type. The wire grid polarizers are made up of gold, aluminum, metal-coated polymer, composite Cu-polythene-Cu with a working frequency range of about 0.1–3 THz (Cho 2018). While the liquid crystal polarizers have a limited working frequency range of only 0.2–1 THz (Ji et al. 2019). The broader frequency range of wire grid polarizers has the advantage of higher extinction ratio with durable strength and acceptance of wide range of angle of light. The detectors form an integral part of THz spectroscopy and are useful in the detection of transmitted and reflected signals. The photoconductive antennas are applied for the strong powerful THz detection. The working principle of the THz detectors can be of either thermal, electronic, or electro-optical based. Different types of detectors used for analyzing the T waves are Schottky mixers, super conductor mixer detectors, Golay cell detectors, pyrometers, and bolometers (Patil et al. 2022). Applications of different types of detectors used for spectroscopy were described by Kasap and Capper (2017). Reconfigurable polarizers and CNT fiber optics polarizers are also used for THz data transmission, spectroscopy, and imaging. Wang et al. (2018c) summarized the applications of carbon nanotube devices such as antennae, emitters, detectors, transistors, amplifiers, and polarizers for generating, modulating, polarizing, and detecting the T waves.

2.3.2 Basics of THz Spectroscopy Systems

The gradual evolution of THz science led to the emergence of various sources and detectors thereby resulting in many new designs of spectrometers. The approaches used for the development of THz techniques are broadly categorized as optical and microwave. The former approach of optical methods is commonly used in coherent radiation sources like quantum cascade lasers (QCLs), optically pumped molecular lasers, and X-ray free electron lasers. The frequency multiplication of a stable reference synthesizer is the conventional approach employed in designing microwave oscillators. The primary components of THz systems based on frequency multiplication comprise of frequency multipliers, amplifiers, and mixers. The combination of frequency multipliers and mixers based on quantum semiconductor superlattices introduces various precise spectral components (Malhotra et al. 2018). The frequency multipliers based on backward wave oscillators (BWOs) with frequency range of 667–1100 GHz in integration with quantum semiconductor harmonic superlattice mixers results in efficient THz frequency conversion (Vaks 2012). Later a THz device with phase lock loop frequency control for BWOs (900 GHz) on the 55th harmonic signal from synthesizer was developed. With the emergence of novel applications of transient microwave components; and solid-state sources led to outcome of working methods of THz spectroscopy that can be operated in frequency phase switching and fast sweeping modes (Vaks et al. 2014a). It is based on the interaction of phase-switched pulses with resonantly absorbing molecules. This causes periodic induction and decay of molecular polarization. As a result, the nonstationary signals that are arising during this process of decay of molecular polarization are recorded at the receiving end of spectrometer. Thus, the magnitude and the shape of the obtained signal are analyzed to determine the concentration of the components under study with higher accuracy.

The THz spectrometer with phase switching system utilizes BWOs as THz source. A high-voltage power supply is fed to these oscillators and the resulting output voltage determines the oscillation frequency (Vaks et al. 2014b). An input signal from BWO is allowed to pass through the analytical cell containing the material under study. The transmitted signal from the analyte cell is computed by a Schottky diode detector. The obtained signal is amplified at receiver end followed by digitization using an analog-to-digital convertor. The weak spectral signals are chosen against the background noise using a digitalized storage device that completes the addition and real-time averaging of input signals. On the other hand, the signal controlling the BWO frequency can be applied either at the reference input of high voltage or at the BWO anode connected with the low resistance device housing in case of spectrometers operating in fast frequency sweeping mode (Liebermeister et al. 2019). A phase lock loop frequency control is used for achieving the accuracy in fixing the instantaneous values of source frequency of radiation should not exceed 10^{-6}. The phase lock loop frequency control for BWO can be operated using highly stable reference frequency synthesizer. In addition, a high-speed analog-to-digital convertor (16 bit) is used for sweeping the signal to have a control over the BWO frequency sweeping. The receiver end of fast frequency sweeping mode spectrometer consists of a detector that comprises of waveguide camera with Schottky diode, detector bias circuit equipped with low-noise preamplifier, and low-pass filter (Vaks et al. 2014a). The intensity of the spectral lines can be determined through rectifying the measured DC voltage on detector at the low-pass filter input. The signal emerged from preamplifier output reaches the analog to digital convertor and then to storage where the received signals are added and processed in real-time. Then the obtained spectral data are transferred to computer for further data averaging. The sensitivity of spectroscopy measurements can be increased by increasing the signal-to-noise ratio through accumulation of the coherent THz signals (Peng et al. 2020). Both the frequency phase switching and fast sweeping modes of THz transient spectrometers are highly useful in analytical measurements and identification of chemical components in various fields including material science, biology, medicine, security screening, and other high-tech applications mainly because of their high sensitivity and spectral resolution.

2.3.3 Classification Based on THz Source Technology

Unlike for the rest of the electromagnetic radiations on either side of the THz width, inexpensive compact high intensity THz sources are not readily available. This holds up the research progress on exploring applications of THz radiation for material sensing and characterization. It is well-known fact that we cannot see the THz radiation; however, we can feel the warmth as the THz spectra share the considerable part of region of far-infrared waves (Lee 2009). Yet this region of electromagnetic radiation remains as the least explored region mainly due to the difficulties in accessing this spectrum. To overcome this technical issue, many research works are actively in progress for the development of efficient and compact THz sources. This section explains the various types of available THz sources that are being used in spectroscopic and imaging systems. Broadly the THz radiation can be generated by the following sources: electronic-based sources, optical-based sources, and laser-based sources, each of which has different methods of generation of THz radiation (Figure 2.2). The electronic-based devices can produce THz radiation at frequencies up to 0.5 THz that includes resonant tunneling diodes, Gunn diodes, bipolar or field effect transistors (Yu 2021). Generally, these types of electronic devices exert a negative differential conductivity due to the properties of band structure that are limited by transition time behavior to a 0.5 THz upper limit. The output power from device is inversely related with the squared frequency. The next common strategy employed to produce THz radiation is the use of lasers. The direct THz generation is highly tedious because the laser operation in the THz region is restricted by density of quantum states (Smye et al. 2001). Since laser-based sources of THz radiation work based on the transition between the different energy states. Even though the molecular gas lasers are in existence, they are large and complex to handle. The potential use of compact solid-state THz sources is also restricted

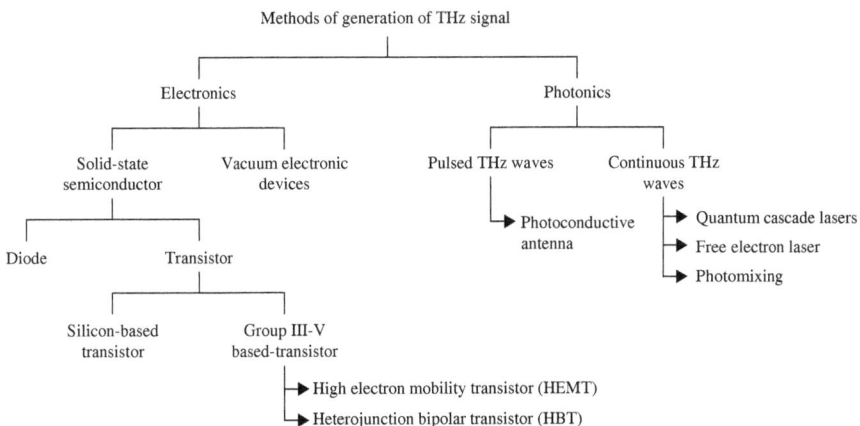

Figure 2.2   Different methods of generation of THz signal.

by the energy density, cost, power consumption, stability, and working range. A two-stage approach for the THz generation is adopted to overcome the difficulty of the direct THz generation using lasers. In such approach, laser source like Cr: Li SAF pumped by a diode or Ti:Sapphire laser pumped by argon laser are commonly employed to generate short pulses of <100 fs (Fortier and Baumann 2019). The generated near-infrared radiation is then directed to photoconductive structures or nonlinear optical crystals. Some of the photoconductive structures that are illuminated by optical femtosecond (fs) pulses are resonant dipoles, large aperture photoconductive antennae, Hertzian dipoles, and dipoles with dielectric lenses (Lavadiya and Sorathiya 2021). The energy conversion efficiency of the optical to THz pulse is lower for most of the techniques except the one that employs femtosecond pulses to illuminate the optoelectronic materials. Most of the THz studies have considered the benefit of the high Q of approximately 10^6 connected with the spectral lines of low-pressure gases (Mittleman 2013). In this regard, most of the laboratory THz spectroscopic systems and corresponding applications require high spectral purity. Strategies to high Q source problem have been addressed with the advancements in nonlinear frequency multiplication techniques and the development of cooled THz detectors, the extended application of electron beam oscillators, use of solid-state oscillators, and basic solid-state oscillators to high frequencies, generation of microwave sidebands on the far-infrared sources, optical heterodyne down conversion, and the demodulation of femtosecond pulse (Consolino et al. 2017). Since most of the THz systems employ cryogenic detectors, the THz systems can be categorized based on the source technology. This includes harmonic generation, mixing of optical sources, electron beam sources, tunable sideband sources, and femtosecond sources. This section provides a brief discussion about various sources and techniques employed for the generation of THz radiation.

2.3.3.1 Harmonic Generation and Frequency Multiplication

Harmonic generation of THz radiation was the base for the early exploration of THz realm of electromagnetic spectrum. The combination of harmonic generation along with the sensitive detection provides a wide coverage of THz spectrum, high resolution, sensitivity, frequency control, simplicity, and overall reliability of system (Picque and Hansch 2019). The early stage of development of THz systems uses cross waveguide harmonic generators that are embedded with nonlinear diode in the microwave structures. These kinds of arrangements are specially designed for the spectroscopic measurements with feature of tuning the back shorts. These multipliers not only have a broad band across the entire waveguide input but also facilitate the simultaneous generation of harmonics. The absorption characteristics of the linear molecules do not possess exactly harmonic relationships. Consequently, the multiple outputs are decoupled and further analyzed by scanning the multiplier's input frequency. This can be done through closely spaced subharmonics of the different absorption frequencies (Picque and Hansch 2019). Similar observations with spectral matching are optimized for narrow spectral range and are operated in low-order multiplication. This kind of THz source remains as the choice for several applications that requires local oscillator power to receivers especially for remote sensing and monitoring. The increased detector sensitivity of the absorption spectroscopy is generally considered for compensating the reduced source power. Thus, the application of sensitive cryogenic detectors in THz systems yields a significant advance in THz spectroscopy measurements (Crowe et al. 2004). The relativity, flexibility, and simplicity of the THz systems pave a way for the wide range of its adoption for spectroscopic and imaging applications. The harmonic generation of THz systems is useful in reading out the spectra of small chemical species including H_2O, CO, O_2, O_3, NO, N_2O, HCN, HNO_3, NH_3, SO_2, HCl, H_2S, CH_3OH, and many others (Peiponen et al. 2012). However, the THz spectral resolution is limited by Doppler broadening with a Q value of ~10^6. Despite this Q range, it is quite adequate for most THz applications, high THz spectral resolution is often considered to be significant (Nagarajan et al. 2017). With modern advancements in the microfabrication methods and the design of broadband circuits, the capabilities of the frequency multipliers are enhanced. This enhances the reliability of the use of these devices in broader range of applications. Most of the recent THz systems often use this approach in addition to THz generation by QCLs.

2.3.3.2 Mixing of Optical Sources

This approach of generation of T waves working on optical mixing of sources has been in practice for a long time. Although the generation of T waves based on this approach is straightforward as the name implies, the practical implementation depends on several factors. This includes the calibration of laser sources, frequency stability, and the efficiency of the mixing process (Pasquazi et al. 2018). A well-known application of this approach is the THz spectrophotometer that works on the basis of the different frequency mixing of the two stabilized CO_2 lasers (Shumyatsky and Alfano 2011). Sometimes, tunable lasers in the range of near-infrared and visible regions are also used for generation of THz radiation as sources based on differential frequency mixing. The THz sources of this kind include photoconductive mixers where the radiating antennas are combined with semiconductor switches. A wide range of frequency control and measurements are quite possible with this kind of THz source since they are operated at higher frequencies of two- to threefolds of magnitude higher than the generated THz radiation. The operation of photomixers is closely related to the frequency multipliers except for the fact that photomixers convert the frequency difference of the two lasers into the submillimeter THz (Kittlaus et al. 2021). These photomixers are operationally more broadband than the frequency multipliers but with a less power at the given frequency. The less operational power is due to the fact that their frequencies are determined by the two larger frequencies. Often the generated frequencies are difficult to calibrate as they have less spectral stability and purity (Consolino et al. 2017). Despite these issues, this kind of THz source best suits for spectroscopy applications working at broader resonances as in case of most solid materials.

2.3.3.3 Electron Beam Sources

The development of solid-state devices especially the electron beam sources had remarkable role in evolution of THz spectroscopy. The use of BWOs in THz systems had increased the broadband power that reflects the capabilities of solid-state THz sources. The BWOs are quite useful in the spectral range of 100 and 1000 GHz (Paoloni et al. 2021). Hence, most of the commercial systems are based on BWOs. The purity of spectrum and remarkable power of BWO sources reduces Doppler limit by making use of saturation spectroscopy. The fast-scan submillimeter spectroscopy technique (FASSST) is based on BWO solid-state electron beam sources. The key characteristic feature of FASSST is the short-term purity of BWO which can be voltage tuned over 50% range of frequency containing $\sim 10^5$ spectral resolution elements (Albert and Lucia 2001). The resulted high resolution THz system can be replaced by optical spectroscopy. Although the longer wavelength decreases the optical precision, greater accuracy in frequency can be achieved. Recently the applications of high frequency and high-power frontiers are realized with the development of vacuum electron beam devices (Sominskii et al. 2018). The use of plasma sources and radar has driven the research potential of THz systems with the development of gyrotrons (Idehara et al. 2020). Among all the other sources, the electron beam sources are key players that had remarkable growth in the design of THz systems with enormous applications in spectroscopy, imaging, remote sensing, and communications.

2.3.3.4 Tunable Sideband Sources

The invention of the gas discharge and optical-based infrared lasers remains as powerful sources for the THz generation. The applications of this type of THz sources are limited because they are essentially fixed frequency sources. Recently, the THz systems relying on mixed tunable microwave sources were developed that considerably increased its spectroscopic applications (Kittlaus et al. 2021). With varying operational conditions, the frequency of the far-infrared lasers would also vary in the range of 1–5 MHz. The provision of stable laser frequency is quite adequate for achieving good accuracy in spectroscopic measurements. The electronic methods of generation of tunable sidebands are an alternative to its generation using far-infrared lasers. The electronic methods are based on tuning of the molecular resonances to a fixed operational frequency using large electric or magnetic fields (Ferguson and Zhang 2002). Since most of the analytical compounds of interest possess dipolar moment, the use of tunable sideband sources based on laser electric resonance (LER) is more general. However, the applications based on laser magnetic resonance (LMR) are also quite common in generation of THz radiation based on tunable sideband sources. This is mainly because of the large tunability of the LMR that are well suited for the characterization of free radical and ions than LER possessing smaller tunability. These THz sources exert exquisite sensitivity because of their fixed frequency and more powerful radiation. However, the greater complexity of analysis of obtained spectra is the resultant characteristic of this method (Tanaka et al. 2011). The applications involving the assessment of the Van der Walls bonds in Ar-HCl complex employ LER sources.

2.3.3.5 Femtosecond Sources

The above-discussed sources of THz generation rely on the production of current pulses represented in time scale as the reciprocal of the generated frequency. This explains the use of femtosecond lasers acting as the driver for the production of current pulses; henceforth it results in generation of THz radiation (Tonouchi 2020). This type of source follows two ways of generation of T waves: use of femtosecond pulse to further generate pulses in a broad spectrum followed by Fourier Transform techniques as employed in Fourier transform infrared (FTIR) spectroscopy and another approach is the use of continuous femtosecond pulses resulting in the production of high-resolution frequencies which can be modulated by scanning the frequency of the input (Baxter and Guglietta 2011). Both these approaches employ photoconductive switch to demodulate the optical pulse to yield THz radiation. The significance and working of femtosecond lasers in THz systems are further discussed in Section 2.4 of this chapter.

2.4 Generation and Detection of THz Radiation

The two main types of the THz spectroscopy systems used for generating T waves are pulsed wave and continuous wave THz systems (Figure 2.3). The former type of the pulsed wave THz systems make use of the electromagnetic pulses (picoseconds) in a broader frequency range of 100 GHz to THz, while the continuous wave systems use THz beam in a single frequency at one instant (Mathanker et al. 2013). The excitation of photoconductive antenna or a laser-based nonlinear crystal (single femtosecond laser) can be used to generate broadband THz radiation. On the other hand, two femtosecond lasers are employed for the THz generation of continuous wave in narrower range. Accordingly, either a photoconductive detector or electro-optical sampling (EOS) module is used for detecting the generated T waves (Bacon et al. 2020). A schematic diagram of different sources of generation of T waves is presented in Figure 2.4. A brief overview of the various methods of generation and detection of T waves based on system components used are presented in this section.

(a)

(b)

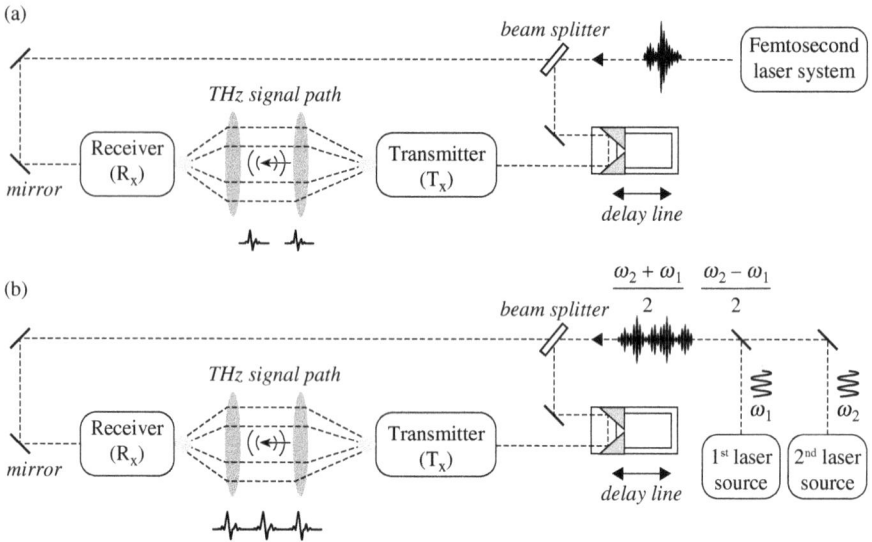

Figure 2.3 THz spectroscopy system: (a) pulsed generation and (b) continuous wave generation.

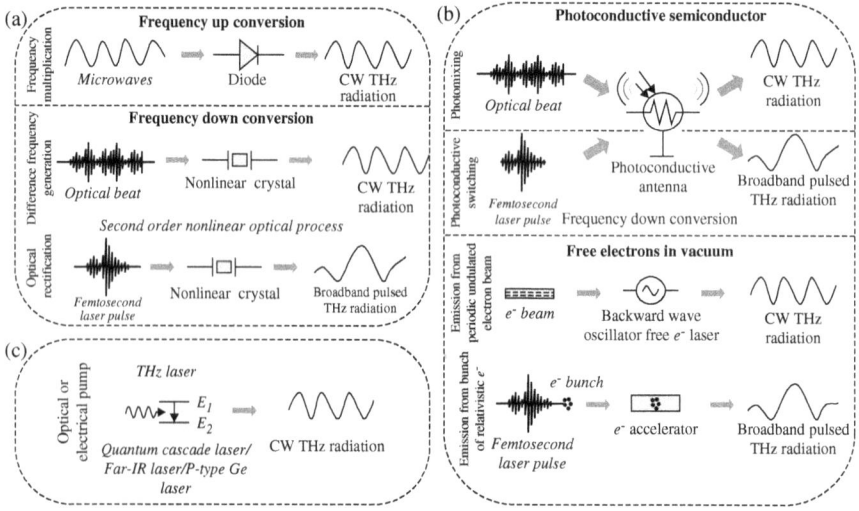

Figure 2.4 Schematic figure showing various sources of THz generation: (a) THz generation in nonlinear media, (b) THz generation from accelerating electrons, and (c) THz emission from laser source.

2.4.1 Photoconductive THz Generation and Detection

The generation of T waves using photoconductive switch involves the impingement of the ultrafast laser pulses (femtosecond) onto the surface of the biased photoconductor (pulsed THz systems). When the metallic electrodes are excited by the femtosecond laser, it generates the THz pulses provided at the bias field in the photoconductive gap. The generated THz pulses are directed to the substrate through the THz dipole antenna (Zhang and Xu 2010a). This antenna resembles a metallic bridge that connects the electrodes. A substrate lens attached to the structure helps in the collimation of the THz beam. The antenna along with the THz generator detects the THz pulses through photoconductive sampling. The photoconductive antenna is also used for the detection of the THz radiation. The two metallic electrodes are connected to the T wave sensors for detection of signals rather than connecting to the power supply for generation of T waves. The required electric field for driving the photocurrent in the THz detector is provided by the incoming THz pulses. Other generation methods of T waves include the optical rectification in a nonlinear crystal, use of semiconductor quantum structures, surging current at the semiconductor surface, and high temperature superconducting metallic bridge (Fujita et al. 2019).

2.4.2 Broadband and Narrow Band THz Generation and Detection

In a broadband coherent T wave generation and detection system, the T waves are generated by excitation of the THz emitter by a laser pulse system. The generated laser pulse is directed toward the sample using an off axis paraboloidal mirror (OPM). Then the transmitted signal from the sample is also passed through the THz detector using another OPM (Takai et al. 2016). A beam splitter used for splitting the laser pulse excited by THz emitter is used to gate the THz detector. Thus, the generated photocurrent by the THz detector can be measured as a THz signal. On the other hand, spatial overlapping of laser beams in a beam carrier would generate a narrow band THz radiation. Later, the overlapped laser beams are focused either on to the surface of the photoconductive antenna or nonlinear crystal. In a continuous wave THz generation system (CW-THz), the lasers must be selected based on the process requirements in order to match the frequency difference in the desired THz range (Wang et al. 2018a). Apart from the difference that exists in the frequency range, functionally CW-THz transmission systems are similar to that of the time domain THz generation and detection. However, the CW-THz systems are more accurate and efficient, compact, and less expensive than conventional time-domain THz systems.

2.4.3 Optical Rectification and Electro-Optical Sampling Methods

The THz pulse generation using optical rectification works by differential frequency generation (DFG) method. The principle of optical rectification relies on the second-order nonlinear optical effect (Li and Li 2020). The THz radiation is produced from the electro-optical (EO) crystal through optical rectification while the detection of the generated THz beam involves the use of the EOS. The EOS is a reciprocal of optical rectification process utilized for THz generation. The detection in this system involves the modification of refractive index of the EO nonlinear crystal by the incident THz radiation. This results in the phase retardation of the linear polarized light and the extent of the retardation helps in the determination of the THz field strength of pulses. The emitted beam because of the interaction of the incident radiation with the sample is combined by the beam splitter. This combined THz signal can pass through the sensor and the polarizer. The Wollaston polarizer is commonly used in optical rectification systems for converting the phase retardation of the THz beam into an intensity modulation (Runge et al. 2020). The intensity-modulated THz signal is detected by a photodiode connected to the structure and the difference signal is applied to the lock-in amplifier to control the pulse processing and to optimize the system performance. In a similar approach, optical rectification is used for the broadband THz generation and detection using a nonlinear crystal. A free-space electro-optical (FSEO) sampling unit is also utilized to detect THz radiation instead of photoconductive antenna (Mathanker et al. 2013). The system can be equipped with either a single laser as in case of time domain THz generation or two lasers as in case of continuous wave THz generation.

2.4.4 Nonlinear Generation and Detection of THz Pulses

In contrast to the broadband THz generation, the nonlinear generation of THz involves the use of FSEO sampling unit for THz detection instead of a photoconductive antenna. Here the generation of THz and its detection is based on optical rectification using nonlinear crystals (Mathanker et al. 2013). With this major difference, the THz system comprises of lasers for its generation. It can be used either as a single laser in case of THz-TDS or as two lasers in case of CW-THz generation.

2.5 Different Forms of THz Spectroscopy

Two different modes of operation of THz spectroscopy are reflection and transmission. The high frequency THz components can be attenuated in the transmission mode while the reflection mode has low attenuation (Mathanker et al. 2013). Compared to the transmission optics, the alignment of the reflection optics THz system is difficult. In the early 1990s, most of the research works on THz spectroscopy were focused on time-domain spectroscopy. Only in recent times, other types of THz spectroscopy like time-resolved spectroscopy, THz gas spectroscopy and attenuated total reflection spectroscopy have emerged. The operation principle of each form of THz spectroscopy is described in this section.

2.5.1 THz Time-Domain Spectroscopy

The use of substrate lenses in the THz systems made it possible to direct the generated picosecond pulses in a specific direction as required. Further, the collimation of THz beams with the use of optics has subsequently led to the development of the THz time-domain spectroscopy (THz-TDS) (Figure 2.5) (Wang et al. 2017b). Thus, THz-TDS is one of the most used THz spectroscopy systems that helps in the measurements of the sample properties in THz frequency range. THz-TDS systems are flexible to be operated in reflection, transmission, and attenuated reflection modes (Figure 2.6). Generally, the working of THz-TDS involves the measurement of reference and sample signals of electromagnetic pulses. The THz pulses are usually shorter than a picosecond. The modifications in the shapes of THz pulses reflect the changes undergone with sample's optical properties. Then the detected pulses undergo Fourier transformation that helps in the calculation of the refractive index and coefficient of absorption (Mohan et al. 2019). Based on obtained dielectric properties, the sample properties like thickness can be easily measured. Compared to FTIR spectroscopy that measures only the intensity of the

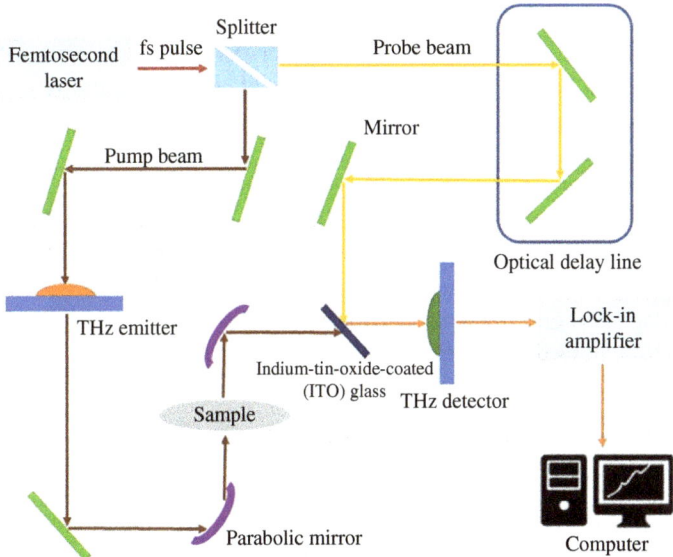

Figure 2.5   Typical experimental setup of THz-TDS system. (Source: Wang et al. 2017b.)

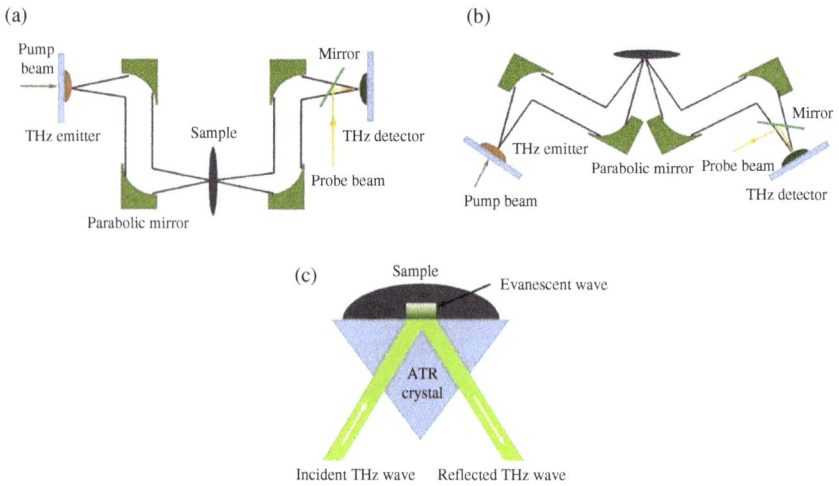

Figure 2.6 Different modes of detection of THz-TDS: (a) transmission, (b) reflection, and (c) attenuated total reflection. (Source: Wang et al. 2017b.)

signal, THz-TDS measures the transient electric fields of the phase and amplitude of the signal (Petrov et al. 2016). This clearly shows the superior sensitivity and dynamic working range of THz-TDS systems than other spectroscopy techniques. The THz-TDS systems can give better output with frequencies below 3 THz while above 10 THz works well for FTIR spectroscopy (Mathanker et al. 2013). The optical information about the samples is analyzed by measuring the dielectric permittivity and conductivity. The propagation of the electromagnetic fields in the matter can be depicted through the coefficient of absorption and refractive index of the sample while the conductivity represents samples with free charge carriers (Ghann and Uddin 2017). THz-TDS allows the measurement of the sample static properties that covers a wide range of applications in diverse fields like molecular spectroscopy, bioassay, sampling, and quality control. The nature of fatty acid composition and the purity of oils have been analyzed using THz-TDS (Karaliunas et al. 2018). Results showed a significant difference in the absorption spectra of pure and degraded oils. The associated nonuniform optical properties were mainly due to oil hydrolysis that determines the extent of the degradation of oils.

2.5.2 *Time-Resolved THz Spectroscopy (TRTS)*

The principle of TRTS (also known as transient THz spectroscopy) involves the evaluation of the photoconductivity in a noncontact manner using electrical probe with a temporal resolution of more than 200 fs. The components of TRTS are nearly same as that of THz-TDS with the exception of one additional pulse laser beam in case of TRTS for splitting the transmitted beam into three parts

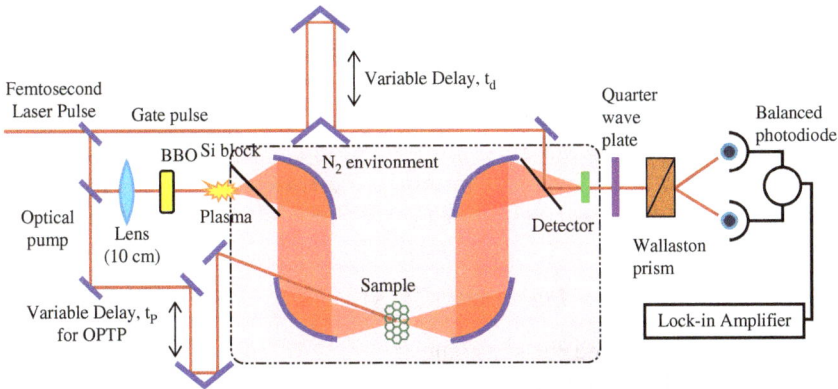

Figure 2.7 Typical experimental setup of time-resolved THz spectroscopy. (Source: Kar et al. 2014.)

(Patil et al. 2022). The additional pump beam is attached to the generation and detection modules. The laser output in TRTS is divided into three beams: generation beam, sampling beam, and the pump beam (Figure 2.7) (Kar et al. 2014). As like the nonlinear THz generation and detection systems, TRTS uses FSEO sampling unit which works on the basis of DFG (Mathanker et al. 2013). Recently, the TRTS has gained a great scope in photo-physics especially in bulk and organic semiconductors, dielectrics at micro and nano levels. The optoelectronic properties of the 2D van der Waal materials have been summarized by Han et al. (2020). In this report, TRTS has been used for investigation of carrier dynamics and transport properties like the exciton formation and relaxation processes of the semiconductor materials. A decreased THz reflection was observed with photoexcitation during the measurements of charge carrier mobilities of thin films with metal substrates using TRTS (Hempel et al. 2017).

2.5.3 THz Emission Spectroscopy (TES)

The TES is a variant of THz spectroscopy systems that employ magnified laser and the detector in transmitters near field. The TES measures the amplitude and shape of the transient electric field discharged from the sample and the analysis is done without focusing the optics for avoiding the artifacts (Patil et al. 2022). The carrier charge dynamics and the emission spectrum of semiconductor gallium nitride (GaN) were measured using TES (Yamahara et al. 2020). The applied non-destructive analytical method for charge determination involves the use of polarity of the amplitude of THz signals to determine the carrier type of the GaN films. The TES is not only used for measuring surface analysis but also used for determination of inter-surface characterization of semiconductor materials (Huang et al. 2019b). The integration of TES with the imaging system is commonly referred to as laser

THz emission microscopy (LTEM), mainly used for characterization of electrical charge carriers and the optical excitation of the materials. When excited with the femtosecond laser, the resulting THz waveforms are useful in analyzing the nature of the electronic materials (Tonouchi 2019). Thus, LTEM has emerged as a novel tool for the determination of electronic state of the materials in material science. Another variation of TES is the spintronic THz emission spectroscopy (STEM) widely applied for near-field THz biosensing and imaging (Guo et al. 2020a).

2.5.4 THz Gas Spectroscopy (TGS)

THz spectroscopy is used for analyzing the mixtures of gases and is referred to as TGS. The detector employed in TGS can differentiate the each gas from a gaseous mixture with absolute specificity, while the conventional gas quantification techniques such as mass spectrometry (MS), gas chromatography (GC), and infrared (IR) spectroscopy have limited specificity in analyzing the gaseous mixtures (Mathanker et al. 2013). The major advantages of using THz spectroscopy in quantification of gas samples are its selectivity and reliability of the measurements. The absorption spectra of gaseous mixtures have been analyzed using TGS at 245 GHz (Li et al. 2020e). In this study, independent component analysis (ICA) was used for recovery of spectral data in predicting the concentrations of individual gas components in the mixture (CH_3OH and CH_3DO). A high-resolution rotational THz spectroscopy was applied for analysis of the polar gas molecules especially in analysis of volatile organic compounds. Hindle et al. (2018) reported a study on the use of THz spectroscopy to detect and monitor hydrogen sulfide (H_2S) during the storage of packed Atlantic salmon. Results showed that volatile compounds like ammonia, ethanol, and mercaptan were considered as suitable indicators with high sensitivity levels while the sensitivity of other compounds like acetone, butanone, and dimethyl sulfide could be improved with two orders of magnitude for its detection limit.

2.5.5 THz Attenuated Total Reflection Spectroscopy (THz-ATRS)

The THz-ATRS is an alternative to the transmission and reflective spectroscopy. In THz-ATRS, the crystal is selected in such a way that the optical properties must match the sample's reflection characteristics. Recently, the applicability of THz-ATRS has been increasing because it allows for the comparison of different operating modes of the THz spectroscopy in selecting the appropriate mode for the material of interest (Wang et al. 2017b). The polar plot of the measured transmission, reflection, and ATR modes can be represented with the x- and y-axis showing the real and imaginary part of the measured sample property, respectively. Based on the optical properties, it was reported that weakly absorbing liquids with low dispersion can be characterized well in the transmission mode while high absorbing liquids like water can be best characterized in ATR mode (Mathanker et al. 2013). The spectral frequency range in the ATR mode is well spread that helps in the ease of processing. On the other hand, the high frequency data in the reflection mode are tedious to work out. In case of transmission, the film thickness of the sample (water) must be <100 μm. It was reported that THz-ATRS is suited well for measuring the dielectric constants of water and sucrose. Thus, the THz-ATRS are very useful in determining the proper operation modes of THz spectroscopy for the materials of interest under study. The advancements of the surface plasmon sensing systems make THz-ATRS to be superior technique in terms of modulation, sensing, analyzing, and imaging (Huang et al. 2020b). A meta-surface enhanced sensing system was developed based on ATRS (Zhong et al. 2020c). Using the developed sensor, the sucrose solution of concentrations from 1 mol/L up to 0.03125 mol/L was detected with a sensitivity of fourfold greater than conventional ATR systems. Thus, this study demonstrates the sensitivity improvement of existing systems with the use of metamaterials for detection of ultrasensitive aqueous samples.

2.5.6 Comparison of Sensitivity in Different Modes of THz Spectroscopy

The three common operation modes of THz spectroscopy are transmission, reflection, and ATR modes (Wang et al. 2017b). Samples with refractive indices lower than that of the prism material have significantly higher sensitivity while operating in ATR mode than that of transmission and reflection modes. This can be explained by comparing the polar plots of weakly absorbing and strongly absorbing THz materials. Jepsen et al. (2011) clearly demonstrated the sensitivity of different operational modes of THz spectroscopy by comparing the polar plots of liquid heptane (weakly absorbing) and liquid water (strongly absorbing). The different plots gave indication of regimes that are useful with different techniques. Thus, the weakly absorbing liquids like heptane characterized with low dispersion are conveniently assessed under transmission mode. The reflection and ATR modes failed in case of weakly absorbing liquids because the optical properties over entire spectral range resulted as a single point in polar plot as depicted by Jepsen et al. (2011). On the other hand, liquids with strong THz absorption like water are more appropriate to be characterized by THz-TDS operated at ATR mode. This was due to distinct depiction of ATR data over the spectral range while the high frequency data of reflection modes were complex to resolve. In addition, this study indicates that maximum thickness of water film should fall around 100 μm for transmission measurements. Thus, the optical properties of the analytes under study are main factors that determine the sensitivity of the operational modes of THz spectroscopy.

2.6 Spectral Data Acquisition and Analysis

The resultant output of TDS is a measure of the magnitude of the electric vector of the THz broadband pulse with respect to time. The delay in time for receiving the signal is directly related to the rate of travel of THz radiation as a distance between the sample compartment and the detector. The time of the arrival of the laser can be computed from the delay line position (Mamrashev et al. 2019). Thus, the computed TDS is characterized by the large intensity burst known as main peak with subsequent amplitude decay of the THz signal. Since the pulse velocity is greatly decreased by the material's refractive index, the optical path of the probe beam is shorter than the pump beam prior to main peak. Before the detection of the main peak, the signal is a measure of background noise. The probe beam path is increased by scanning of the delay line and the received signal at the detector gives to the intensity of generated THz beam at subsequent times (Baxter and Guglietta 2011). The spectroscopic measurements are based on the coherent broadband THz radiation. Hence, the output signal of distributed frequencies resembles a collection of superimposed cosine waves with constructive interference at main peak. With this output, both the THz reference and the sample signals are Fourier transformed to convert time domain data into frequency domain. The resulting frequency domain plot is known as the single-beam spectrum since the reference and the samples are measured individually instead of simultaneous measurement as in case of dual-beam spectrometer. Finally, the absorbance spectrum $A(v)$ (Equation 2.1) can be determined by performing log_{10} of the ratio of the single beam spectrum of sample to reference (Smith and Arnold 2011). The signal averaging is done to enhance the signal-to-noise ratio which in turn helps in reducing the random noise.

$$A(v) = -\log_{10}\left[\frac{I(v)_{sample}}{I(v)_{reference}}\right] \qquad (2.1)$$

To enhance the final spectral quality, the raw TDS signals are preprocessed prior to fast Fourier transformation. Apodization and zero-filling/padding are common signal preprocessing methods used (Sitnikov et al. 2019). The apodization method is generally performed on time domain signal prior to the frequency domain transformation. This process is defined as the weighing of signal which is performed by applying windowing function. The apodization function excludes the artifacts and reduces background noise thereby improves the quality of spectrum. This step of signal preprocessing requires utmost care as sometimes extreme apodization may lead to signal distortion and loss of spectral features. Various kinds of apodization functions available for improving final quality of frequency domain spectrum are boxcar, Hamming, Happ-Genzel, triangular, Norton-Beer, and Blackman-Harris functions (Smith and Arnold 2011). In which the simplest apodization function is the boxcar function (also referred as rectangular or Dirichlet function) where the raw TDS data are multiplied by one so that fast Fourier transform could be applied on the raw unweighted signal. While other apodization functions have diverse pattern of multiplication that helps in enhancement of different regions of TDS and the rest of them are attenuated in magnitude. Often it considers unit value near the main peak and gradually gets decreased toward zero when moved nearer to extremities of the sampled time window. The rapidity with which apodization reaches a minimum value is termed as severity. The function that approaches toward minimum value near to main peak results in a smooth frequency domain spectrum (Galvao et al. 2007). However, such function tends to decline and broaden the narrow absorption bands. Hence, the oscillation of such signal remains distinct at greater distances from the main peak. Although absorbance peak width is increased, the spectral features are smoothed through apodization. In addition, the asymmetric characteristics of the TDS signal can be optimized using apodization function. In which the function can be applied asymmetrically with each end reach a similar value resulting in calculation of function despite of the equal length of pre-peak portion as that of post-peak portion. Thus, the application of stronger apodization function to raw TDS data greatly helps in reducing the high frequency noise during processing of the signal (Zhai et al. 2020).

The second common method of preprocessing of signal is zero-padding (also known as zero-filling or Fourier interpolation. This method is applied when there is a need for more data points within a given spectral range (Sitnikov et al. 2019). Here, the added points do not contain specific chemical information; despite it denotes the interpolation between the actual and relevant points. This eventually results in the spectral smoothening. The process requires addition of zeros to the end-of-time domain spectra before applying fast Fourier transformation. The smoothening effect can be evident in a spectrum by an increase in the number of points with a decrease in point spacing (Smith and Arnold 2011). However, the resolution of spectrum was not improved by adding data points as the modified signal does not contain any true chemical information. Thus, zero-padding is mostly applied to improve peak fitting through simplifying the shape of the spectral features that are captured with spectrometer of normal resolution. Therefore, experimental procedure of material characterization starts with preparation of sample, reference spectrum acquisition, sample spectrum acquisition, signal preprocessing (apodization and zero-filling), and single beam/double beam absorbance spectrum calculation.

2.7 Challenges, Limitations, and Current Research Advancements

The THz science is emerging as a promising technology, especially for wireless communication systems that are complementary to available technology of receivers and transmitters working at high frequency and microwave ranges. This is mainly due to faster data transfer speed that is 100 folds by THz than conventionally available technology (Lepeshov et al. 2017). Because of few hurdles with some of the materials like metals and water exhibiting strong THz absorbance its broader applicability in material science and engineering is limited. However, the characteristic absorbance of T waves by water is very useful in characterizing the materials in bioscience. Until recently the coherent generation of T waves was quite tedious. With gradual exploration in semiconductors and understanding of science of interaction of optical short pulses with semiconductors and nonlinear components has opened up new innovations and compact low power operating THz transmitters have been developed (Berry et al. 2013). The most common method of generation of low power T waves is based on the optical pump conversion involving the use of semiconductor materials with its characteristic surface conductivity. It comprises of two or more conductive electrodes separated by certain gap which has been placed onto a semiconductor surface. Such structure is known as photoconductive antenna if pumped by optical femtosecond pulses. On the other hand, the semiconductor surface is known as THz photomixers when the two laser sources are operated for continuous pumping at closer wavelength with a difference exist only in THz spectral range (Burford and El-Shenawee 2017). The placed electrodes over the semiconductor surface are biased by applied external voltage. The THz pulses are generated upon excitation at the electrode gap in case of femtosecond laser with sharp increase in the concentration of the charge carriers over a short time. Here, the duration of the generated pulse and its spectrum are measured by the charge carrier's lifetime (Lepeshov et al. 2017). On the other hand, the concentration of the charge carriers varies with frequency difference of pump wavelengths resulting in the CW-THz emission. The efficiency of these sources is highly restricted by the amount of optical energy absorbed at the photoconductive antenna gap. Although the energy absorption is small, the high values refractive index of semiconductors at optical frequencies leads to high coefficient of reflection. Henceforth, the performance of conventional photoconductive antennas is reduced by low drift velocities of the optically induced charge carriers in semiconductor substrate and the material breakdown threshold level (Jooshesh et al. 2014).

With recent progress in material science, it is proved that the use of optical nanoantenna (NA) into the photoconductive gap significantly improved THz pulse generation efficiency (Krasnok et al. 2013). Thus, the structure contains an array of optical NAs in addition to the photoconductive THz antenna and is termed as THz optical hybrid photoconductive antenna. Hence, the efficiency of these hybrid photoconductive antennas is usually higher by at least onefold than conventionally used THz antenna. In addition to enhancement of pump field in semiconductor substrate, the plasmonic NAs increase thermal stability of photoconductive antenna (Lepeshov et al. 2018). This is because of large thermal conductivity of metals present in the substrate. Considering THz detectors, the application of hybrid antenna with optical resonant nanoparticles at the photoconductive gap enhances the resolution of near-field imaging (Lepeshov et al. 2017). Another approach is the use of resonant plasmonic structures at the THz frequencies in detectors. The application of plasmonic resonant structures in spectrometers promisingly increases the sensitivity of the THz measurements.

Another major challenge existing with the progression of THz components is lack of availability of efficient and compact sources operating at room temperatures. This includes lack of tunable continuous wave sources with narrow bandwidth required for high resolution spectral and imaging measurements (Tonouchi 2007). Considering this limitation, the semiconductor-based sources are quite favorable in terms of compactness, efficiency, and economy. In this regard, QCLs are used as CW-THz sources; however, to operate at low temperatures, it requires additional cryogenic cooling unit. The highest working temperatures without externally applied magnetic field are 163 K and 186 K at 1.8 THz and 3.9 THz, respectively (Ferguson and Zhang 2002; Kumar et al. 2009). To overcome this issue, photomixing in ultrafast photoconductive semiconductor-based materials is used in THz systems because photomixers work best with high resolution tunability at low frequency (Preu et al. 2011). The performance efficiency of the photomixers depends on the semiconductor material's internal efficiency as well as radiating power of metallic planar antenna. In the early 2000s, available semiconductors and antenna suffered from low efficiency that in turn significantly affected the resulting power of the CW photomixers (Brown 2003). Therefore, the past decade witnessed active research on the optimization of materials used as photomixer, semiconductors, and antenna designs for THz systems. However, very less attention has been paid for improving the photomixers efficiency. Focusing on this aspect, the interdigitated electrode structures are used in active sites for improving the photon to current conversion efficiency (Mikulics et al. 2008). However, the optimization process may lead to loss between either of photocarrier generation intensity or electrode finger separation. Hence, the transit time reduction for photogenerated carriers is required to achieve higher efficiency which in turn obligates reduction in the finger electrode separation. This scheme allows only small portion of photoconductive area to be exposed to the optical laser that leads to less generation of photocarriers. Therefore, a novel active region design of structures could evade the existing limitation and aid in improving the efficiency of CW-THz photomixers.

Recent research on plasmonic metamaterials makes to realize the distinct opportunities in modulating THz domain of the electromagnetic spectrum. The development of novel THz components like perfect absorber, spoof surface plasmon polaritons, collimation of THz QCLs, and frequency iterated filters and modulators are some of the notable examples (Zhang et al. 2020b). The plasmonic electrode structures are promisingly used for significant enhancement of the THz pulsed emitter used in femtosecond laser that works under THz realm. However, the application of plasmonic metamaterials for enhancing THz system performance is more critical, and hence more research should be carried out in developing desired CW-THz sources. Tanoto et al. (2013) reported on the use of highly efficient CW-THz photoconductive antenna based on photomixer equipped with nano-spaced electrodes in active region of semiconductor. This structural tip-to-tip design of nano gap electrode enhances strong THz field that acts as nanoantenna in radiating the T waves. Results showed that the output intensity with the developed structure was about twofold greater than the photomixer with interdigitated electrodes. Hence, a notable improvement was observed on the THz emission bandwidth. Thus, this kind of novel design of photomixers would result in efficient CW-THz sources resulting in compact structures of THz systems for use in spectroscopic and imaging applications.

2.8 Summary

Although THz spectroscopy is at its early stage of development than other spectroscopic methods, it has a great scope to be applied for vast array of quantitative and qualitative measurements across diverse industrial sectors with several applications. The current research progress led to emergence of diverse detection methods, source technology, and optical configurations that take THz spectroscopy to the next level of practical implementation. The time-resolved feature of the THz systems results in retrieving unique spectral data that are hardly possible with CW instrumentation which are often applied in material characterization, quality control, and analysis. This chapter detailed about physics of T waves, design components and instrumentation, source technology, data acquisition, and processing providing better understanding of the science of THz and provides insights on the future perspectives in system design and enhancement of performance efficiency of THz spectroscopy. The data processing of THz systems is applied to raw data followed by Fourier transformation that helps in achieving enhanced signal-to-noise ratio thereby leads to precise spectral characterization using THz spectroscopy. Since the spectral features are highly influenced by the extent of hydration and material properties, more rigorous methods are needed. As like any other analytical methods, the challenges in analyzing solid samples must be overcome with research progress on development of methods with high radiant power. The acquisition of data is affected by scattering and reflection effects which could be reduced by adopting appropriate sample preparation and processing methods. Thus, this chapter outlines the potential prospects of THz radiation in spectroscopic applications and remains as valuable source for gaining better knowledge before moving to real-time applications in subsequent chapters of this book.

Chapter 3

THz Imaging

3.1 Introduction

Terahertz (THz) waves are an integral part of the electromagnetic spectrum that resides in between the far infrared (IR) and the microwave realms with frequency ranging from 0.1 to 10 THz having a wavelength of 0.03–3 mm (Nikitkina et al. 2021). The interactions of THz radiation with the chemical matter molecules such as dipole transitions, hydrogen bonds, and van der Waals forces correspond to weaker interactions (Zhong et al. 2020a). The major applications of THz waves are their use for spectral imaging. The spectroscopic applications of THz employ a femtosecond (fs) laser pulses as in case of THz time-domain spectroscopy (THz-TDS) for the production and detection of time-resolved THz electrical field and capture the spectral information of the analytes using Fourier transformation (Mathanker et al. 2013). Thus, the structural and physical properties of the samples are analyzed and identified based on their characteristic frequencies. This is because the energy levels of macromolecules (molecular vibrations and rotations) lie on the THz band that enables the characterization of chemical molecules through THz spectroscopy (Ferguson and Zhang 2002). While THz imaging employs THz waves to irradiate the samples resulting in the acquisition of the data based on transmission through and reflection from the object. Therefore, THz imaging is not only used for the construction of the geometrical images of the sample but also used to analyze the chemical composition through extracting the optical information of samples under THz region (He et al. 2021). Because of this characteristic feature, THz imaging is receiving a greater research attention. The THz imaging is an emerging quality assessment technique with its applications in safety checks, quality monitoring and control especially through noninvasive image processing procedures (Afsah-Hejri et al. 2019). One of the unique features of the THz waves is its ability to penetrate through packaging materials that provides information about the nature of substances inside packages like detection of harmful explosive materials based on spectral information of the materials present inside (Ren et al. 2019b).

In the agri-food sector, the quality and security of the raw and processed food products remain the top priority among the producers and consumers since it is closely related to the human health and wellness (Wang et al. 2018a). Some of the common food safety threats that are encountered in agri-food processing industries are presence of pesticide residues, foreign bodies, additives, adulterants, and heavy metal traces. The group of foreign bodies includes contamination of food products with stone, peel, seeds, papers, plastics, metal pieces, bones, glass pieces, and so on. These foreign bodies may accidentally enter processed foods leading to severe life threats. Thus, to ensure safe delivery of foods to consumers, it is quite important to check the final quality of products by the manufacturers before dispatching. Hence, it is quite necessary to monitor the final quality of processed foods using fast, accurate, and reliable qualitative or quantitative methods that could distinguish the foreign bodies from the foods (Feng and Otani 2021). Various imaging techniques such as X-ray imaging, electrostatic techniques, ultrasonic imaging, thermal imaging, fluorescence imaging, and IR imaging are currently used for non-destructive assessments of biological materials (Vadivambal and Jayas 2016). All these techniques have their own advantages and disadvantages in terms of compatibility with samples and versatility of imaging systems. However, no single technique can meet all the detection requirements. For instance, the identification of low-density foreign objects in a food matrix is difficult with X-ray imaging and also the use of ionizing radiations of X-rays can be harmful to analysts and cause damage to the samples (Wang et al. 2020b). Similarly, it is nearly impossible to distinguish the samples with similar characteristics such as color, shape, and volume using machine vision systems. The lower penetration depth of the IR radiation limits its uses in the agri-food sector. Thus, exploration of novel emerging non-destructive imaging method is the pressing need for research and development community and industry to overcome the drawbacks of conventional imaging approaches.

The application of THz radiation in imaging is a novel emerging technique that helps in the characterization of agri-food materials. The THz waves also are referred to as "T waves" and have the characteristic of both the submillimeter waves and the fingerprint spectra of the IR waves. Further, the THz energy is far lower than X-rays in the range of 0.41–41 meV (Jiang et al. 2016). Hence, the THz waves do not damage samples due to photoionization and are relatively safe to operators. Thus, the characteristic features of THz waves such as fingerprinting, penetrability, and safety have resulted in a great range of spectroscopic and imaging applications in the agri-food sector (Sun and Liu 2020b). This chapter mainly focuses on the principles of THz imaging, THz imaging systems and design components, different THz imaging techniques, and tomographic imaging methods based on the THz waves. Further, specific applications of the THz imaging in the pharmaceutical field, which is similar in scope to the agri-food sector, are highlighted. Various advancements and ongoing research progress in THz imaging are also briefly presented. Finally, this chapter provides valuable information for future research directions exploring the applications of THz waves for noninvasive contact-free detection and quality assessment of agri-food materials.

3.2 THz Imaging – System Components and Operating Principle

The THz imaging technique relies on potential use of the THz waves for the detection and identification of foreign bodies, harmful materials like concealed explosives, illicit drugs, and other chemical and biological agents by exploiting its characteristic properties of reflection or transmission spectra in the THz range (Kundu and Pragti 2022). In addition to the information as obtained by optical images, THz imaging also provides the spectral information in relation to depth. The THz waves are very capable of penetrating through dry, nonmetallic, and nonpolar materials like paper, cardboard, plastics, and organic substances (Yan et al. 2006). The absorption is governed by the optical photons in case of dielectric materials which in turn depend on the material's optical polarity and phonon resonance. While the metals are not transparent to THz waves and polar liquids like water strongly absorbs the THz waves. Due to these spectral properties, the THz waves are envisaged to have a broad range of biomedical imaging applications. This includes characterization of skin, muscles, fat, and other tissues to produce *in-vivo* images for diagnosis (Samanta et al. 2022). However, the THz imaging applications in the agri-food sector are in their infant stage and have a long way to go. Food scientists and engineers are putting forth great efforts in exploring the applications of THz imaging, especially for design of system components. This section describes the principle of THz imaging with consideration to system components and other accessories.

The THz imaging system uses the same concept as that of TDS with an additional imaging system for image acquisition and data processing (Figure 3.1). A

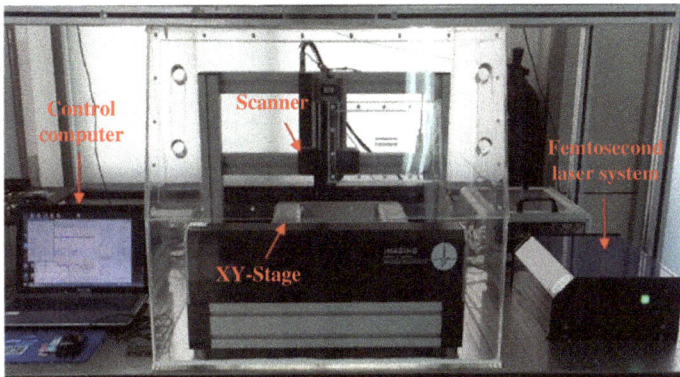

Figure 3.1   Photograph of fiber-coupled time-domain spectrometer. (Source: Wang et al. 2019b.)

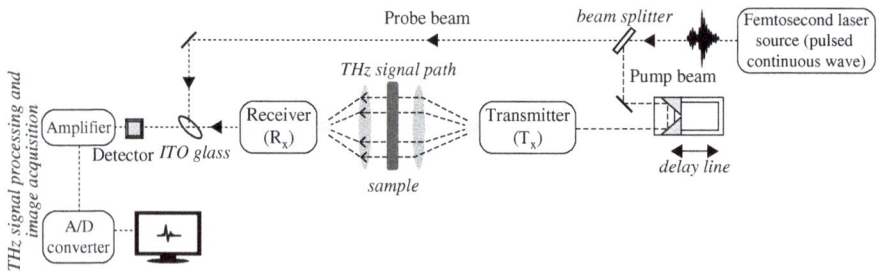

Figure 3.2 Scheme of a typical THz imaging system.

typical transmission THz imaging system consists of a THz source, THz emitter, parabolic mirror, lens, splitter, THz detector, preamplifier, A/D convertor, digital signal processor (DSP), and display unit (Zhang et al. 2021b). The capturing of THz images starts with the placement of samples on XY moving stage (Figure 3.2). Then the beam is directed to sample using a set of parabolic mirrors followed by transmission of THz radiation through the sample. The beam transmission is based on the sample thickness. The resulted amplitude and phase are obtained using different sets of paraboloids and then directed to optically gated THz dipole detector. The transmitted beam is amplified using preamplifier where the signals are digitized and processed by DSP during its flight for each pixel (Qin et al. 2013). Thus, the image of the sample could be captured using transmittance-based THz imaging system. A similar kind of setup as that of transmittance system is used for reflection imaging with exceptions of few modifications to the THz beam path (Figure 3.3). Although transmission-based THz scanning system provides

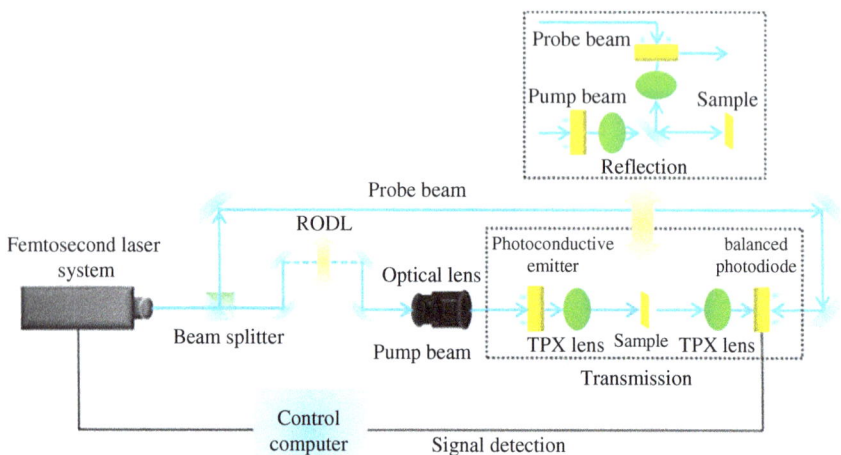

Figure 3.3 THz-TD imaging setup in transmission and reflection operation modes. (Source: Wang et al. 2019b.)

information of the time domain spectra through pixel-by-pixel scanning, the process takes a long time and resulted image has poor spatial resolution due to long THz wavelength. With advancements in THz spectroscopic and imaging systems, the problem of poor spatial resolution and time-consuming procedures are being solved with the emergence of continuous and near-field imaging systems (Wan et al. 2020).

Based on the operating principle, the THz systems are broadly classified as THz pulsed and continuous wave (CW) systems. The pulsed systems are also termed as time domain (TD) systems. The four main components of TD systems are ultrafast pulsed laser which is used as primary THz source that is capable to emit sub 100 fs pulses, THz emitter, a time delay stage, and THz detector (Zappia et al. 2021). The two major technologies that are commonly employed for the THz pulse generation and detection are photoconductive antenna and electro-optic crystal (EOC) (Zhang et al. 2021d). Even though the approaches based on array of detectors are in practice, one emitter and one detector are used in a typical THz system. The generated ultrafast laser beam from THz source can be of two types pump and probe beams. The THz pulses produced by the pump beam affect the emitter while the THz pulses from probe beam are applied to gate the detector. The THz pulses from probe beam are focused onto the sample using mirrors or lenses. Then the THz waves ejected from the sample after the interaction is collimated and refocused on the THz detector. The offsets of the pump beam and probe beam are carried out using the delay stage which permits the THz temporal sampling of signal. On the other side, the CW-THz systems are also referred to as frequency domain THz system that relies on the difference in the wavelength of two laser beams that lie in THz region. Semiconductive photomixers are used for excitation and detection of THz radiation (Yang et al. 2016b). The THz radiation is discharged into the free space when the photomixer is attached to antenna. The emitted CW-THz signal is then collimated and guided as like that of the pulsed THz systems. The TD systems are comparatively complex, heavy, and expensive than CW systems. However, the TD systems can work in a wider bandwidth of 0.1–6 THz that collects the information in the form of signals as a function of time (Zappia et al. 2021). Further, the TD system allows to reconstruct the gathered information about the sample through the use of sophisticated data processing procedures. In contrast, the CW-THz systems use a narrow band that gathers only the signal's intensity. These signals are stored in a form of matrix and are transformed into a raster image. Thus, the CW systems are more effective in generation of real-time THz images. However, the CW-THz systems are less flexible because of inaccessibility of some features and information about the sample. Notably, the TD-THz systems also include the depth information because of variation with pulse timing. Thus, the measured peak deviation of temporal location represents an optical path length change from emitter to sample and then from sample to detector. The depth information of samples can also be measured in CW systems but only through the application of additional detectors or by repeated scanning procedures (Karpowicz et al. 2005). In such case, the scattered wave can be measured instead of the specular reflection by angle change between the emitter and the detector. The assessment at different angles by adopting this procedure allows to collect intensity data that accounts for scattering and reflection phenomenon at varied depths. This in turn assists and retracts information about the image features of the measured sample in a layered fashion. Also, both the TD and CW systems can be used in either transmission or reflection modes based on the THz interaction with sample (Wan et al. 2020). The choice of selection of pulsed or CW systems relies on the end applications of the research or application interest. The TD-THz systems are mostly preferred for quality monitoring in the agri-food sectors due to noninvasive procedures for detection of foreign bodies. This is mainly because the TD systems provide an absolute information and aid in drawing a 3D assessment of the samples (Zappia et al. 2021). However, the CW systems are very useful in obtaining a 2D image and are good enough for applications like in-line monitoring of packaging materials. A detailed discussion on various forms of THz imaging systems is provided in the following sections of this chapter.

3.3 THz Imaging Techniques

Imaging using THz involves analyzing the transmitted and reflected spectral information about the sample. A typical THz imaging system consists of THz optical system, light source, THz detector, image acquisition system, and computer for data interpretation. THz imaging has the following advantages over the traditional imaging systems: THz images have higher spatial resolution that is very useful in material identification and component analysis, THz imaging is noninvasive procedure that is safer to use for biomaterials as it does not damage samples due to photoionization, and THz images also give spectral information describing frequency range and spatial density (Gong et al. 2020). This section describes in detail about various forms of THz imaging techniques.

3.3.1 Pulsed THz Imaging

The discovery of the electromagnetic pulses in the late 20th century marked the emergence of use of THz optoelectronics with generation of pulses having width of picosecond (Auston et al. 1984). Since then, there has been a great progress in the design of system components that are used in pulse generation and detection. The pulsed systems are capable of generation and detection of the transient electromagnetic pulses. The pulsed imaging technique is used for transforming spatial information of the THz signals into electrical signals through detection of digital pixels (Wang et al. 2018a). Later the transformed image is then translated into images through appropriate image processing. The THz image pixel is taken in TD with high signal-to-noise ratio (SNR). The fundamental of THz pulsed imaging method is the TD imaging (D'Arco et al. 2020). Based on TD imaging, the other methods such as THz near-field imaging and THz real-time imaging have been developed with improved imaging performance.

3.3.1.1 THz Time-Domain Imaging

The advent of TDS inspired the emergence and the growth of THz imaging techniques. The first obtained THz image was based on THz-TD imaging (THz-TDI) system that consisted of a typical THz-TDS with additional imaging facility, i.e., 2D translation stage and image processing unit (Hu and Nuss 1995). The basic components of THz-TDI system consist of fs laser, THz emitter, time delay stage, and detector (Figure 3.4). The process starts with the splitting of the fs laser beam using a beam splitter at a specific energy ratio into pump and probe beams. In which the pump beam is directed to produce THz ultrafast pulses by excitation using photoconductive antenna (Wang et al. 2018a). After the sample interaction, the THz pulses from the detector module are integrated with probe beam using indium-tin-oxide glass and directed to THz detector. The detector of THz system comprised of a zinc telluride (ZnTe) nonlinear crystal, a quarter wave plate, photodiodes, and a Wollaston polarizer that allows the realization of instantaneous recording of the THz electrical field through electro-optical sampling (EOS) method (Runge et al. 2020). The THz temporal profile is assessed with the movement of the time delay stage at varied time offsets between the pump and probe beams. This results in deriving THz pixel of the sample. Thus, THz image is acquired by pixel-by-pixel scanning of the sample. The components like preamplifier, A/D converter, and a display screen are employed to further process the signal. In addition, some of the THz systems are equipped with movable source and a detector that move synchronously with a fixed sample in a raster scanning approach (Ok et al. 2019).

Based on the properties of samples and geometric constraints, the detection module has different configurations that allow to acquire THz images in different detection modes such as reflection, transmission, and attenuated total reflection (ATR). The transmission mode is more commonly used and is more apt for imaging samples that are moderate or weak absorbing and weak scattering. On the contrary, the samples with highly absorbing properties are measured under reflection or ATR modes. Specifically, the ATR mode is well suited for strongly absorbing liquids (Wang et al. 2018a). The THz image can be formed by adding the number of pixels through data processing. This is followed by the feature (parameter) extraction process which starts with a single pixel to visualize THz image. The basic feature extraction procedure is briefly described as follows. When the two THz-TD waveforms are propagated through a sample and a reference under a transmission mode, these get recorded. With appropriate Fourier transformation of these waveforms, the amplitude and phase signals are determined, from which the complex dielectric constant of the sample is measured. For instance, considering the weak absorbing sample illuminated at the normal incidence, both refraction index and the

(a)

(b)

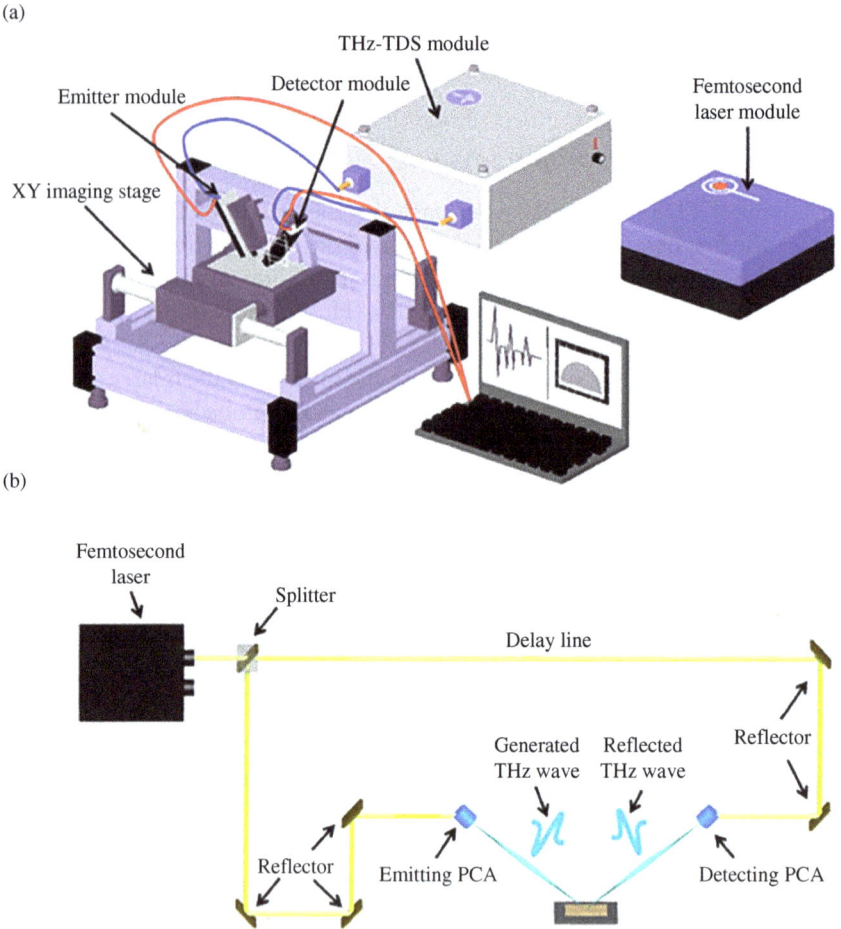

Figure 3.4 THz-TD imaging setup: (a) representation of THz-TD imaging setup and (b) schematic view of integral components of THz-TD imaging system. (Source: Ryu et al. 2016.)

absorption coefficient are calculated using the following equations (Equations 3.1 and 3.2) (Jepsen et al. 2011).

$$n(\omega) = 1 + \frac{\Delta\varphi(\omega)c}{\omega x} \qquad (3.1)$$

$$\alpha(\omega) = \frac{2}{x}\int\left[\frac{4n(\omega)}{\rho(\omega)\big[n(\omega)+1\big]^2}\right] \qquad (3.2)$$

where, $n(\omega)$ is sample's refractive index, $\alpha(\omega)$ is sample's absorption coefficient, $\rho(\omega)$ is amplitude ratio of reference and sample, $\Delta\varphi(\omega)$ is phase difference between the reference and sample, ω is angular frequency, x is sample thickness, and c is speed of the light in vacuum. In a similar way, the data analysis and feature extraction methods are followed for other detection modes. Likewise, the TD and frequency domain of all the pixels of the sample can be obtained, from which the THz image can be obtained from different portions of the retrieved data that give the overall characteristics of the sample. Thus, the sample composition and chemical distribution are identified from frequency domain information based on parameters like coefficient of absorption, amplitude, and dielectric constant at particular frequency (Qin et al. 2013). The refractive index change results in variation in THz phase pulse from which the sample is profiled.

Advantageously the image could be visualized based on TD information that would be otherwise impossible with the conventional optical imaging techniques where it records only the information related to intensity. Thus, the depth information of each and every pixel can be measured based on reflection through time of flight imaging approach (Xu et al. 2021). In a similar way, the other characteristic parameters such as peak area and peak-to-peak amplitude ratio are also used for visualizing the THz images. In brief, TDI is an extended form of the typical THz-TDS system equipped with components that acquire information from a single point to a 3D space. It is possible to obtain time, space, and frequency domains by gathering information from entire pixel elements as an electrical signal and reconstructing them into an image through its characteristic parameters (Chen et al. 2018). The THz-TDI is an attractive noninvasive approach for obtaining information about the sample composition and chemical distribution making use of features of THz radiation.

3.3.1.2 THz Real-Time Imaging

A typical TDI process is time consuming because the system scans object in a pixel-by-pixel manner through focusing THz beams. To overcome this limitation and to extend its practical applications, faster scanning can be performed which yields multichannel assessments using parallel elements for THz wave generation and their detection (Karpowicz et al. 2005). However, the higher cost of infrastructure for faster scanning procedure limits its practical applications. In late 20th century, a 2D electro-optic imaging based on THz real-time imaging system was developed (Wu et al. 1996). This system restricted relative movement between the THz beam and sample that allowed to save time to a large extent during coherent detection. A real-time THz imaging operated in transmission mode resembles same as that of THz-TDS system with fewer differences as explained. In the pump channel of the real-time imaging system, a concave and convex lens positioned in front of the time delay stage for expanding pump beam (Wang et al. 2018a). Also, two polyethylene lenses are used in real-time imaging for focusing THz beams to pass through the sample instead of pairs of parabolic mirrors that are commonly used in THz-TDS

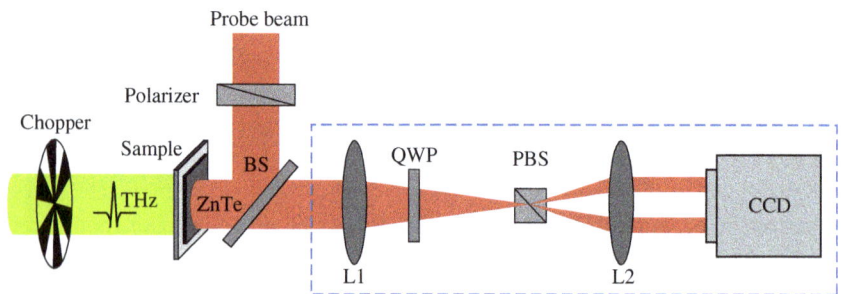

Figure 3.5 **THz real-time imaging setup with balanced electro-optic system (QWP – quarter wave plate, PBS – polarized beam splitter, BS – beam splitter, and L$_1$ and L$_2$ are two achromatic lenses). (Source: Wang et al. 2010b.)**

system. In a similar manner, concave and convex lenses are applied to enlarge the probe beam within the probe channel. Considering the construction of THz image, a THz imaging system utilizes 2D EOS with a charge-coupled device (CCD) camera that is technically varied from common electro-optic unit adopted in THz-TDS system (Wiegand et al. 2010). The transmittance axis of one polarizer is placed in perpendicular to another polarizer in front of the CCD camera within the probe channel. When the polarized linear probe beam is allowed as a sequence to pass through ZnTe crystal and polarizer, the THz beam along with sample information causes a change in polarization. This induced polarization change is converted into intensity change and gets recorded by the CCD camera. Thus, the stored information gives the instant details of entire pixels instead of scanning them pixel-by-pixel. This pixel information reveals the characteristics of the sample (Figure 3.5). Only the THz signal from the sample is detected by eliminating the background noise during the measurement of signals. Also, the measured signal is linear with the electric field during real-time imaging. It also allows the measurement of THz temporal profile through relative movement of time delay stage (Wang et al. 2018a).

In contrast to THz-TDI, the real-time THz imaging does not follow 2D scanning of the sample which saves lots of time during imaging process. In addition, the scan information in temporal dimension can be reserved without the loss of TD information (Wang et al. 2010b). However, the size of the sample's image from a single measurement during THz real-time imaging is relatively small due to limitations with intensity of THz source. Thus, in order to improve working performance and promote the practical applications, more research works on the development of a high-power THz source and well-optimized optical configurations are required.

3.3.1.3 THz Far-Field Imaging

In far-field imaging, a pulsed beam source is used for scanning the objects and the transmitted beam is recorded using a detector that is placed away from the sample (Figure 3.6). This results in a complete spectrum of information about the sample

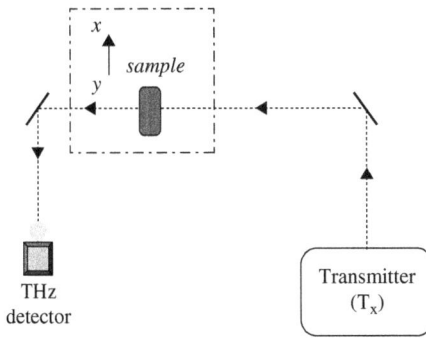

Figure 3.6   A simple configuration of far-field THz imaging system.

at each scanned position (Mathanker et al. 2013). The image of the scanned object can be obtained by measuring the absorption strength of the characteristic THz frequencies assigned to the resulted pixels of the spatial data (Ghann and Uddin 2017). Thus, the image of the object under study can be scanned through THz far-field imaging systems at a particular band frequency. Here, spatial resolution is restricted by diffraction in the orders of the THz wavelength of beams used. Hence, the spatial resolution is represented as a definite spot of THz focus used during the analysis. Huang et al. (2019a) reported on far-field super resolution imaging in microwave and THz regimes. In this study, surface plasmon polaritons (SSP) were used as sensitive probe for the extraction of spatial harmonics in near-field mode. Later, the information was sent to the far field for the sampling of broadband spatial spectrum of the target. From the collected data, subwavelength image can be obtained by inverse Fourier transformation of the sampled spatial spectrum. Thus, the far-field subwavelength imaging based on spatial spectrum sampling has greater imaging ability that could further be improved through optimization of the SSP structures. The reduced resolution of the obtained image is the main hindrance of far-field imaging (Zhang and Xu 2010a). In order to improve the resolution of the scanned image, alternative techniques such as THz near-field imaging are commonly employed in agri-food industry.

3.3.1.4 THz Near-Field Imaging

The imaging speed and the spatial resolution are the two significant factors that determine performance of THz imaging system. In general, the THz source used in the THz systems has a wavelength in the range of 0.1–3 mm. Due to the diffraction limit, the resulted smallest resolvable feature of THz image is not less than its wavelength (Chan et al. 2007). This limitation affects spatial resolution of THz image only up to 0.1 mm. The resulted poor resolution restricts the THz imaging applications especially for non-destructive assessments where longer wavelength sources are commonly used. To overcome the diffraction limits, near-field

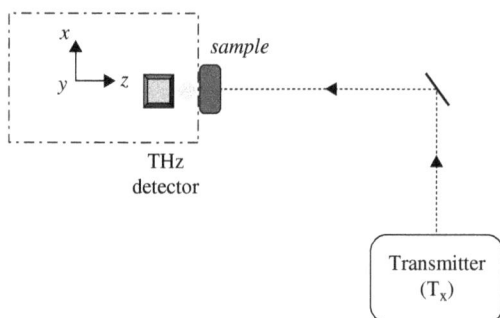

Figure 3.7   A simple configuration of near-field THz imaging system.

imaging approach is introduced for improving spatial resolution of the THz images (Mathanker et al. 2013). A near-field probe system is used for raster scanning of the objects in the near-field imaging. A detector module is attached immediately next to the sample that assists in the detection of the transmitted THz beam with a relatively higher resolution (Figure 3.7). Thus, the limited resolution of the far-field imaging could be overviewed using near-field imaging by applying measurements of transmitted THz waves under a sample proximity (Wang et al. 2018a). So, the spatial resolution of scanned object is determined by probe interaction with THz field and not by analyzing the THz spots dimension. Some of common THz near-field techniques that have been in practice are tip-based near-field imaging, aperture-based near-field imaging, and highly focused beam techniques. A typical schematic diagram of THz quasi near-field real-time imaging is represented in Figure 3.8.

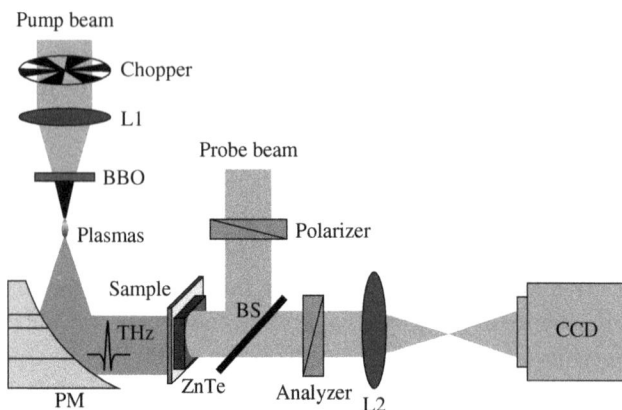

Figure 3.8   Scheme of THz quasi real-time imaging system with near-field configuration (BS – beam splitter, BBO – β-barium borate crystal, and L_1 and L_2 are two convex lenses). (Source: Wang et al. 2009)

There are two kinds of near-field setups in consideration with the type of components used. The first kind of THz near-field imaging setup is based on TDI with a hollow metallic aperture placed in front of sample (Mueckstein et al. 2011). The aperture resembles a metallic narrow tube with subwavelength opening and subwavelength distance to the sample. This allows an interaction of THz waves with samples before it gets diffracted. Thus, the near-field images are obtained by focusing the THz beams through aperture thereby scanning the samples in 2D. Hence, the size of the aperture is used for determining the image resolution rather than the wavelength of THz source. This enables capturing of THz images with subwavelength spatial resolution. Here the aperture is considered as a metallic waveguide with a well-known cut-off. Therefore, a desired spectrum obtained through near-field THz imaging should neglect the cut-off point. Efforts were made to improve the spatial resolution at the desired wavelength through optimizing the aperture shape and size to yield a 3 μm resolution (Mitrofanov et al. 2017; Wang et al. 2018a). An aperture-based dynamic near-field THz imaging was demonstrated by allowing the beam optical gating onto a semiconductor substrate. The second setup of near-field imaging is based on utilizing THz near-field detector as an integral component (Adam 2011). Thus, a typical near-field THz microscopy relies on THz coherent pulse emission and detection by photoconductive antenna. Here, the THz electrical field is directly detected using a photo-gated receiver antenna instead of collimating and focusing THz beams as in case of THz-TDS system. A photo-gated antenna is in sliding contact with the sample. By adopting this approach, the spatial image resolution of the images can be enhanced to ~25 μm at 600 μm wavelength. Other than these two methods, sometimes sharp metallic tips are used for THz near-field imaging that improves spatial resolution by 100 nm (Wang et al. 2018a).

The working of the typical TDI systems could be improved through adopting the near-field imaging approach resulting in higher resolution. The near-field imaging allows to obtain images of minute sample targets. Hence, THz imaging at higher spatial resolution is considered as a better alternative tool for imaging samples in biomedical research. This kind of imaging by near-field is apt for both the pulsed and CW-THz imaging (Kyoung et al. 2009). A near-field THz imaging was used for noninvasive detection of hidden objects using single element detector (Stantchev et al. 2016). Results confirmed the possibility of the detection of surface fissures and deformations under subwavelength resolution in range of micrometer size. Thus, THz near-field imaging has a great scope in *ex-vivo* imaging of materials.

3.3.2 Continuous Wave THz Imaging

The CW imaging is another form of imaging where the femtosecond laser pulse source is replaced with the photomixers (Figure 3.9). The concept of CW imaging was first demonstrated based on the photomixing of lasers (Kleine-Ostmann et al. 2001). With immense research growth, the CW-THz has been widely adopted for imaging due to its low cost, high power, and high spectral resolution. The CW-THz

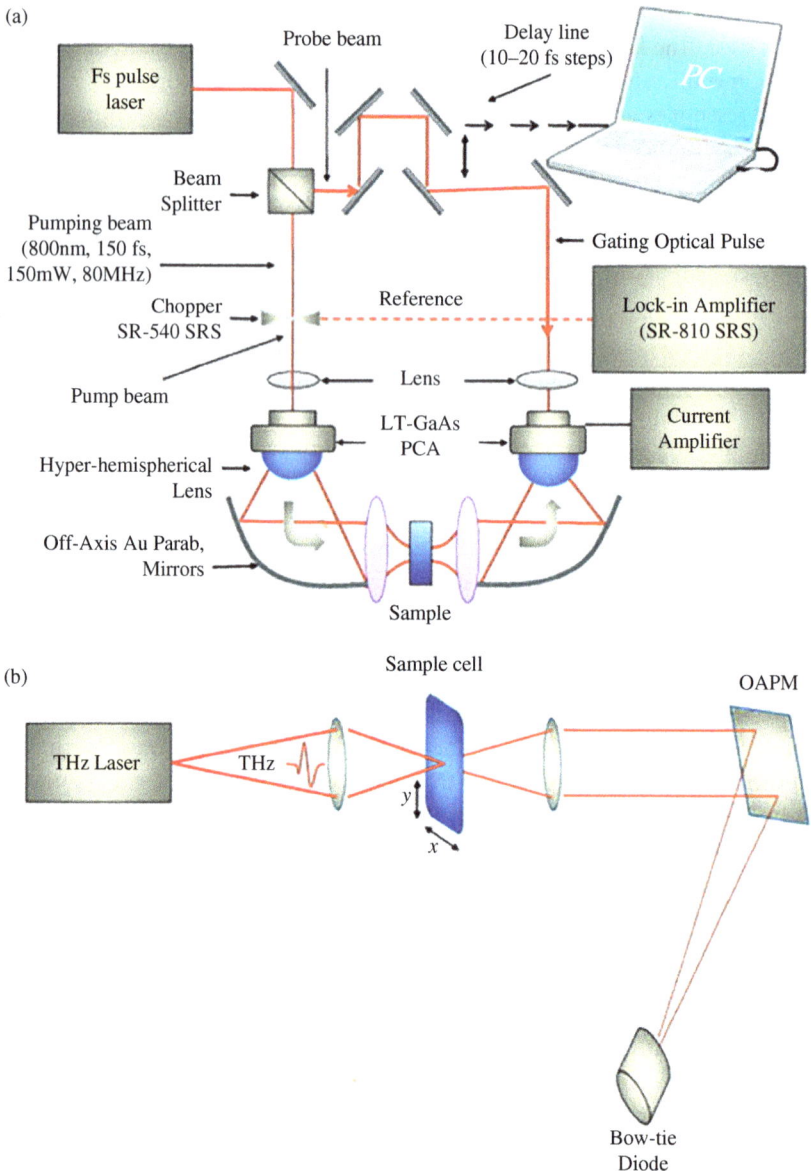

Figure 3.9 Schematic diagram of THz imaging system: (a) pulsed TD THz imaging setup and (b) CW-THz imaging setup. (Source: Wahaia et al. 2011.)

system captures image at different local points on the sample's surface through assessing the retrieved THz intensity information. This is followed by the storage of information in matrix form and subsequent integration to obtain the sample's 2D image (Karaliunas et al. 2018). The former techniques of near- and far-field imaging are more suitable for THz pulsed imaging. Often, the time-domain imaging requires scanning in three-dimension (3D), comprising of one temporal and two spatial dimensions (Feng and Otani 2021). The 3D scanning consumes more time and in cases where the spectral information of the object is not required, the THz imaging is done by fixing temporal delay through recording signal amplitude. Thus, the CW-THz imaging is also applied to obtain similar kinds of images through monitoring the transmitted and reflected signal intensity at the considered THz frequency (Mathanker et al. 2013).

Comparatively, the CW imaging is relatively less expensive with faster spectral acquisition than the pulsed wave imaging systems. Lee et al. (2012) worked on the use of CW imaging for detection of high- and low-density objects in instant noodles. This study demonstrated a noninvasive approach for accurate detection of the cricket and maggot pieces that are generally not visible under X-rays. However, fine details about the high-density objects such as granite and aluminum pieces cannot be observed using CW-THz imaging due to diffraction of waves. In order to improve the spatial resolution and SNR ratio, a horn-type antenna was introduced into the THz imaging system (Kim et al. 2012). The addition of antenna increased the transmitted power by sixfold that improved the accuracy in the detection of stones, metal pieces, paper, and insects in the flour and instant noodles. The developed method has a higher sensitivity with a spatial resolution of the captured image up to 500 μm. In general, the THz radiation could be detected either coherently or incoherently with both the pulsed and CW-THz imaging systems. In a coherent detection, THz amplitude and phase information of electrical field is measured. On the other hand, only the intensity information is measured in an incoherent detection technique.

3.3.2.1 Coherent Detection of CW-THz

A typical CW imaging setup facilitated with coherent detection comprises of two independent diode lasers that emit the two laser beams. These lasers are used to produce CW radiation by photomixing principle in a photoconductive device (Safian et al. 2019). The tuning of the difference in the wavelengths of two lasers determines the final wavelength of the emitted CW-THz radiation. Meanwhile, the THz waves are detected by another photoconductive device based on second photomixing using the two CW lasers. This series of CW-THz generation and detection are considered to be up-conversion of THz beam with subsequent homodyne detection (Wang et al. 2018a). A CW-THz image can be obtained by working on the time delay stage and sample in one temporal and two spatial dimensions, respectively. Thus, the THz phase of the emitted radiation could be obtained through moving time delay stage. A rapid phase modulation technique (100 kHz) was adopted in a CW-THz imaging system for achieving a higher imaging speed. This has resulted in coherent detection of THz waves at a scanning speed of ~3-fold magnitude which is relatively faster than other available mechanical approaches (Sinyukov et al. 2008).

A pulsed source TDI system possesses a broad spectral band in a single pulse. Contrary to pulsed system, the CW imaging system operates at the main wavelength and intensity with coherent detection. Therefore, the phase details could be obtained at the operating wavelength by a single measurement. Hence, the bandwidth of the THz spectrum in a CW imaging is directly influenced by the tunable wavelength of CW lasers (Liebermeister et al. 2019). The multiple measurements with optimal tuning of the associated wavelength are done for obtaining desired broadband spectral information. However, this procedure is more time consuming because a four-dimensional (4D) scan includes a 1D frequency domain, time domain, and 2D spatial domains. This 4D scanning procedure is essentially needed in a CW imaging for obtaining the same amount of features as that obtained from the THz-TDI. As an additional feature, the CW imaging possesses good spectral resolution at the MHz avoiding use of expensive fs lasers whilst ensuring not to lose the capability of obtaining the amplitude and the phase information across a narrow band (Wang et al. 2018a). Hence, the CW imaging with coherent detection can be applied to determine the amplitude and phase of the signal at required wavelength. Notably, the pulsed THz imaging systems are better suited for applications where amplitude and phase information are required over a broader bandwidth.

3.3.2.2 Incoherent Detection of CW-THz

The incoherent CW detection is a better alternative than other imaging techniques for applications where only the intensity information alone is required. The first CW system with incoherent detection was demonstrated with the use of a bolometer as THz detector with a low SNR (Kleine-Ostmann et al. 2001). Later with the advancements in THz research, novel designed efficient THz sources and detectors were assessed for improving the performance of the CW-THz incoherent detection systems. The CW beam is discharged from a Gunn diode oscillator in a typical incoherent CW imaging system and detection by a Golay cell or Schottky diode. Here, the image is generated through pixel-by-pixel scanning of the sample. Unlike other imaging systems based on synchronous laser, the source and detectors used in CW-THz incoherent imaging system are independent of each other (Karpowicz et al. 2005). Since the detection is incoherent, the time delay stage can be avoided. Researchers have used high-power sources like quantum cascade lasers and backward wave oscillators in a CW-THz imaging systems (Rothbart et al. 2013; Delfanazari et al. 2020; Wang et al. 2020b). Regardless of system configuration, the CW-THz incoherent imaging system is quite simple than other imaging systems. A typical CW incoherent imaging with reflection geometry is used for obtaining the intensity information at single frequency because of limited spectral tunability of the Gunn diode oscillator. Considering broadband spectra and phase, a CW incoherent imaging seems to be advantageous in terms of high spectral resolution, high power, high imaging speed, compactness, and low cost (Wang et al. 2018a). In addition, it is possible to select the CW source wavelength at a single frequency which helps in avoiding the water absorption lines thereby obtaining an image at a stand-off distance.

3.3.3 Summary of Imaging Systems

The main differences among the THz imaging techniques are in their ability and methods used for obtaining the information from time and frequency domains. The difference exists with the type of THz source and corresponding system configuration. All the information can be obtained by a single measurement in case of pulsed THz imaging (Karpowicz et al. 2005) but multiple measurements with tunable wavelength are needed for CW imaging with coherent detection to gather the same information. On the other hand, the CW imaging with incoherent detection allows to measure only the intensity at one frequency. The THz near-field imaging and real-time imaging are used for improving the system performance in relation to spatial resolution and imaging speed, respectively (Wang et al. 2018a). In consideration with simplicity and cost, pulsed imaging systems are more advantageous for basic research as these provide complete information about sample. Based on the collected analytical information from the pulsed systems, the other information related to target identification is extracted as guide to select key parameter for CW-THz imaging suitable for practical needs and applications.

3.4 Tomographic Imaging

Although the THz imaging is a well-demonstrated noninvasive analytical approach, the obtained images are often 2D. The 3D imaging in transmission configuration is often restricted with thin samples due to sample's absorption added to direction of propagation. Hence, the tomographic procedures are used for obtaining the 3D images in THz range (Weijs et al. 2020). The science of tomography includes the approaches used to obtain the cross-sectional images that help to know more about the internal details of the objects. With the developments in THz imaging, several 3D imaging with THz waves have been demonstrated. Some of the common tomographic techniques based on THz waves used to obtain 3D information are described in this section.

3.4.1 THz Diffraction Tomography

In diffraction tomography, the spatial dispersion of the sample's refractive index is determined by measuring diffracted field. The relation between the THz wave distribution and the sample's refractive index with respect to position is defined by the Maxwell's equation (Zhang and Xu 2010b). The linearization of inverse scattering follows multiple hypotheses as applied in common ultrasound tomography is the basis for THz diffraction tomography. A probe beam is allowed to interact with the sample in diffraction tomography and the scattered waves from the target are used for building a 3D image of the sample. This makes it distinct from the computed tomography (CT) where the amplitude signal is utilized for direct transmission of signal through the sample (Gbur and Wolf 2001). The diffraction tomographic signal holds the same information as that of CT where the structural details of the target in 3D are analyzed by the diffracted distribution. The THz diffraction tomography is more commonly used for fine-structured complex samples where diffraction predominates the measurements. The first demonstration of THz tomography was to image a polyethylene (PE) cylinder with the image reconstruction followed the Born linearized wave equation (Ferguson et al. 2002). In another study, the working components of THz tomography system were demonstrated. The system comprised of mode-locked fs laser used for generating THz waves through optical rectification from ZnTe crystal and signal detection by CCD camera (Wang and Zhang 2004). The reconstruction based on theory of the electromagnetic wave scattering was proposed. This setup was used for imaging three rectangular PE cylinders using THz pattern at different projection angles. The major potential of the diffraction tomography is its ability to get useful information of the refractive index. On comparing the image quality as a function of the THz frequency, the THz tomography gives poor reconstructed image due to issues with reconstruction algorithms and interpretation. Often reconstruction is limited by a low SNR of THz field measurements (Guillet et al. 2014). Therefore, these techniques assume that the target is dispersion less. This assumption totally removes the advantage of the THz techniques, especially in the extraction of the spectral data. The image acquisition speed of the diffraction THz tomography is faster than the THz-CT. The diffraction THz tomography is based on the reconstruction of inverse of refractive index of the sample.

3.4.2 THz Tomosynthesis

Tomosynthesis (TS) is a common procedure involving the reconstruction of the slices of 3D object from limited projections. Hence, the reconstruction of tomographic images is based on the projections within the small angles (usually between −50° and +50°), and it does not essentially need the measurements in all directions (Catapano and Soldovieri 2019). Therefore, the THz-TS found most of its applications in the medical imaging, e.g., detection and diagnosis of breast cancer. The principle of acquisition systems of the THz-TS is explained as follows. The projections are also referred to as radiographs captured at different views by moving the THz source along a linear trajectory (Vandewal et al. 2013). A detector plane positioned in opposition to the plan of emitter determines the attenuation of signal transmission. By following these procedures, a sequence of radiographs is captured from different positions. Advantageously, this technique could be able to gather the desired layers as against out-of-focus layers. Like CT, this procedure utilizes absorption images and various projections from different points of view. However, TS relies on less count at the considered angles that can be viewed as CT image with lost information (Guillet et al. 2014). Due to common features between the TS and tomography, most of the developments for tomographic reconstruction can be conveniently applied for the data acquisition in TS. Over the years, different ranges of new methods were introduced since the reconstruction is done in accordance with XY plane instead of XZ plane. However, these approaches follow iterative algorithms such as simultaneous art (SART) and ordered subset expectation maximization (OSEM) (Recur et al. 2011). Specific algorithms are followed for superimposing and transforming all the acquired radiographs through point-by-point superposition to recover XY plane for a specific depth which yields a single focal plane. It is quite important to shift the radiographs before starting the superposition to obtain all the Z slices. Thus, shifting depth measures the focal depth in TS.

3.4.3 THz Time of Flight

The pulsed imaging has a distinct attribute that involves 3D mapping of the target using time of flight (TOF) of the reflected pulses. In this approach, the THz pulses are directed toward sample, and the amplitude and the phase of the reflected beam is measured. The temporal status of the reflected pulses shows the interface along direction of beam propagation (Xu et al. 2021). The depth of the target is measured based on the TOF difference from the pixel to pixel which gives 3D profile of target. The 3D image of the floppy disk was created by using reflection mode TD setup (Mittleman et al. 1997). The TOF approach of THz imaging is based on the following assumptions: samples have less dispersion and diffraction, weak reflection and hence multiple reflections are avoided, and refractive index is consistent within each layer of the object. This approach was widely applied for 3D imaging of nontransparent materials such as metallic ions, SD card, razor blade, and drugs based on pulsed THz radiation as well as in the identification of defects in foam materials (Guillet et al. 2014). A tomographic THz system based on TD and 2D EOS with high-speed complementary semiconductor camera was used for detection of defects in materials. Further, a high-depth resolution of order of magnitude of 1 μm can be achieved. The TOF resolution is influenced by the incident THz pulse width which can be improved further when the generated pulses are short. The distinct reflected pulse becomes difficult when the THz pulse is not perfect resulting in variation with time. This would result in occurrence of ghost interfaces in the tomographic image. This issue of ghost image can be rectified through adopting appropriate signal processing methods say for instance the use of employing deconvolution algorithms (Niu et al. 2019). This approach found its applications with reflection imaging where transmission is hindered and for the nontransparent materials that cannot be imaged under transmission mode (Zhong et al. 2005). In all these applications, the large bandwidth is beneficial that enables accessing different wavelengths in each layer. It helps in extracting the depth information and the quasi-3D image of the object. Based on this principle, a powerful imaging modality of T-ray reflection CT was demonstrated (Pearce et al. 2005). This method allows edge mapping of cross-sections of the sample which can be assessed from various reflection measurements at varied angles.

3.4.4 3D THz Holography

The THz 3D holography emerged from the advent of radar and the optical holography technologies. The main hypothesis relies on separation between the scattering centers is much larger than the THz pulse width. The differentiation of scattered waves of varied order of scattering allows to implement holographic approach with THz beam. The intensity and the phase distributions are determined for obtaining THz holograms (He et al. 2020). The interference pattern between the sample and reference is noted. The stored data contain information more than the focused image. Then the hologram reconstruction based on Fourier optics aids to view actual 3D image of the sample. Ruffin et al. (2001) reported about time reversal imaging which was termed as 2D holographic imaging in TD. The THz holography provides images of high fidelity with targets of scattered points located at well-defined distinct planes. However, it contains notable count of relevant deficiencies. Hence, it is considered as insignificant to more complex targets, and it does not give accurate refractive index of the reconstructed target (Guillet et al. 2014). Thus, this approach greatly depends on numerous scattering measurements and diffraction effects that significantly influence the image quality. Since the THz holography image depends on incident pulse, it limits its applicability in extraction of spectroscopic information about the target. A CW in-line digital holographic multi-plane system was reported for achieving a 3D sectional image of sample (Figure 3.10) (Huang et al. 2017). The result of proposed method confirms that the in-line digital holographic multi-plane imaging is an efficient tool to acquire multi-plane information of sample in THz range.

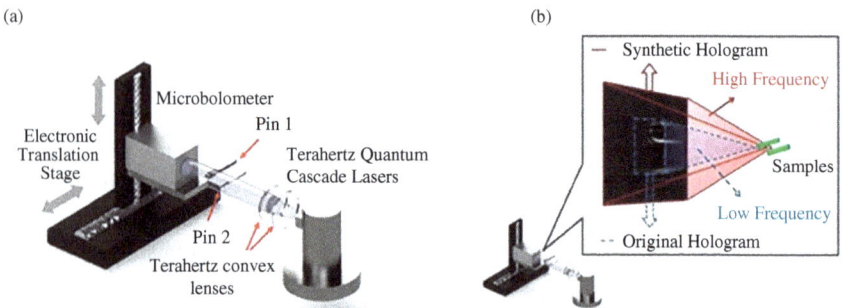

Figure 3.10 Schematic diagram of THz holographic imaging system: (a) schematic layout of the CW in-line THz holographic system and (b) schematic representation of synthetic aperture in holographic system. (Source: Huang et al. 2017.)

3.4.5 Fresnel Lenses-Based THz Approach

The diffractive Fresnel zone plates also termed as Fresnel lenses are used for focusing the light instead of traditional refractive lenses. Often the uses of Fresnel lenses in imaging systems have advantages in terms of weight and size over the common refractive lenses. However, diffraction Fresnel zone plates can only be used with narrowband applications because of frequency-dependent focal length (Kim et al. 2018). It is possible to image various samples positioned along the beam propagation path using a Fresnel lens in consideration with formation of image based on THz radiation. This allows broadband illumination tomographic imaging of sample. The THz focal depth determines the third dimension Z image resolution (Guillet et al. 2014). Hence, there is no limitation for narrowband applications. This tomographic approach is well suited for cases where the rotation of sample is impossible.

3.4.6 Synthetic Aperture Processing

A set of optics are used for focusing light into a narrow beam in conventional imaging technique which aids in achieving minimum dimensions called beam spot. A desired beam spot is adequate which influences the resolution that is optimum at its midpoint and declines in the vicinities. Here, the Rayleigh length is taken as a reference dimension for maximum thickness of sample (Toh et al. 2017). Eventually, the measured sample area must be positioned where the beam spot is located. The signal detection could be maximized by allowing the beam normal to the sample's surface. When the thickness of sample is either equal or greater than the Rayleigh length, the image depth is not correctly captured. In such cases, the synthetic aperture (SA) measurements are used for imaging the samples. In contrast to the conventional imaging systems of point-to-point measurements, the SA systems use an unfocused or diverging beam for data collection (Matsui and Kidera 2020). The target to be measured is irradiated by the sensor from different scanning positions to provide a wider beam for analysis. The energies obtained from all adjacent positions are much lower than that used in imaging techniques. Hence, a coherent integration of obtained signals is applied for increasing the SNR of the final image of the target. The energy compaction process is termed as SA processing which is usually carried out using time or frequency domain providing a computationally challenging result (Cristofani et al. 2014). The algorithms in SA are based on the integration of coherent raw data and matched filter where exists a lack of correlation between the noise and the signals. This greatly reduces the influence of noise during reconstruction of the SA image. The image from SA processing provides a constant and range independent resolution which is very useful in measuring the sample thickness. Therefore, it is feasible to measure the maximum sample thickness which is no longer hindered by Rayleigh length. However, the working parameters of sensor in relation to the sample under inspection must be carefully considered.

Sometimes, a wider beam is used to overcome the limitation of the mechanical performance of the scanner during complex scanning path or small scanning curvatures (Guillet et al. 2014). Thus, several successful demonstrations of THz imaging based on synthetic phased array with improved spatial resolution were reported in literature (O'Hara and Grischkowsky 2002, 2004; Zhang and Buma 2009).

3.4.7 Time Reversal THz Imaging

The time reversal imaging is a novel indirect approach based on pulsed THz. By taking advantage of the time reversal symmetry of the Maxwell's equation, the basic reconstruction algorithm was derived based on TD Huygens Fresnel diffraction equation (Ruffin et al. 2002). This approach allows to reform the 1D, 2D, and 3D images based on diffracted radiation at numerous different angles. The image resolution is provided by the spatial distinction of the two points on the plane of object that results in emergence of THz pulses with a notable time difference which is captured by the detector resulting in 674 μm resolution (Guillet et al. 2014). This was significantly less than the average wavelength in other approaches. This type of imaging was also used for phase contrast and reflection imaging. Efforts were made for increasing the numerical aperture through waveguide approach as adapted in ultrasound imaging and new algorithm during image reconstruction to enhance the time reversal THz imaging (Wan et al. 2020).

3.4.8 THz Computed Tomography

X-ray CT was introduced as a biomedical imaging for diagnosis of ailments that captures specific tomographic slices of specific region of organs. A real 3D image of the object is reconstructed from 2D X-ray images (Wang and Zhang 2004). This approach has slowly grabbed industrial interest for applications related to identification of defects and in quality control. However, the harmful effects of X-rays greatly suppress its application in other allied industrial sectors. The THz CT imaging was first demonstrated in the year 2002 (Ferguson et al. 2002). Compared to X-ray imaging, the THz CT imaging allows measurement of the amplitude and the phase of the object. A recent report demonstrated the applicability for spectral analysis of the materials (Wang et al. 2022b). The basic limitation of THz CT imaging is slowing down the absorption phenomenon that greatly limits thickness to be imaged. It is envisaged that most of the experiments are conducted with real samples under optimal wavelength, resolution, and its ability to capture the object that supports potential applications of THz imaging. Research efforts were made in overcoming the long acquisition time with THz-TDS. However, the complexity involved and cost of operation of THz sources limits its applications. Another limitation encountered is the nature of the sample that should possess a transparency sufficient and suitable enough in the THz bandwidth (Guillet et al. 2014). The THz CT imaging still requires a lot of research on new algorithm patterns and reconstruction strategies as the propagation of THz beam is far from X-rays. For which a broader knowledge on all other existing image processing techniques must be understood to fully explore the design and system developments of THz CT imaging.

3.5 Imaging Analysis

The various applications of THz imaging in the agri-food sector have been detailed in the Chapters 5 to 8. Hence, this section on image analysis is restricted to THz imaging applications for pharmaceutical solids. The THz imaging applications in the agri-food sector are an extension of applications from pharmaceutical field. Nevertheless, it is worth mentioning that THz imaging applications in the agri-food sector are not well explored and have a long way to go for implementation as commercial applications.

3.5.1 Characterization of Pharmaceutical Solids

3.5.1.1 Tablet Microstructure

The most exploited and well-known THz imaging application is the determination of the thickness of a sample. Hence, THz imaging is most commonly used in pharmaceutical field, especially for assessing the thickness of coatings on tablets (Figure 3.11) (Shen 2011). In addition to chemical imaging, THz imaging allows characterization of the microstructures of samples in determining the tablet density and hardness (Lu et al. 2020). The amplitude of the reflected THz signal from the sample surface relates to the refractive index. The received intensity from the

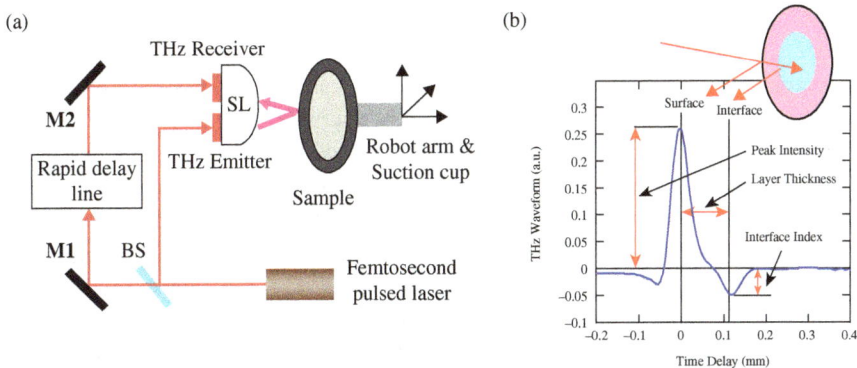

Figure 3.11   Pharmaceutical quantification of coating thickness using THz pulsed imaging method. (a) THz pulsed imaging and (b) THz waveform measured from a single-layer coated tablet. (Source: Shen 2011.)

surface of a sample helps in determining its density. Thus, density of flat-faced tablet can be determined based on calibrating the amplitude change by either univariate or multivariate models in consideration with the refractive index of spectral component (Palermo et al. 2008). Based on the surface geometries (flat or biconvex), the applicability of THz pulsed imaging in assessing the variations in refractive index in association with changes in tablet density was assessed in relation to varied compression pressures (May et al. 2013). A close correlation was observed between the refractive index of tablet's surface and its bulk density. Similar correlation was derived between the crushing force (hardness) of tablet with surface refractive index. In another study, the relative density of the static or moving powder (silicified microcrystalline cellulose and lactose) in an in-line process was measured using THz waves (Stranzinger et al. 2019). Results of density maps clearly showed the presence of local density differences in the powdered bed of both the samples. A precise resolution of the spatial distribution of the relative density was achieved with this approach that facilitates the in-line sensing of powder characteristics of samples upon compaction.

3.5.1.2 Cracks and Delamination

Apart from examining thickness of coatings, density, and surface thickness of tablets, it is possible to assess the underlayers below the tablet's surface. It was reported that the sample up to 2 mm below the surface of the tri-layered tablet was assessed using a standard THz pulsed imaging approach (Zeitler et al. 2007a). Despite the ability of THz imaging for detection of cracks and delamination having been proved, these applications are not well adopted. This is mainly because of the requirements of optimization of commercial optic THz systems to resolve thin structures at or near the object's surface with best resolution. Hence, the focal length is relatively short resulting in strong convergence of THz beam. With the advancements of THz imaging systems, the 3D SA techniques were used for radar imaging at THz frequency range (Toh et al. 2017). With further developments of imaging reconstruction algorithms, the applications of THz pulsed imaging have broadened in resolving the deep microstructures of pharmaceutical solids. However, these applications are currently restricted by diffraction and scattering effects. In another study, risk of crack formation and delamination of coatings in bi-layered tablets was reported (Niwa et al. 2013). The large deviation in the refractive index of air and tablet matrix results in a strong reflection if there is any defect in layers (like delamination). The associated reflection amplitude at the air-tablet interface aids in faster detection of cracks on tablet's surface and risk of delamination. By adopting a similar approach of THz imaging, the multi-delamination and thickness of the glass fiber-reinforced polymer composite plate were analyzed using fiber-coupled THz-TDS system (Han and Kang 2018). The obtained B and C scans from imaging technique confirmed the identification of locations of hidden multi-delamination in samples. Further, the calculated and measured thickness of sample had close consistency with error of <3.7% that was clearly visualized from B-scan image. It was inferred that TD visualization was well suited for determining thickness and location of delamination in reflection mode of scanning while the frequency domain visualization was suited for assorting the sizes of samples. These results showed that larger-sized defects are efficiently visualized at lower frequencies while smaller-sized defects can be visualized at higher frequencies.

3.5.1.3 Porosity

Several other applications dealing with the determination of porous matrix and optical properties using TDS were reported (Zeitler and Gladden 2009; Markl et al. 2018; Bawuah et al. 2020). Such applications are quite informative as these result in extracting a large amount of physical information about the sample based on transmission measurements. Using THz imaging it is feasible to measure the material's porosity, sample weight, and quantification of potential drug (active ingredient) content in a pharmaceutical tablet with a single measurement. Further, it allows to quantify the volume of the tablet that helps in getting an understanding of the bulk behavior of sample. The potential features of high penetration power and short measurement duration in range of milliseconds of THz radiation would complement the destructive techniques of current practice for determining the thickness, weight, and hardness measurements during the tablet compaction process (Markl et al. 2017). The extraction of mechanical properties of pharmaceutical solids directly from the transmission measurements during THz imaging was demonstrated (Peiponen et al. 2015). Based on measured tablet porosity, the Young's modulus was determined using power law relationship. It was reported that the porosity of the pharmaceutical biconvex microcrystalline cellulose compacts was linearly correlated with THz pulse delay (Bawuah et al. 2016). Based on this relationship, it is possible to determine the porosity of tablets through construction of calibration line between the porosity and THz pulse delay. A detailed review on porosity and microporous structural detection of pharmaceutical solids was presented (Lu et al. 2020). In this review, the porosity measurements based on the refractive index such as approaches of zero-porosity approximation (ZPA), anisotropic Bruggeman effective medium approximation (AB-EMA), and traditional Bruggeman effective medium approximation (TB-EMA) were explained.

3.5.1.4 Disintegration Testing

The high temporal resolution and high contrast during THz pulsed imaging enhance the applications related to analysis with deep penetration of pharmaceutical solids. It was feasible to use THz imaging to determine rapid disintegration of oral dosage formulations (Al-Sharabi et al. 2020). Thus, the diffusion of mass transport of tablet matrix into dissolution medium can be monitored by 1D imaging experiments. Here, the THz pulse exhibits characteristic reflection from surface of tablet as well as from the penetrating liquid front (Yassin et al. 2015). The acquired THz data were analyzed for extracting the information about the penetration front profile within the tablet and to monitor the swelling upon contact with the dissolution medium. Through THz imaging it is possible to assess the effects of variations in formulations like lubricants, and fillers on the microstructural properties like compaction characteristics, porosity, and particle size for basic understanding of process during the tablet disintegration (Quodbach and Kleinebudde 2016). Further, the mathematical modeling of these processes allows for rational design of disintegration behavior of dosages.

3.5.2 Polymeric Film Analysis

3.5.2.1 Coating Uniformity

Out of all applications of THz imaging, the assessment of uniformity and the integrity of polymeric films during quality inspection is the most explored application. The major advantages of THz imaging over other conventional quality inspection techniques in analyzing the coating uniformity of films are non-destructive technique, provides excellent contrast between different layers of inner structure, operates at broader wavelength in the range of mm to µm, and rapid measurements (Zeitler 2016). The samples are drawn at regular intervals from the coating pan for the assessment of coating uniformity. The drawn samples are analyzed using THz pulsed imaging technique which results in coating map showing the spatial distribution of thickness over sample's surface. The information of mapped distribution of thickness coatings is summarized by histogram plots which help in assessing whether during the process a uniform coating is achieved or not (Ho et al. 2009). This approach of THz pulsed imaging is perfect quality monitoring probe for process control. In a study reported on comparison of THz imaging and near-infrared (NIR) imaging, it was observed that the THz measurements did not cause any rise in temperature of samples (Maurer and Leuenberger 2009). Hence, THz imaging is advantageous over NIR as application of heat during NIR measurements alters structure of coating. It is also feasible to detect defects and cracks in coatings using THz imaging. Further, this technique is also useful in analyzing thick coatings that are difficult to examine with other sensing modalities (Vynckier et al. 2015). It was proven that THz imaging is not only used for measuring the thickness (>400 µm) of the sample coating but also helps in the detection of air gaps due to delamination between the structures and thickness quantification.

3.5.2.2 Functional Coatings

The active coating content of pharmaceutical solids is usually analyzed using high-performance liquid chromatography (HPLC) in combination with dissolution testing. This approach is destructive and time-consuming. Although it provides precise and accurate results, it is limited with the amount of samples tested from a particular batch due to associated cost and time. To overcome this limitation, a study was reported on development of noninvasive approach to quantify the active pharmaceutical ingredient of the functional coatings using THz imaging (Brock et al. 2012). The THz imaging took less analysis time than HPLC. Hence, THz imaging is an excellent alternative tool for assessing performance efficiency of coating process. It is also possible to determine conditions to achieve a good intra-tablet coating uniformity by keeping control over the uniformity of layers. In a report on assessing the effect of process conditions on dissolution properties of coatings, the results of THz coating were directly comparable to the virtual images of coatings measured using X-ray microtomography (Niwa and Hiraishi 2014). In addition, the THz imaging is also useful in analyzing the diffusion membrane in coatings of sustained release. The coating thickness and porosity are the crucial process parameters in determining the releasing profile of active ingredient. It was demonstrated that the THz imaging has a comparable predictive power as that of dissolution testing (Ho et al. 2008). The reflection peak from core-coating interface is resolved based on multivariate calibration. A quantitative model is used for predicting the mean dissolution time of the sustained release based on reflection data. The THz imaging is also employed for better understanding of the changes in the microstructure of solid formulations during scaleup operations. It was also reported that both THz electric field peak strength (TEFPS) and the THz coating thickness measurements were well correlated with the mean dissolution time for batches of samples coated under varied process conditions (Ho et al. 2008). It was inferred that pilot scale sample had thinner and denser coating layer despite of its slower release than lab-scale process. This was mainly because of the magnitude of the mechanical force associated with different scaling processes.

3.5.2.3 Sensor Calibration

The possibility of measurements of absolute thickness of coatings with less calibration requirements in determining the refractive index of samples made THz imaging an ideal reference for spectroscopic in-line sensor calibration such as NIR and Raman fiber-coupled probes which are more sturdy and low cost than THz pulsed imaging (Alves-Lima et al. 2020). The coating thickness of the samples is not directly measured using NIR and Raman sensor. In such cases, the thickness of the coating can be measured based on spectral change which are unique for each material (coating/core). The obtained intensity is further used to build indirect chemometric models for calibrating spectral signature against polymer weight gain based on known training data sets (Zeitler 2016). In contrast, the measurements are made directly using THz imaging that has significantly reduced sample count required for analysis and time required for model development. However, THz imaging is hindered by sample thickness. The experimental results of optical coherence tomography and pulsed imaging have been compared for the quantitative

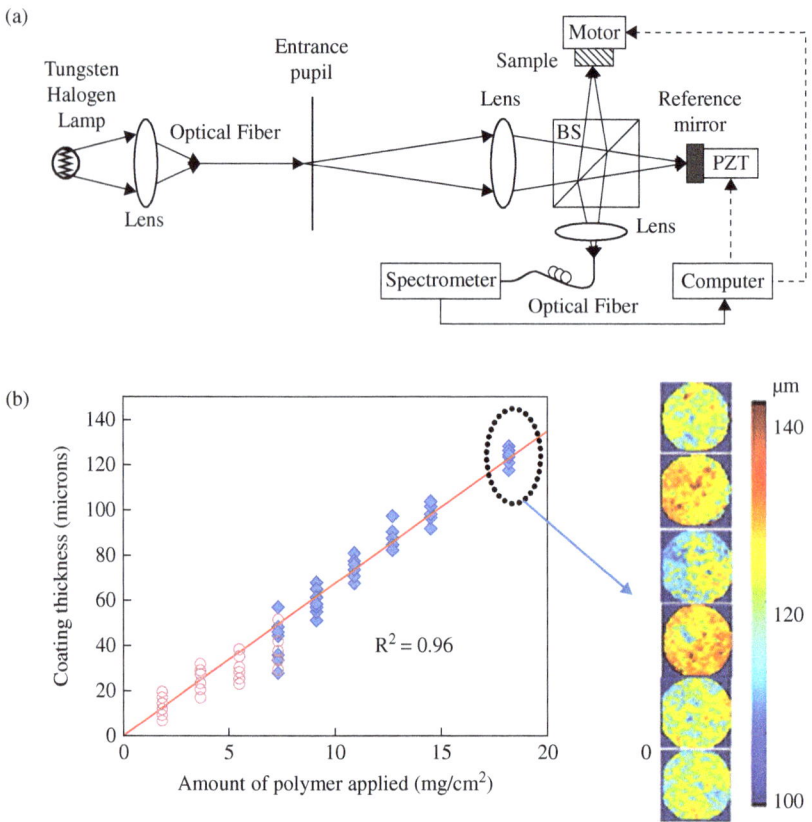

Figure 3.12   THz pulsed imaging combined with optical coherence tomography: (a) scheme of spectral domain optical coherence tomography (BS – 50:50 beam splitter and PZT – piezoelectric transducer) and (b) correlation between coating thickness and polymer amount in tablet. (Source: Zhong et al. 2011.)

measurements of tablet coatings of 10–140 μm thickness (Figure 3.12) (Zhong et al. 2011). Results inferred that both THz pulsed imaging and optical coherence tomography were complementary techniques that could be promisingly used for quantitative characterization of thickness coatings of solid dosages. Therefore, the coating thickness of thin samples is easily measured using optical coherence tomography. It was reported that THz imaging in combination with optical coherence tomography has great scope to focus entire range of coating thickness for calibration (Lin et al. 2017).

3.5.2.4 *In-Line Sensing*

The tablet coating thickness of the randomly moving individual samples can be measured using THz pulsed imaging sensor (Alves-Lima et al. 2020). The sensor collects a continuous array of reflected waveforms with faster acquisition rates (<10 ms). The waveforms corresponding to reflection peaks from samples obtained from THz imaging sensors are determined using selection algorithm. It is feasible to achieve a meaningful statistics of distribution of tablet coating thickness in real time using THz-TDI reflection waveforms (May et al. 2011). The THz in-line measurements retrieved during coating of 150 kg of tablets showed that the distribution of inter-tablet coating thickness among the samples within coating pan was broader than off-line sample assessment that was obtained from coating pan at the same time (Lin et al. 2015; Zeitler 2016). The THz in-line sensor readings were measured through a prototype sensor as this technology is still not commercially available. However, the above findings clearly prove efficiency and usefulness of THz technology in understanding the complex process of film coating of polymers that are currently measured based on empirical control strategies (Zhong et al. 2020b).

3.5.3 Chemical Imaging

The THz imaging can be promisingly used for determining the uniformity of active ingredient in transcutaneous patches (Sakamoto et al. 2009). The THz imaging was successfully used for mapping the distributions of different ingredients in formulations and their homogeneity. The same concept is applied for understanding the phase transitions in tablets. The drug theophylline tablets were prepared from a mixture of microcrystalline cellulose and magnesium stearate and the study demonstrated how the induced moisture got transformed from anhydrous to hydrate form and vice versa (Hisazumi et al. 2012). In this study, the transformation of chemical forms of moisture was monitored using THz imaging technique. It was also reported that the spectral information of samples was extracted from deep inner layers within tablets using reflection imaging even if the tablet was entirely covered or coated. Thus, THz imaging along with THz pulsed spectroscopy helps in analysis of crystalline properties as well as the non-invasive chemical recognition of the 3D object (Alves-Lima et al. 2020). In the 3D spectral dataset of the sample, the first two axes describe the spatial dimensions in horizontal and vertical directions, while the third axis represents the depth information as signal time delay. Sometimes, the fourth dimension of the dataset is generated by performing Fourier transformation which helps in analysis of the tablets with inner buried structures of different chemical composition. These applications confirm and pave a way for a range of different exciting applications of THz in pharmaceutical fields. The recent advancements in production technology especially functionalized multilayered capsules/solid dosages put forth the need and demand for new technologies to analyze the process (Arshad et al. 2021). Considering this, the THz imaging would remain as novel chemical analytical technology of future in identifying and recognizing the chemical species using THz spectral data.

3.6 Recent Advancements, Challenges, and Limitations

The THz imaging provides tremendous opportunity for the agri-food industry in food quality control, detection of foreign bodies, monitoring of packaging materials, and so on. Recently THz imaging has been emerging as an anticounterfeit monitoring tool that is based on extraction of structural fingerprint of samples. The urgent need for development of quick imaging technique for in-line monitoring acts as drive for advancements of THz imaging. The integration of THz emission spectroscopy (TES) with the imaging system is commonly referred to as laser THz emission microscopy (LTEM), mainly used for characterization of electrical charge carriers and the optical excitation of the materials. When excited with the fs laser, the resulting THz waveforms are useful in analyzing the nature of the electronic materials (Tonouchi 2019). Thus, LTEM has emerged as a novel tool for the determination of electronic state of the materials in material science. Another variation of TES is the spintronic THz emission spectroscopy (STEM) widely applied for near-field THz biosensing and imaging (Guo et al. 2020a). However, the cost associated with the devices used for generation of THz radiation limits its applications. The development of cost-effective THz imaging systems especially for in-line/on-site monitoring is considered by adopting the alternative methods of THz generation. This includes the applicability of photomixing-based systems with the use of low-cost diode lasers or quantum cascade lasers based on CW-THz imaging.

The following future opportunities are summarized in a report on extreme THz science (Zhang et al. 2017c). This report states that the applications of metamaterials for nonlinear THz processes would remain as a promising strategy for strong generation and THz frequency conversion. This kind of artificially engineered metamaterials offers greater flexibility and controls over the phase, amplitude, and polarization of THz radiation. Another promising future opportunity of THz science is the exploitation of THz quantum sensing and imaging. Taking the advantage of utilization of nonlinear optical processes, it is possible to produce a set of entangled photons namely visible and THz photon. The entangled photon pairs find their application in ghost imaging. The spectral information can be computed by interpreting the data from visible photon, which can be detected by a detector without its interaction with target. This approach seems to be beneficial when dealing with visible photon's high detection sensitivity, but it is difficult to realize the entanglement degree between photons through experiments. The challenges associated with multidimensional spectral characterization of complex sample with spatial structure at nano level can be addressed with diffractive coherent imaging by applying synchronized THz pump and X-ray pulses. With active research progress in THz science, the focus of its implementation has shifted from developing cost-effective THz systems toward its applicability. Hence, the limitation of THz technology must be better understood to overcome the technical issues associated with it.

The major limitation of wide-scale commercial applicability of THz imaging is the cost associated with the THz components including sources, detectors, and data acquisition systems. Although research works are being focused on the development of new cost-effective systems, the implementation and process evaluation has long time to put forth for practical applications (Afsah-Hejri et al. 2020). Despite the success of the THz imaging as novel noninvasive tool for quality inspection and monitoring, most of the works have been demonstrated under lab-scale setup. Hence, the adoption of the THz imaging is long way off from the actual implementation to economically beneficial applications. The developments of subregion THz systems are applications specific where the sampled subregion must match with the desired response. A higher SNR is relatively problematic to access certain portion of THz spectrum. However, this limitation can be overcome with improved data acquisition procedures through multiple scanning and averaging (Wang et al. 2022c). Also, applications of THz technology are significantly affected by the extent and amount of water content of samples. One such limitation in the detection of moisture content of samples is that THz spectral imaging is not applicable for high moisture samples with thickness <1 mm. This is mainly because of the high absorptive power of water toward THz radiation. Since most of the agri-food products are characterized with considerable amount of water, the application of THz imaging is limited to the quantity of moisture in biological samples (Yan et al. 2022). In such cases, reflection imaging is another option. Sometimes, it would result in image artifacts with the evident standing waves due to optical path length difference. Another challenge in THz imaging applications during quality monitoring of fresh produce is the effect of variations in the physical attributes like particle size on the refractive index of biomaterial. In addition, the adverse effects of scattering significantly influence of quality of THz measurements. The THz extinction spectra are highly affected by the scattering losses especially for materials where the grain size of the solid sample is comparable to THz wavelength (Osman and Arbab 2019). In relevance to broader applicability of THz imaging applications, the following considerations must be strongly addressed for future practical implementations in agri-food industry.

- The influence of process conditions on data quality, accuracy, precision, and repeatability of THz measurements
- Determining the optimal measurement conditions and operative procedures for different ranges of samples with varied biophysical properties and knowledge about the sample interaction and response with THz signals
- Analysis and development of suitable chemometric methods and selective algorithm for translation of THz signal intensity into actual information about sample
- Integration of appropriate chemometric method with THz spectroscopy and imaging techniques through modeling of process and product variables for better monitoring of quality
- Exploring the compatibility of process integration of THz system with other conventional quality detection approaches for cutting down the cost of operations useful for practical industrial applications

3.7 Summary

THz imaging is an emerging noninvasive tool for sample analysis, especially in quality monitoring and inspection of agri-food products. In addition to the system components of THz-TDS, the imaging system includes special system for image capturing which provides spectral information of samples. The THz imaging approach relies on the analytical processing of the transmitted and reflected THz signals from the sample. Based on the forms of the THz waves, the THz imaging systems are broadly categorized as pulsed and CW-THz imaging. Different forms of THz imaging techniques are detailed in this chapter. The 2D and 3D mapping of chemical information of samples is possible using THz imaging. However, the THz 3D chemical imaging is still under progress which would open a novel way of non-destructive approach for sample analysis. Comparatively, THz imaging is a safe method of imaging, especially for the biological materials. Although conventional technique such as X-ray imaging is simple and cheaper, it has limitation of possibility of sample damage due to photoionization. The attractive feature of THz imaging is the low photon energy and good penetration ability of THz waves that best suit for agri-food samples. Both the spatial and spectral information in the THz frequency range is extracted from the captured THz images. Further, the THz imaging possesses superior resolution and hence used for structural and molecular characterization. In addition to the detailed discussion on THz imaging principle and system components, the recent advancements and limitations are also briefly described. The THz imaging applications are comparable to that of THz spectroscopic applications, especially in the agri-food sector. The applicability and feasibility of THz system developments for imaging applications are well explored in pharmaceutical field with active research progresses over the past years. These advancements assure a similar growth and potential scope of THz imaging for agri-food applications in near future.

Chapter 4

Spectral Measurements, Image Acquisition, and Processing

4.1 Introduction

Research related to THz science has increased significantly in recent times, especially due to advancements in THz sources, emitters, and detectors. The THz spectroscopic and imaging techniques are applied for both basic and applied research in material science (Mittleman 2017). Hence, THz spectroscopic and imaging technology has found many innovative applications in security checks, material characterization, quality control, inspection, and monitoring of foreign bodies, harmful substances, and hidden objects. The unique properties of THz waves such as chemical specificity, spatial resolution, fingerprint spectra, and dielectric penetrability make it distinct from other bands of electromagnetic spectrum (Bandyopadhyay and Sengupta 2022). Since THz science is an emerging analytical technique, much active research works are in progress, especially in biosensing, industrial process management, cultural heritage preservation, agricultural plant growth monitoring, hydration analysis, agri-photonics, biochemical characterization, and identification. Despite this enhanced progress, the THz spectroscopy and imaging techniques suffer from technical limitations. These include low acquisition speed, high cost of system components, limited spectral contrast, and low spatial resolution (Gowen et al. 2012). Unlike other non-destructive quality assessment techniques, THz waves are safer to handle as these do not damage sample due to their non-ionizing nature. Hence, in-depth works on THz science in exploring the effects of interactions of materials and THz waves would promisingly assist in design of compact, flexible, and efficient THz systems.

The basic principle and working of THz spectroscopy and imaging technology is the combination of spectral sensing of sample and its surroundings through scanning under THz frequency range (Jepsen et al. 2011). This initial process of material interaction with THz waves is followed by the recording of the electronic signals as emitted from the sample. This process is referred to as spectral analysis and data acquisition which is an important step in the material characterization. Later, the acquired THz signals with phase, amplitude, or intensity information are preprocessed/processed to extract useful features like absorption coefficient and refractive index (Zeitler 2016). The material characterization or chemical species identification is made from the unique chemical attributes of samples. This helps in the determination of specific needs such as thickness measurements and detection of cracks and foreign substances. The first three chapters of this book discuss about the basic introduction, working concept of THz spectroscopy and imaging, and various forms or types of system configurations. The latter processes of spectral processing, image acquisition, and data analysis are equally important as that of the understanding of the interactions between the THz waves and the sample. Hence, this chapter focuses on this aspect of spectral and data analysis. This chapter provides a clear understanding of how the obtained raw spectral data are processed to obtain useful information for real-time applications.

4.2 THz Spectral Data Analysis

For the simple understanding of the concept of spectra and image acquisition, this section describes the process in context with THz time-domain spectroscopy (THz-TDS) system. The sample positions to be analyzed under THz waves are selected initially for single-point THz spectral acquisition. The THz signal from each desired location is measured. During the measurement of sample, the other process conditions such as humidity around instrument should be set as required based on applications. The obtained THz spectral features are reproducible. From which, the absorption spectra are viewed as distinct peaks (Qu et al. 2018a). The environment space is filled with nitrogen as reference under the transmission mode while the metal-based total reflection mirror acts as a reference operating in the reflection mode. The sample fixed on to the metal plate of sampling stage is allowed to move in a path of trajectory in two-dimensional plane at a desired resolution as required during image acquisition process (Wang et al. 2020a). The length of the scanning time delay line is optimized to remove the effects of the echo pulses. Since the obtained original spectra are based on time domain, the next step is to transform the spectral data. This is usually done by applying the conventional fast

Fourier transformation (FFT) function to collect the sample's frequency domain information as expressed in Equation 4.1 (Zhang and Xu 2010b).

$$E(\omega) = A(\omega)e^{-i\varphi(\omega)} = \int E(t)e^{-i\omega t}\,dt \tag{4.1}$$

where, $E(\omega)$ is the Fourier transform of the experimental data, $A(\omega)$ is the amplitude of the electric field, $E(t)$ is the time domain wave, and $\varphi(\omega)$ is the phase. The absorption coefficient can be measured by the logarithm of ratio of amplitude of reference spectrum to the sample spectrum divided by the sample thickness (Equation 4.2) (Zhang and Xu 2010b).

$$\alpha = \frac{\ln\left(A_r / A_s\right)}{x} \tag{4.2}$$

where, α is the absorption coefficient, A_r is the amplitude of the reference, A_s is the amplitude of the sample, and x is the thickness of the sample.

Based on the amplitude of the reference and sample, the transmittance (T) of the sample is determined by Equation 4.3 (Dorney et al. 2001).

$$T = \left(A_s / A_r\right)^2 \tag{4.3}$$

Later the chemometric methods like principal component analysis (PCA) are applied to analyze the THz transmission spectra for demonstrating the desired end application of THz spectroscopy and imaging system say e.g., the discrimination of foreign bodies from food samples. The PCA is the dimensionality reduction method that transforms the original random vector into new one (Liu et al. 2016b). Here the components are non-interrelated by orthogonal transform. This is followed by size reduction of the multivariate system into low dimensional variable with high precision. The low dimensional system is transformed through appropriate value function to one-dimensional system and is displayed on 2D plane. Thus, the samples are classified from this obtained 2D graph. Using the PCA, the characteristics of the samples are extracted (Hocine et al. 2014). This in turn helps in narrowing down the THz transmission spectral range for designing the predictive models and analyzing the data. Different methods of chemometrics as applicable in THz spectroscopy and imaging technique are described in the subsequent section of this chapter (Section 4.7).

4.3 THz Signals in Time and Frequency Domains

The signals obtained from the THz-TDS assess the magnitude of the electrical vector of the THz pulses as a time function. The time delay in the measured signal represents the traveling rate of the THz radiation passing the sample cell until reaching the detector (Ahi et al. 2015). The measurement timing is modulated based on the arrival of the near-infrared pulses as measured by the delay line position. The recorded TDS signal is characterized as the intensity burst known as the main peak with subsequent delay in the signal amplitude. The optical path of the probe beam is generally smaller than the pump beam prior to the recording of main peak from the sample. The influence of the sample's refractive index would result in reduction in the velocity of the THz pulse (Zeitler 2016). Therefore, the recording of THz signal before the measuring of main peak denotes the background noise since the THz pulses have not reached the detector. So, the scanning of the delay line increases the probe beam path, and the detected THz signal corresponds to the intensity from the sample. Based on the nature of the sample, the THz signal arrives at the detector in subsequent times. Thus, the obtained spectral information of the sample is translated to useful information in context of real-world applications.

A typical THz-TDS signal looks similar to that of asymmetric interferogram from Fourier transform infrared spectroscopy. The THz-TDS plots are obtained when the incident THz radiation is allowed to pass through the sample in different modes like transmission, reflection, and attenuated total reflection (ATR). The THz-TDS plot as obtained from solid sample comprises of sample, reference, and the difference of sample and reference time-domain (TD) (Smith and Arnold 2011). The difference between the sample TD and reference TD indicates the original portion of THz signal as absorbed by the sample. The THz-TDS measurements are based on the coherent broadband THz pulses. Hence, the obtained THz signal is a collection of cosine waves represented as the distribution of the superimposed frequencies with constructive interference observed at the main peak (Damyanov et al. 2020). This obtained raw THz signal is directly subjected to FFT for converting the data from TD into frequency domain (FD). Since the measurements of sample and reference are made independently rather than simultaneously, the FD plot is given by a single beam spectrum. After performing the Fourier transformation, the absorbance spectrum of the sample is determined as given in Equation 4.4:

$$A(v) = -\log_{10}\left[I(v)_s \Big/ I(v)_r \right] \qquad (4.4)$$

where, $A(v)$ is the absorbance spectrum of sample calculated from single-beam spectra, $I(v)_s$ and $I(v)_r$ are the single beam spectra of sample and the reference, respectively. Similar to any other spectroscopic methods, the signal averaging is done with the THz-TDS for enhancing the SNR thereby reducing the random noise (Smith and Arnold 2011).

4.4 Signal Preprocessing

Before the processing of THz data using FFT algorithm, the raw THz-TDS signal is subjected to signal preprocessing to enhance and improve final THz spectral quality. The two most applied preprocessing methods in preprocessing of THz signals are apodization and zero-filling. The experimental procedure of material characterization starts with preparation of the sample, reference spectrum acquisition, sample spectrum acquisition, signal preprocessing (apodization and zero-filling), and single beam/double beam absorbance spectrum calculation. Different process steps involved in THz spectral and image acquisition are summarized in Figure 4.1.

4.4.1 Apodization Function

The first method of signal preprocessing is apodization that is usually carried out with TD signal immediately prior to transforming the signal into FD. This process step involves the weighting of the TD signal performed using one variety of apodization or windowing functions (Yin et al. 2017). The application of apodization function eliminates the artifacts and decreases effects of noise thereby aids in improving the spectral quality. It should also be noted that sometimes the intense apodization causes distortion of the THz spectral features with loss of information when the system resolution reaches the width of the sample spectral parameters.

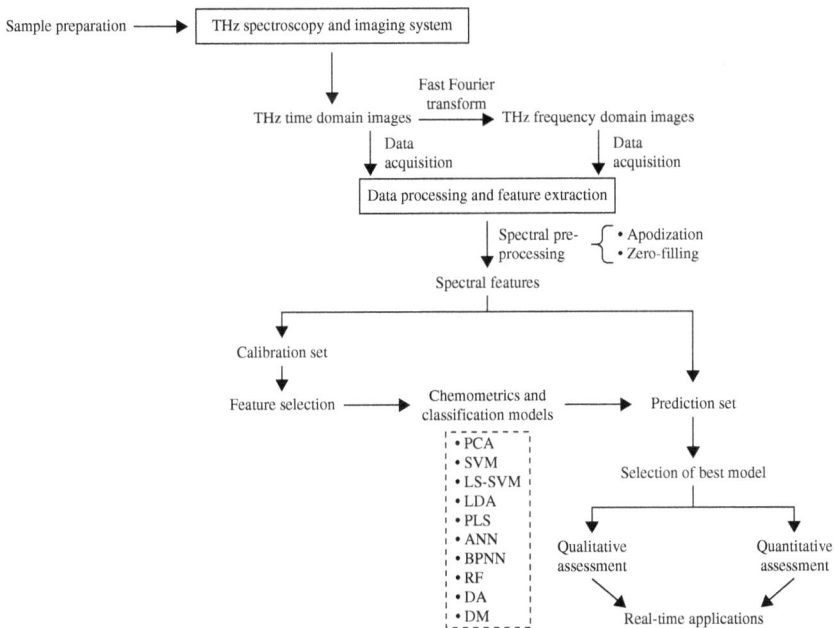

Figure 4.1   Process steps involved in THz spectral and image acquisition.

Different forms of apodization functions are available for improving the FD of THz spectrum. These include Hamming, boxcar, triangular, Happ-Genzel, Norton-Beer, and Blackmann-Harris three-term or four-term (Smith 2012). Among which the simplest function is the boxcar also termed as Dirichlet or rectangular apodization. Here, the unprocessed spectral data are multiplied by one followed by FFT algorithm on unweighted raw TD THz signal. While other functions of the apodization have different multiplication patterns that enable different regions of TD spectrum while the rest are attenuated in magnitude. All these functions assume a unit value at the region nearer to main peaks and show a decreasing trend toward zero in regions nearer to extremities of the window sampled time.

The severity of the applied function on THz spectrum is described in terms of rapidity with which the apodization reaches its minimum value. The apodization functions approaching zero closer to the main peak provides a smoother FD spectrum. However, these functions have inclination to drop and broaden the narrow absorption bands. Although the absorbance peaks are increased during this process of apodization, it helps in smoothening of the spectral features (Lasch 2012). The asymmetry characteristic of THz-TDS allows variation of one other feature to optimize the apodization function. Here, the apodization functions are used in asymmetrical fashion with each end making a symmetrical value where the function is determined since signal's pre-peak length is same as the post-peak. It is notable that high frequency noise of the absorbance spectra could be reduced by employing stronger apodization functions to THz-TDS data.

4.4.2 Zero-Filling Function

The second method employed in THz signal preprocessing is zero-filling. This function is also referred to as zooming or zero-padding or Fourier interpolation. This method is used for obtaining the spectral points within a given range (Sitnikov et al. 2019). The additional data points are the interpolated points between the chemical points without any specific chemical information. As a result, the spectral features get smoothened within the spectrum. This method involves the addition of zeros at the end of TDS before subjecting the raw data to FFT (Smith and Arnold 2011). Thus zero-filling is adopted to improve fitting of peak by elucidating the shape of obtained spectral features that could be observed at nominal resolution of the spectrometer.

4.5 Data Processing and Reconstruction

Different data processing and reconstruction methods used for THz tomographic processes are discussed in this section. These methods are similarly applied for the data processing of THz image processing in general. The THz tomography comprises of two distinct steps of image processing. The first one is the direct method which describes the acquisition of how to get the measured dataset from the observed physical phenomenon. The second method is the inverse model that defines the procedure of reconstruction of volume from the acquired dataset. These models have to be discretized and are not applied directly as they are defined in continuous domain (Recur et al. 2012b). Sometimes, the acquired data possess dynamic and static misalignments that significantly affect the outcome of the acquisition process. The noise associated with image acquisition and the induced approximations by the discretization could lead to occurrence of errors in the reconstructed data. Addressing these limitations, research works recently focused on development of methods to reduce the noise errors and discretization in the acquired data during image reconstruction process. The common reconstruction methods and their optimizations in obtaining the image of target are direct and iterative methods (Goncharsky and Romanov 2017). The direct method describes the image acquisition and model reconstruction in the space domain and the FD based on Radon transform and the Fourier slice theorem, respectively. While the Karcmarz method detailing about the algebraic reconstructions comes under the iterative method.

4.5.1 Direct Method

This section gives a brief note on the acquisition algorithms based on Radon transform and inverse algorithm. These models must be discretized since they are defined in continuous domain. Factors like discretization step in THz tomography process and its limitations are the drawbacks of this direct method of data processing.

4.5.1.1 Measuring Dataset from Physical Phenomenon to Reconstruction

The penetration power of THz waves enables the radiation to travel through the sample. The THz beam gets attenuated based on material composition and the sample density. After the acquisition of two-dimensional projection (named sinogram) as in case of tomography, a mathematical calculus function like retro-projection is used for reconstruction of the 2D slice of the sample (Kuba and Hermann 2008; Katz 2013). Considering an example of 2D centered orthonormal domain, the 1D projection (R_θ) is defined with an angle θ. The projection line along θ is usually based on its position as given by the module ρ which can be represented as function (θ, ρ). The measured dataset relies on the obtained data in θ direction corresponding to the attenuation of THz radiation during its interaction with sample along the projection line. Hence, at a given angle of θ, module set defines the angle projection (R_θ). A sinogram (S) defined as a set of projections is usually obtained around the sample in a complete angle of rotation. The acquired domain is reconstructed based on the values of sinograms using back-projection (Guillet et al. 2014). As stated earlier, the inverse algorithm is the indication of the acquired sample that is described by adopting back projection function of the filtered projections. It should be noted that a single projection is not enough to reconstruct the entire domain. Hence, the original domain is usually reconstructed based on many projections of sinogram in an accurate way.

4.5.1.2 Radon Transforms

The function Radon transform is used to model the image acquisition and reconstruction. This function was named after Johann Radon who first introduced it in 1919. The direct transformation R is a projection line acquisition transforming two-dimensional function say $f(x, y)$ into a 1D projection (Equation 4.5) (Stanley 1983; Toft 1996).

$$R_\theta(\rho) = \int\int_{-\infty}^{\infty} f(x, y)\delta(\rho - x\cos\theta - y\sin\theta) \, dx.dy \qquad (4.5)$$

where, ρ and θ are the radial and angular coordinates of the projection line (θ, ρ), respectively. The $R_\theta(\rho)$ is the total absorption of points traversed by projection line in the domain. The inversion process recovers the original function from the projections. For a sinogram S with multiple infinite projections say $\theta \in [0, \pi]$ and $\rho \in R$, the inverse function can be given as follows (Equation 4.6) (Guillet et al. 2014):

$$f(x, y) = \int_{0}^{\pi}\int_{-\infty}^{\infty} F(|\rho|) R_\theta(\rho)\delta(\rho - x\cos\theta - y\sin\theta) \, d\rho.d\theta \qquad (4.6)$$

where, F is the Fourier transform. This inverse function executes a ramp filtering on each projection to acquire more information. Then the value of the point (x, y) is enumerated from the total of the filtered projections. Hence, this approach is termed as the Filtered Back-Projection.

4.5.1.3 Fourier Slice Theorem

This is another approach used in inverse function where the two-dimensional Fourier transform of the original function is inversed by integrating the information of projection Fourier transforms. The Fourier slice theorem (FST) specifies that the one-dimensional Fourier transform of the projection R_θ of $f(x, y)$ along θ belongs to Fourier line space of f along the same angle θ (Ozanyan 2015). In terms of polar coordinates $(x = r\cos\varphi \ and \ y = r\sin\varphi)$, the function f can be represented as in Equation 4.7:

$$R_\theta(\rho) = \int_{0}^{2\pi}\int_{-\infty}^{\infty} f(r\cos\varphi, r\sin\varphi)\delta(\rho - r\cos(\varphi - \theta)|r| \, dr.d\varphi \qquad (4.7)$$

The 1D Fourier transform F_{1D} of the $R_\theta(\rho)$ is given by Equation 4.8:

$$F_\theta(\nu) = F_{1D}(R_\theta(\rho)) = \int_{-\infty}^{\infty} R_\theta(\rho)e^{-2i\pi\rho\nu} \, d\rho \qquad (4.8)$$

On combining Equations 4.7 and 4.8,

$$F_\theta(v) = \int_0^{2\pi} \iint f(r\cos\varphi, r\sin\varphi)\, \delta(\rho - r\cos(\varphi-\theta))|r|\, e^{-2i\pi\rho v}.\, d\rho.dr.d\varphi \quad (4.9)$$

Since Dirac impulse $\delta(\cdot)$ cannot be null at $\rho = r\cos(\varphi-\theta)$, the Equation 4.9 becomes,

$$F_\theta(v) = \int_0^{2\pi} \int_{-\infty}^{\infty} f(r\cos\varphi, r\sin\varphi)|r|\, e^{-2i\pi\rho v\cos(\varphi-\theta)}\, dr.d\varphi \quad (4.10)$$

Considering 2D Fourier transform $\bar{F}_{2D}(\psi, v)$ of function f in polar coordinates ($X = v\cos\varphi$ and $Y = v\sin\psi$),

$$\bar{F}_{2D}(\psi, v) = \int_\theta^{2\pi} \int_{-\infty}^{\infty} f(r\cos\varphi, r\sin\varphi)|r|\, e^{-2i\pi\rho v\cos(\varphi-\theta)}\, dr.d\varphi \quad (4.11)$$

From Equations 4.10 and 4.11, the equation corresponding to the FST is given by Equation 4.12.

$$F_\theta(v) = F_{1D}\left(R_\theta(\rho)\right) = F_{2D}\left[f\left(r\cos(\varphi), r\sin(\varphi)\right)\right] = \bar{F}_{2D}(\psi, v)_{\psi=0} \quad (4.12)$$

Thus, the two-dimensional Fourier transform of the original function can be derived from the projection data. Then the original function is obtained through inversion.

4.5.1.4 Acquisition Properties

The image acquisition under real condition is comprised of finite projections that are uniformly distributed between the 0 and π (Guillet et al. 2014). The sample count N_ρ is finite and remains constant during acquisition. The angular step $d\theta$ between a set of projections is expressed as π/N_θ. The distance between two projection lines indicates the sampling step on $d\rho$ projections. The sampling ratio among the image and projection is denoted as $d\rho$ during discrete reconstruction of image (I) from $W \times H$ sized pixel and is given by Equation 4.13:

$$d\rho = \frac{\max(WH)}{N_\rho} = 1 \quad (4.13)$$

The image acquisition with finite number of projections N_θ is comprised of samples with sinogram S^*. This S^* is a subset of ideal sinogram S and is represented by 2D image with a size of $N_\theta \times N_\rho$. Each line in 2D image corresponds to acquisition value of respective projections. For instance, the pixel (i_θ, i_ρ) contains an acquisition value $R_\theta(\rho)$ in a range of $0 \le i_\theta < N_\theta$ and $0 \le i_\rho < N_\rho$; where i_θ is the projection index and i_ρ is the module index.

A sinogram obtained from the Shepp-Logan phantom of the continuous model based on analytic definition of objects is used in the reconstruction of the image I which denotes the acquired function (Shepp and Logan 1974; Toft 1996). This process can be done by applying discrete function of the Fourier construction or Radon inversion. The reconstruction of image I during acquisition based on the inverse discrete Radon transform (R^{-1}) is given by Equation 4.14:

$$R^{-1}(i,j) = I(i,j) = \sum_{i_\theta=0}^{N_\theta-1}\sum_{i_\rho=0}^{N_\rho-1} R_\theta^{BFP}(\rho)\, pk(\theta,\rho,i,j) \qquad (4.14)$$

where pk represents pixel kernel. This pixel kernel is generally applied to assess how the projection line intersects pixel. The value of the filtered projection $[R_\theta^{BFP}(\rho)]$ can be found by Equation 4.15:

$$R_\theta^{BFP}(\rho) = \sum_{\rho s}|\nu|\left(\sum_\nu R_\theta(\rho_s)e^{-i2\pi\rho_s\nu}\right)e^{i2\pi\rho\nu} \qquad (4.15)$$

To proceed with the Radon inversion, the Fourier reconstruction should be discretized for getting an image from the projection Fourier transforms. Considering this, the simplification done in Equation 4.9 is not possible over here. This is because of the slight deviation in the location of polar point is not positioned at the exact pixel center (Toft 1996).

$$F_\theta(\nu) = \bar{F}_{2D}(\psi,\nu)_{\psi=0}$$

$$= \sum_{\phi=0}^{N_\theta}\sum_{\rho=0}^{N_\rho}\left[\sum_{r=0}^{N_\rho} I(r\cos\phi, r\sin\phi)|r|e^{-2i\pi\rho\cos(\phi-\theta)}\right]\Delta(\rho - r\cos(\phi-\theta)) \qquad (4.16)$$

The sum between the square brackets belongs to one-dimensional Fourier transform. While the double sum outside the brackets gives the polar points that are useful in computing the pixel $\bar{F}_{2D}(\psi,\upsilon)$ (Equation 4.16). Thus, the obtained discretization output is an interpolation from the polar to cartesian grid that corresponds to Fourier space of image (Guillet et al. 2014).

4.5.2 Iterative Reconstructions

In contrast to the direct approach, the iteration process has the advantage of correction of induced errors by one or set of projections based on image state under reconstruction and the available data on other projections.

4.5.2.1 Algebraic Methods

This method of iterative approach was introduced in 1970s and has been known as the Algebraic Reconstruction Technique (ART) (Gordon et al. 1970; Gordon

1974). This iterative algorithm is used to approach the linear equation solution system based on Karmarz theorem by updating each pixel. Here the pixel correction is performed based on the measured error of the initial and recomputed sinogram from previous iteration of image. The projection value of line in a discrete geometry is given by a linear array of pixels (i, j) weighed by a pixel kernel (Equation 4.17).

$$R_\theta(\rho) = \sum_{i=0}^{W-1} \sum_{j=0}^{H-1} \rho k(\theta, \rho, i, j) I(i, j) \tag{4.17}$$

The sinogram R derived by vector $N_\theta \times N_\rho$ and the image I to reconstruct is described by $W \times H$ vector. The overall pixel contribution in the projection lines is given by Equation 4.18.

$$R = AI \tag{4.18}$$

where, A is the weight matrix defined by $N_\theta \times N_\rho$ rows and $W \times H$ columns. The inversion of Equation 4.18 retrieves the original vector I from the data of projection R (Equation 4.19).

$$I = A^T R \tag{4.19}$$

The main goal of algebraic methods is the equation system resolution. By knowing the sizes of image and acquisition, the algebraic systems become complicated and too complex to be solved by matrix inversion approach. In addition, the linear approximations and the noise of projection do not consider the exact resolution. The Karmarz theorem derives the solution by vector estimation at I^k at the kth iteration. Each pixel $I^k(i, j)$ is improved by comparing the measured $R_\theta(\rho)$ and computed $R_\theta^k(\rho)$ values from I^{k-1}. On iterating, it converges to a solution through error minimization between the R and R^k thereby leads to an absolute solution I. This method requires initial imaging as a result of direct method. However, the results suffer from artifacts as given by the initial analysis (Herman and Meyer 1993; Herman 1995). Hence, a uniform image with every pixel valued with average initial sinogram is generally selected.

Several forms of ART techniques have been developed for data reconstruction over the years that are used in image processing and are listed as follows:

- Multivariate Algebraic Reconstruction Technique (MART): The pixel is computed as the ratio of R and R^k (Lent 1976).
- Adaptive Algebraic Reconstruction Technique (AART): This method employs an adaptive adjustment of relaxation attribute at every reconstruction pace (Lu and Yin 2004).

■ Simultaneous Iterative Reconstruction Technique (SIRT): Every pixel is improved at once from the line of projection of all the projections (Gilbert 1972).

■ Simultaneous Algebraic Reconstruction Technique (SART): This method is an agreement in combination of the ART and SIRT (Guillet et al. 2014). It utilizes all the projection lines from a single projection in updating the image. The pixel error is averaged from the sinogram and subsequently optimized by projection access scheme.

4.5.2.2 Stochastic Methods

The reconstruction techniques based on statistic elucidations formalize the problem as to what is the most likely image I knowing the observed R (Shepp and Vardi 1982). On iterating, the algorithms lead to increase the probability, $P(I \mid R)$ that is the probability of getting I based on the projection of R. Bayes theorem can be defined as follows (Equation 4.20).

$$P(I|R) = P(R \mid I) \times P(I) / P(R) \tag{4.20}$$

where, $P(I|R)$ is the value to be optimized, $P(I)$ is the priori of the I, and $P(R)$ is the priori of projection which is equal to 1 as the projections are known. $P(R \mid I)$ is the probability having R projections according to the I which is the likelihood of the projections. By considering the solution space with limit $P(I) = 1$.

$$P(I|R) = P(R \mid I) \tag{4.21}$$

The projection likelihood can be increased by knowing the image I to maximize the probability of getting I knowing the R (Equation 4.21). It can be done by decreasing the distance between the observed and the calculated projections. This process is like to the algebraic method since minimization is done by iterative updates in k of I. The maximum likelihood expectation maximization (MLEM) algorithm defines the likelihood by modeling R^k as Poisson distribution and logarithmic version of this algorithm is also used for exploiting this model (Shepp and Vardi 1982). The aim is to maximize the mathematical expression through partial derivation along I when it is considered as null. The maximization of likelihood seems to consist of minimization of its partial. Here individual pixel is independently used to universally reduce the partial derivative. Initially, the MLEM algorithm determines the expectation of likelihood depending on the set of projections of the I under reformation. This pace is followed by computing the expectation maximum through canceling the partial derivative according to I through iterative pixel update. Thus, the maximization considers the projection set where the MLEM algorithm becomes identical to simultaneous multiplicative algebraic reconstruction technique (SMART). This can sometimes be denoted as simultaneous multiplicative iterative reconstruction technique (SMIRT). Another probabilistic algorithm based on Gaussian distribution is introduced. This model brings out many conjugate gradient imaging reconstructions alike to the MLEM in additive form (Kaufman 1993; Mumcuoglu et al. 1994). Several optimizations based on separation of the projection sets under different orders have been proposed due to impede convergence of the MLEM (Recur et al. 2011). This ordered subset reconstruction approach is applied for applications in biomedical field requiring high convergence speed and quality accuracy.

4.5.3 Propagation Beam Assessment and Modeling

The THz ray follows a Gaussian distribution during propagation and depends on properties of THz waves as well as the lenses used to focus beam or beam shape during continuous imaging (Guillet et al. 2014). The Gaussian beam radius from the beam axis takes a minimal value at the beam waist. The radius of the Gaussian beam $w(z)$ at an operated wavelength λ and the z position from the beam midriff is given by Equation 4.22.

$$w(z) = w_0 \sqrt{1 - \left(\frac{z}{z_r}\right)^2} \qquad (4.22)$$

where, w_0 is the radius of Gaussian beam axis at its minimum value in the beam waist, and z_r is the Rayleigh range given by $\pi w_0^2 / \lambda$. And the distribution of energy at the cross-section is given by Equation 4.23.

$$E(r, z) = E_0 \left(w_0 / w(z)\right)^2 e^{\frac{-2r^2}{w^2(z)}} \qquad (4.23)$$

where, E_0 is the energy at the center of the beam midriff and r is the beam axis distance.

Following a Gaussian distribution, the energy declines from beam center to edge. By knowing the properties of source using Equations 4.22 and 4.23, the observed Gaussian is computed to simulate the source. The THz source gets altered to Gaussian propagation when the lens focuses the beam based on the impact of the lens. Altogether the Gaussian beam is indicated as Gaussian beam convolution that models both the source and lens. The overall model can be estimated using nonlinear least square three-dimensional Newton-Gauss algorithm in cases when the properties of lens are not known. Thus, the Gaussian beam modeling has been applied as convolution filter in iterative technique such as SART and OSEM. This greatly helps in improving the accuracy and quality of the reconstructed image (Recur et al. 2011, 2012a). As a rigorous quantitative approach, the Gaussian beam modeling provides insights in improving the 3D THz-CT imaging.

4.6 Data Transformation

A crucial decision-making step in chemometrics for a particular set of data depends on the raw spectral data organization. The matrix of the obtained spectral results must be defined and structured before subjecting to further processing using software. In general, the common method of building spectral data matrix for chemometrics involves arrangement of mean values and response variables in rows and columns, respectively. This step must be preceded by the so-called spectral preprocessing. It is adequate to balance the statistical importance of the response variables by adopting desired spectral preprocessing methods such as mean centering, variance centering, autoscaling, filtering, normalization, standard normal variate, multiplicative scatter correction, derivatives, smoothing, detrend, orthogonal signal correction, Fourier transforms, and wavelet transforms (Adegbenjo et al. 2020). The smoothing method is often used for spectroscopic data while autoscaling is adopted for continuous quantitative responses. The transformation of data should be done to ensure the distribution to be symmetrical thereby giving each response variable the same weight and equal priority during analysis. A proper preprocessing of data is a must otherwise it would result in extra variation in the data set (Engel et al. 2013). Thus, spectral preprocessing of data is a critical step that must be carefully done and has a direct influence over the outcomes of the chemometrics and accuracy of decision-making as well.

Autoscaling is one of the data preprocessing techniques. Regardless of the distribution of data, the results are transformed by centering through subtracting of the mean of each attribute. This is followed by scaling through dividing the attributes by their standard deviation. Thus, the experimental results are transformed into z scores (Equation 4.24) (Zielinski et al. 2014).

$$Z_{ij} = \frac{X_{ij} - \bar{X}_j}{s_j} \tag{4.24}$$

where, Z_{ij} is the standardized value of response variable, X_{ij} is the original value for the sample i of measured attribute j, $\bar{X}_j$ is the mean of variable j, and s_j is the standard deviation of the response variable j.

The next method of spectral preprocessing is variance centering which is an adjustment of the data set that equilibrates the variance of individual response variable. This method results in decreasing the extent of effects of each response in data set. This procedure involves dividing the observation by standard deviation of entire column of response variable (Equation 4.25) (Karaman 2017).

$$Z_{ij} = \frac{X_{ij}}{s_j} \tag{4.25}$$

Mean centering is a preprocessing method that is applied for calibration of multivariate models. This method involves determining the mean of the response variable and the obtained individual result is subtracted by their corresponding mean value (Equation 4.26) (Cunha et al. 2017). As far as the spectral data are concerned, the removal of the mean from each spectrum results in differences among the samples with significant enhancement in the concentration as well as spectral response. This leads to an outcome of a calibration model rendering more accurate predictions.

$$Z_{ij} = x_{ij} - \overline{x}_{ij} \tag{4.26}$$

The method of smoothing is used for preprocessing of spectral data for reducing noise and elimination of the narrow peaks. Sometimes, differentiation function can be applied to extract related information and in correcting the baseline. The Savitzky-Golay algorithm is the most used technique for smoothing and differentiation. This algorithm is a local polynomial nonparametric regression where the spectral data for smaller intervals of wavelength are fitted by polynomial equation (Jahani et al. 2018). These fitted values provide better estimate than the measured ones as some noises are removed during this process. For each spectral point j with x_j value, the weighted total of the neighboring values is determined. These calculated weights evaluate the post-operation whether smoothing should be carried out or a derivative is calculated (Zielinski et al. 2014). The number of the adjacent neighbors and polynomial degree controls the smoothing strength. The extent of smoothing depends on the count of adjacent neighbors that are applied for computing the polynomial fit to the spectral results. Thus, these are some of the common preprocessing methods used for spectral data. Since there is no proper established guideline when to use which preprocessing method for certain applications, it is necessary to test and compare these methods for selecting the best possible method for data set for certain food safety applications.

4.7 Chemometrics

In recent years, the processing of acquired THz spectral data is made easier through the integration with chemometric methods. The implementation of chemometric methods with spectroscopy greatly reduces the amount of process variables thus facilitates analytical examination of the samples. The spectral data processing using the chemometric methods speeds up analysis time and improves the process efficiency. The application of chemometrics in THz technology helps in removing the undesirable and irrelevant optical features. The chemometric techniques are usually applied during preprocessing and in the multivariate data assessment (Wang et al. 2017a). The common chemometric methods that are combined with THz spectroscopy include PCA, support vector machine (SVM), linear discriminant analysis (LDA), least-squares support vector machine (LS-SVM), partial least squares (PLS), quadratic discriminate analysis (QDA), principal component regression (PCR), probabilistic neural network (PNN), artificial neural network (ANN), back propagation neural network (BPNN), and multiple linear regression (MLR) to name a few (Hussain et al. 2019; Gong et al. 2020). For realizing the real-time applications of THz spectroscopy and imaging like discrimination and classification of foreign materials and harmful substances from desired good quality materials, a model is usually developed from the extracted spectra based on multivariate analysis. The PCA and PLS methods are applied for eliminating the unnecessary information and in reducing the image dimensionality. Thus, chemometrics is an efficient tool used in multivariate analysis for developing models for various tasks such as classification, prediction, and identification in food safety and quality evaluation applications.

The modern analytical techniques applied for THz fingerprinting involve generation and analysis of large amount of spectral data. The generated signals from THz spectroscopy and imaging systems are not easy to understand thus often require simplification of data. The obtained spectral data are subjected to complex statistical analysis to reveal the actual information of sample and for human understanding. The modulation of the multivariate data and extraction of chemical information involves use of chemometrics (Badia-Melis et al. 2015). The chemometric models are conveniently coupled with analytical measurements like THz spectroscopy and imaging for quality assessment of agri-food products. The classification models based on chemometrics are used when there is a large amount of data with high number of variables to reduce data dimensionality. The chemometric methods are classified as exploratory analysis (unsupervised models), classification or discriminant analysis (supervised models), and prediction models or regression analysis (Medina et al. 2019). A general outline of classification of multivariate data analysis is presented in Figure 4.2. This section describes some of the common chemometrics classification models used in THz spectroscopy and imaging systems.

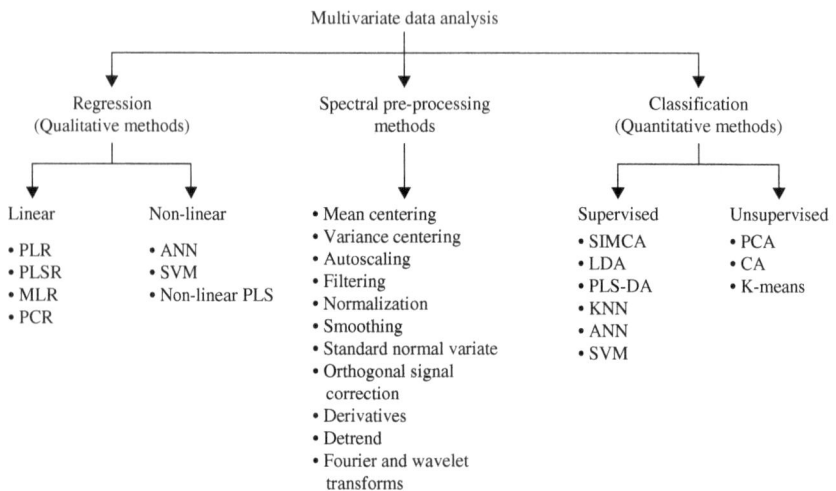

Figure 4.2 Classification of multivariate data analysis.

4.7.1 Classification Models

4.7.1.1 Principal Component Analysis

The PCA is the most employed unsupervised classification model. For initial exploration of visualizing spectral data and to identify the outliers, the PCA is applied. The PCA could provide a valuable information demonstrating its potential capability to distinguish the samples (Liu et al. 2016b). As stated earlier, PCA method is helpful in reducing the image dimension of spectral data through orthogonal transformation. Here, the correlated variables get transformed into variables that are linearly uncorrelated. During the transformation process, the spectral data are compressed by removing noise and other redundant information. The transformed data reveal and reflect the information of the original data that are known as principal components (Gong et al. 2020). The variance used in judging the principal component gives the sample's original information. In other words, the higher variance of the principal component yields greater the extracted information (Tao 2016). The focus of this approach is simplification of the mathematical model and thereby improving data analysis efficiency.

4.7.1.2 Support Vector Machine

The SVM is another statistical approach used in solving the problems related to classification and nonlinear mapping. This method was first introduced by Cortes and Vapnik which has been widely adopted for classification and regression tasks. It is a machine-learning approach based on theory of statistical learning (Shao et al. 2012). The classification is based on the Vapnik-Chervonenkis dimension principle and structural risk minimization. This method of classification is more appropriate and effective in small batch of samples, high dimension, nonlinear, and local minimal problems.

4.7.1.3 Least-Squares Support Vector Machine

The LS-SVM is an augmentation of SVM where it solves the multivariate problem along with effective elimination of noise and reduction of dimensional data. The LS-SVM is based on structural risk minimization principle in training process that helps in reducing the computation complexities (Gong et al. 2020). It also results in increasing the algorithm speed relative to SVM method. Due to these additional features, LS-SVM is preferred than simply SVM classification model. A radial basis function (RBF) along with Gaussian function is considered as kernel function for reducing training procedure. This results in good performance of model under general smoothness assumptions. The LS-SVM requires two crucial parameters $\left(\gamma, \sigma^2\right)$ and the optimal value of these factors are used in developing a model for calibration with higher accuracy based on leave-one-out cross validation (Liu et al. 2016b).

4.7.1.4 Linear Discriminant Analysis

The LDA is a supervised pattern recognition algorithm that retrieves the discriminant function based on intra-class variance at its minimum and inter-class variance at its maximum principle (Gong et al. 2020). Based on utilizing this discriminate function, the spectral data are categorized. This method of classification is widely applied for applications in medical diagnosis, geological survey, market prediction, and forestry pest prediction to name a few.

4.7.1.5 Partial Least Squares

The PLS method of classification was first applied in social science field. Until 1990s, this method was not applied as a chemometric method in multivariate regression (Gong et al. 2020). The PLS is mostly applied to reduce complexity between the input and the output data in determining maximum data correlation of input and output latent variables (Wang et al. 2017b). This method is based on relation of covariance among the input and the output data. The SVM is more competent and conveniently applied for high-dimensional data.

4.7.1.6 Principal Component Regression

The PCR is a form of regression analysis based on consideration of the principal components as independent factors rather than considering the original attributes. The principal components are the linear combination of original factors obtained by the PCA method (Suryanarayana and Mistry 2016). The PCR models can be developed using the principal components as inputs for the prediction and comparing the output with the multiple linear regression algorithms. Consideration of principal components in PCR improves the model prediction and greatly reduces the complexity of model through elimination of multicollinearity.

4.7.1.7 Artificial Neural Network

The ANN is a digitalized statistical method used for simulating the structural function of biological brain using computer. This method is mostly applied for problems dealing with nonlinearity. In contrast to other chemometrics, the ANN could realize the unsupervised learning and increases the information processing efficiency (Pfeiffer and Pfeil 2018). Due to this characteristic feature, ANN is broadly adopted in different fields such as psychology, bioinformatics, computational studies, microelectronics, and neurophysiology.

4.7.1.8 Back Propagation Neural Network

This classification model is a multifaceted feed-forward neural network and comes under ANN. It is comprised of input, output, and a convert layers. Each layer at least contains one node, i.e., neuron. As like any other classification model, the BPNN is used for classification, prediction, and decision-making using spectral data. The external and internal validations are applied for evaluating the model of prediction. The goodness of fit and robustness belongs to the internal model validation while the predictability is the characteristic aspect of the external model validation. The assumption is that output data (Y) is linearly related with input data (X) (Gong et al. 2020).

It is well-known fact that the sample absorption and concentration have positive relation in THz spectrum. According to Beer-Lambert law, the linear equation among X and Y is as follows (Equation 4.27):

$$Y = \beta_0 + \beta_1\beta_2 + \beta_2\beta_2 + \ldots. + \beta_k X_k + \mu \qquad (4.27)$$

In a matrix form, the $Y_1, Y_2, \ldots Y_n$ are represented as given in Equation 4.28.

$$\begin{bmatrix} Y_1 \\ Y_2 \\ \vdots \\ Y_n \end{bmatrix} = \begin{bmatrix} 1 & X_{11}X_{21}\ldots X_{k1} \\ 1 & X_{12}X_{22}\ldots X_{k2} \\ & \vdots\vdots\vdots\vdots\vdots \\ 1 & X_{1n}X_{2n}\ldots X_{kn} \end{bmatrix} \begin{bmatrix} \beta_0 \\ \beta_1 \\ \vdots \\ \beta_k \end{bmatrix} + \begin{bmatrix} \mu_1 \\ \mu_2 \\ \vdots \\ \mu_n \end{bmatrix} \qquad (4.28)$$

Considering, $\begin{bmatrix} Y_1 \\ Y_2 \\ \vdots \\ Y_n \end{bmatrix} = Y_{nX1}$ is the actual concentration assessed by the standard ana-

lytics for n groups of samples, $\begin{bmatrix} 1 & X_{11}X_{21}\ldots X_{k1} \\ 1 & X_{12}X_{22}\ldots X_{k2} \\ & \vdots\vdots\vdots\vdots\vdots \\ 1 & X_{1n}X_{2n}\ldots X_{kn} \end{bmatrix} = X_{nX(k+1)}$ is the group of spectral

data of n samples, $\begin{bmatrix} \beta_0 \\ \beta_1 \\ \vdots \\ \beta_k \end{bmatrix} = \beta_{(k+1)X1}$ is the absorption factor, and $\begin{bmatrix} \mu_1 \\ \mu_2 \\ \vdots \\ \mu_n \end{bmatrix} = \mu_{nX1}$ is the

measurement error.

The unknown absorption factors $\beta_0, \beta_1, \ldots., \beta_k$ are determined using the obtained spectral values and the actual sample concentration. By replacing the unknown parameter $(\beta_0, \beta_1, \ldots., \beta_k)$ of the regression function with the estimated value $(\hat{\beta}_0, \hat{\beta}_1, \ldots, \hat{\beta}_k)$ of absorption factor, the multiple linear regression equation can be derived (Equation 4.29) as:

$$\hat{Y}_i = \hat{\beta}_0 + \hat{\beta}_1 X_{1i} + \hat{\beta}_2 X_{2i} + \ldots\ldots + \hat{\beta}_k X_{kn} \tag{4.29}$$

where, $\hat{Y}_i$ is the estimated concentration of the sample in the range of $= 1, 2, \ldots, n$.

$$e_i = Y_i - \hat{Y}_i \tag{4.30}$$

where e_i is the residual which is the deviation between the sample's estimated concentration Y_i and the actual concentration $\hat{Y}_i$ (Equation 4.30).

Thus, the BPNN is a nonlinear neural network that can be used for solving multiplex problems more accurately than linear methods. For PCA-BPNN model, the PCA was performed initially for extracting the overall information of the whole spectral regions followed by the BPNN. Only a few principal components from PCA analysis were applied to input layer for BPNN. The leave-one-out cross validation approach was adopted during calibration (Liu et al. 2016b). They examined several architectural networks by varying the neurons count in the hidden layer with differed initial weights. The optimum parameters of the hidden nodes, errors, and iteration times were found based on least prediction error.

4.7.1.9 Random Forest

The RF is one of the successful classifier algorithms used more often in recent days. This model is based on ensemble learning algorithm that is very efficient on the multivariate problems, especially in analysis of chemical components in spectroscopy measurements (Liu et al. 2013). The RF is an unpruned group of classification and regression tree (CART) that includes an additional layer of randomness to bagging. It starts with delineating a bootstrap sample using bagging approach followed by selection of best splitting among the randomly chosen subset instead of all descriptor variables at each node. This kind of peculiar pattern recognition with superior performance is mostly used in cheminformatics and bioinformatics fields. The RF is an ensemble of B trees as $\{T_1(x), \ldots, T_B(x)\}$ where $x \in \{x_1, x_2, \ldots, x_m\}$. Here, $x \in \{x_1, x_2, \ldots, x_m\}$ is an m dimensional vector of variables of classified object and m dimensional vector of features obtained from THz images of samples (Liu et al. 2016b). The ensemble produces B outputs $\{\hat{y} = T_1(x), \ldots, \hat{y}_B = T_B(x)\}$ in which $\hat{y}_B$ is the predicted observation of the classified object by the Bth tree where $B \in \{1, 2, \ldots, B\}$. A final prediction $\hat{y}$ is produced from aggregation of the outputs of all the trees. In case of classification problems, $\hat{y}$ is the predicted class by most trees and the average of individual tree prediction in case of regression problems.

4.7.1.10 Discriminant Analysis and Distance Match

The chemometrics reflects the importance of the application of statistical and mathematical methods in extracting valuable spectral information from physical and chemical data. The discriminant analysis (DA) is applied to extract classification information of the samples into the optimal discriminant vector space (Tang et al. 2005). It ensures substantial inter-class distance as the smallest intra-class distance in new subspace. The most calculated distances are the Mahalanobis distance (MD) (Equation 4.31) and the Euclidean distance (ED). The former one has a broader scope of applications because of considering the correlation among the different attributes and scale invariant (Yin et al. 2021).

$$D_M(X) = \sqrt{(x-u)^T \sum S^{-1}(x-u)} \qquad (4.31)$$

where, $D_M(X)$ is the Mahalanobis distance, $u = (u_1, \ldots, u_p)^T$ is the vector of mean values of independent variables, i.e., mean matrix, $x = (x_1, \ldots, x_p)^T$ is the vector of data, i.e., covariance matrix, and S^{-1} is the inverse covariance matrix of the independent variables.

The DM method is used for identifying the samples with slight difference in contaminants. When the matching value is the lowest in any specific category that denotes that spectrum matches the best with this category (Equation 4.32) (Yin et al. 2021).

$$Z_{ei} = \frac{e_i}{S_e} = \frac{y_i - \hat{y}_i}{S_e} \qquad (4.32)$$

where, Z_{ei} is the matching value, e_i is the residual spectrum, S_e is the standard deviation spectrum, y_i is the unknown spectrum, and $\hat{y}_i$ is the average spectrum.

Likewise, other algorithms that are used for processing of spectral data includes genetic algorithm (GA), decision trees (DT), cluster analysis algorithm (CA), simple linear regression, partial least squares discriminant algorithm, and weighted linear discriminant analysis.

4.7.2 Performance Evaluation of Prediction Models

The performance of the prediction model can be assessed by determining the correlation coefficients of the prediction set (R_p), calibration set (R_c), and cross validation set (R_{cv}), root mean square errors of prediction $(RMSEP)$, calibration $(RMSEC)$, and cross validation $(RMSECV)$ (Chen et al. 2020). The used classification model having outstanding prediction must have a high correlation coefficient (R) (Equation 4.33), low root means square error $(RMSE)$ (Equation 4.34), and a small difference between the $RMSEC$ and $RMSEP$ or $RMSECV$. Sometimes, the prediction to deviation ratio (RPD) and error range ratio (RER) are calculated to assess the prediction capability of the calibrated models (Equations 4.35 and 4.36). In general, the considered calibration model would be successful when *RPD* is > 3 and *RER* > 10 (Wang et al. 2017a).

$$Correlation\ coefficient\ (R) = \frac{\sum_{i=1}^{n} (y_{ir} - \bar{y}_{ir})(y_{ip} - \bar{y}_{ip})}{\sqrt{\sum_{i=1}^{n} (y_{ir} - \bar{y}_{ir})^2 (y_{ip} - \bar{y}_{ip})^2}} \qquad (4.33)$$

$$Root\ means\ square\ error\ (RMSE) = \sqrt{\frac{\sum_{i=1}^{n} (y_{ir} - y_{ip})^2}{n}} \qquad (4.34)$$

$$Ratio\ prediction\ to\ deviation\ (RPD) = \frac{SD}{SEP} \qquad (4.35)$$

$$Ratio\ error\ range\ (RER) = \frac{y_{max} - y_{min}}{SEP} \qquad (4.36)$$

where, y_{ir} is the value of reference of the ith sample, y_{ip} is the predicted observation of the ith sample, $\bar{y}_{ir}$ is the average of y_{ir} and $\bar{y}_{ip}$ is the average of y_{ip}, n is the sample number in the training set, y_{min} and y_{max} are the minimum and maximum in reference of the cross validation samples, SD is standard deviation in reference of cross validation samples, and SEP is the standard error of the predicted set.

4.7.3 Software Used for Chemometrics

The complexity of spectral data is high due to large number of data sets and associated estimations. The successful application of any chemometrics depends on the trained high-end technical software. Many different statistical packages are currently available of paid and free versions for chemometric calculations. Some of the commercial software used for chemometrics are XLSTAT, Statistica, Pirouette, and Unscrambler X. While Inria and R are the cost-free licensed versions that are still under development for user-friendly graphical interface with command line programming (Zielinski et al. 2014). There are some specific toolboxes with graphical interface facilitating the use of these software programs. Chemoface is a user-friendly free graphical interface used in chemometric analysis in solving number of supervised and unsupervised problems, experimental designs, and response surface methodology (Mohammadpour et al. 2020).

The Waikato Environment for Knowledge Analysis (WEKA) is usually applied for preprocessing and classification of the spectral data. It performs regression analysis, association rules, clustering, and multivariate visualization of data (Singhal and Jena 2013). The limitation with this software is its combined features of mathematical and statistical techniques that arise the need to combine WEKA with other software for analyzing data. Similarly, the use of Statistica software also requires the purchase of certain other multivariate tools for data analysis. While Action software is the free statistical package which is the first setup to utilize R platform along with Microsoft Excel. It can be used to perform basic descriptive statistical and inferential data analysis as well as unsupervised classification algorithm like cluster analysis (Zielinski et al. 2014). The drawback of the Action software is its application limited to only hierarchical cluster analysis (HCA) while other classification algorithms like PCA, DA, and PLS-DA are not available. On the other hand, R, the often-used comprehensive arithmetic tool and statistical software, offers a range of different functions like linear, nonlinear, descriptive statistics, regression analysis, cluster analysis, and classification models (Bouveyron et al. 2019). As an additional feature, the graphics outputs from R have a high definition. Although it is very useful, R requires some computational skills of programming that limits its application. Though all these software seem to be simple, these require certain level of computational knowledge and operational skills to achieve the correct results with higher accuracy for a particular application.

4.8 Data Processing and Feature Extraction

In addition to the spectral information of the sample, data also contain other irrelevant information like noise, baseline drift, and information overlap. These irrelevant spectral data significantly affect the accuracy of model. In such cases, the pretreatment (selection of variable) should be done to eliminate this irrelevant spectral information. In practical situations, subjecting the spectral data to pretreatments yields better results with decreased RMSEP values (Ciccoritti et al. 2019). The common pretreatment methods used for spectral techniques include smoothing, baseline, derivative, multiplicative scatter correction (MSC), standard normal variate (SNV), and detrend. Some of the other variable selection approaches are uninformative variables (UVE), genetic algorithm (GA), successive projections algorithm (SPA), competitive adaptive reweighted sampling (CARS), and interval partial least squares (iPLS) (He et al. 2021). Broadly, the chemometrics are categorized as qualitative and quantitative methods. This section gives an overview of various chemometric methods that come under qualitative and quantitative analyses. The detection of adulteration or identification of contaminants is based on qualitative and quantitative methods (Wang et al. 2020a). The former method of qualitative analysis classifies different degrees and types of adulterants. On the other hand, the quantitative method helps in determining the proportion of the adulterant. Various combinations of chemometrics with THz spectroscopy and imaging systems in agri-food sector are listed in Table 4.1.

Table 4.1 Recent Works on Evaluation of Agri-Food Products Using Chemometric Methods Combined with THz Spectroscopy and Imaging Systems

Sample	Analytical Measurement	Chemometrics	Data Pretreatment	Reference
Orange extract	Hesperidin and naringin	PLSR	Normalization, SNV, MSC, first and second derivatives	Feng et al. (2022)
Soybean oil	Capsaicin	LS-SVM, BPNN, and PLS	BPNN and GA	Xia et al. (2022)
Wheat	Qualitative analysis of wheat based on producing area	PLS-DA, BPNN, and LS-SVM	SG, MSC, mean centering, and SNV	Shen et al. (2022b)
Soybeans	Detection of protein content	PLS, PCA-RBFNN, and ABC-SVR	Auto-scaling, SNV, MSC, first derivative, second derivative, MSC + second derivative, SNV + second derivative, and DOSC	Wei et al. (2021)
Peanut oil	Evaluation of peroxide value	iPLS, BPNN, SVR, and PLSR	GA and PCA	Liu et al. (2021)
Coffee beans	Geographical origin of coffee beans	CNN, LDA, SVM	GA and PCA	Yang et al. (2021a)
Soybean oil	Detection of aflatoxin B_1 (AFB$_1$)	LS-SVM, BPNN, RF, and PLS	PCA and t-SNE	Liu et al. (2019b)
Corn	Corn variety discrimination	LDA and SVM	PCA	Yang et al. (2021b)

Table 4.1 (*Continued*) Recent Works on Evaluation of Agri-Food Products Using Chemometric Methods Combined with THz Spectroscopy and Imaging Systems

Extra-virgin olive oil	Discrimination of geographical origin of extra-virgin olive oil	LS-SVM, BPNN, and RF	GA and PCA	Liu et al. (2018a)
Flavanols – myricetin, quercetin, and kaempferol	Qualitative and quantitative analysis of flavanols	KNN, ELM, RF, PLSR, and LS-SVM	Smoothing	Yan et al. (2018)
Rice powder	Quantitative analysis of imidacloprid	PLS, SVR, biPLS, and iPLS	AsLS	Chen et al. (2015b)
Rice kernels	Differentiate transgenic rice	DA and PCA	Derivative	Xu et al. (2015)
Soybean seeds	Differentiate transgenic soybean	PCA, LS-SVM, and PCA-BPNN	First and second derivative	Liu et al. (2016a)
Yellow fox millet	Qualitative and quantitative analysis of ternary mixture of L-Glutamic acid, L-Glutamine, and L-Tyrosine	PLS and SVM	MSC, S-G smoothing, first derivative, and wavelet transform	Zhang et al. (2017d)

(Continued)

Table 4.1 (*Continued*) Recent Works on Evaluation of Agri-Food Products Using Chemometric Methods Combined with THz Spectroscopy and Imaging Systems

Cotton seeds	Differentiate transgenic cotton seeds	SVM, MPGA, DT, KNN, and DA	PCA	Qin et al. (2017a)
Maize	Identification of transgenic ingredient in maize	SVM and selection kernels (linear, polynomial, and radial basis function)	PCA	Lian et al. (2017)

ABC-SVR = Artificial bee colony algorithm support vector regression; AsLS = asymmetric least square; biPLS = backward interval partial least squares; BPNN = back propagation neural network; CNN = convolutional neural network; DA = discriminant analysis; DOSC = direct orthogonal signal correction; DT = decision tree; ELM = extreme learning machine; GA = genetic algorithm; iPLS = interval partial least square; KNN = k-nearest neighbor; LDA = linear discriminant analysis; LS-SVM = least squares-support vector machine; MPGA = multi-population genetic algorithm; MSC = multiplicative scatter correction; PCA = principal component analysis; PCA-BPNN = principal component analysis-back propagation neural network; PCA-RBFNN = principal component analysis-radial basis function neural network; PLS = partial least squares; PLS-DA = partial least squares-discriminant analysis; PLSR = partial least squares regression; RF = random forest; S-G = Savitzky-Golay; SNV = standard normal variate; SVM = support vector machine; SVR = support vector regression; t-SNE = t-distributed stochastic neighbor embedding.

4.8.1 Qualitative Assessment

Some of the commonly applied machine learning algorithms for qualitative assessment are PCA, LDA, PLS-DA, ANN, SVM, soft independent modeling by class analogy (SIMCA), and k-nearest neighbor (KNN) (Zielinski et al. 2014). The PCA is a simpler method used to reduce dimension of the data and uses fewer new variables for explaining the variance of the original data. Also, PCA is an unsupervised pattern recognition method. Many studies were reported on cluster analysis that was conducted based on PCA (Patras et al. 2011; Eroglu et al. 2015; Qi et al. 2017). However, the use of unsupervised classification model has many limitations that significantly affect the performance. Hence, supervised machine learning algorithm is a branch of the multivariate statistics. Here the unknown samples are classified and discriminated based on the familiar classification algorithm by employing a set of discriminant functions (Medina et al. 2019). The performance of supervised discrimination methods varies based on applications and different analytical spectral techniques. For a certain application, more than one classification model are employed to determine the optimal method that best suits for specific application. Other models like canonical discriminant analysis (CDA), partial least squares density modeling (PLS-DM), DT, HCA, Simple Trees, and Bagged Tree are also used as classification models (He et al. 2021). However, most of works on classification models were done under laboratory scale with small batch of samples. Hence, more research works are required to explore the robust performance and universal application in real-time scenarios.

4.8.2 Quantitative Assessment

The classification models based on regression analysis are used for quantitative assessment applications. The regression analysis gives a relation between the dependent and independent variables. The PLS classification model is the commonly used regression algorithm to determine the adulterants. The acquired data and the adulterant content had a linear relationship under PLS analysis (He et al. 2021). Other than PLS, models like ANN and PCR are also used in quantitative assessments of contaminants. Some researchers developed linear regression functions with features like spectral intensity and reference value of adulteration content (Khan et al. 2015; Casado-Gavalda et al. 2017; He et al. 2021). Like qualitative assessments, the models of quantitative assessments were applied to acquired data under laboratory experiments. This remains as a major limitation for applications to large-scale samples of real-time applications. Also, the samples used in laboratory situations were very small. Hence, results of the small batches of samples are not reliable for real-world applications.

4.9 Considerations and Current Perspectives

Chemometrics is a statistical and mathematical modeling approach applied for the analytical spectral data under THz range (in case of THz spectroscopy and imaging). The developed models are used for obtaining relevant information in establishing food safety and authenticity. The advantages of chemometrics over normal statistical models are its simplicity, robustness, and reliable decision-making (Stefanuto et al. 2021). The synergistic approach of THz spectroscopy and imaging methods in combination with chemometrics yields a promising result for developing of protocols for food authentication and traceability of agri-food commodities. The decision-making in food safety and quality considerations in real-time applications is a multiplex task. For such complicated process, the univariate models like analysis of variance (ANOVA) with multiple comparison means tests such as Duncan multiple range test, Tukey test, and Fisher test are not sufficient to handle and understand data structure related to food safety and quality consideration. Hence, more sophisticated statistical methods are required in establishing the associations and interactions between the variables of test samples to characterize the materials. These multivariate statistical methods are known as chemometrics. It provides useful information about samples through classification indices even for samples with a large set of response variables (Granato et al. 2018). These chemometric models find their applications in processing of spectral data through mathematical and statistical tools in deriving the empirical models. The developed mathematical empirical models are used in predicting complex analytical data that are otherwise difficult to measure directly based on simple response variables.

Although the chemometrics are very useful in obtaining the chemical information about the sample based on THz spectral data, there are certain hurdles to fully explore their applications for commercial industrial scale in food sector. The adoption of chemometrics in routine operations in food industries requires personnel training to analyze and interpret the statistical results and sometimes there is disinterest of statistical analysis due to complexity and expense of statistical package in implementing in-line processing at industrial level (Zielinski et al. 2014). Additionally, it often requires constant statistical updates and add-on components. These constraints remarkably limit application and use of chemometrics in agri-food industries, despite the implementation of chemometrics along with THz spectroscopy and imaging has a great scope in detection of adulterants, contaminants, and other foreign bodies. Also, it helps in food authentication by ensuring the origin and provenance of commodity. To explore its potential applications and effective implementation at commercial scale, more research studies at laboratory and pilot levels are required in streamlining the appropriate method of spectral data collection, preprocessing, and classification that must be unique for specific needs of agri-food industry. Thus, the chemometrics is an important assessment tool that has a promising scope in quality inspection, monitoring, and evaluation of agri-food products.

4.10 Summary

This chapter discusses about the image acquisition, data processing, and feature extraction. The success of THz systems being utilized in real-time applications depends not only on spectroscopy and image systems but also on data processing of the collected spectral signals. Various preprocessing methods and classification models are detailed. A thorough fundamental and theoretical knowledge on statistical and mathematical modeling is needed for proper selection of appropriate chemometric methods considering their advantages and disadvantages. The practical use of chemometrics requires basic understanding of principle of each classification model and meaning of each individual input parameter along with critical evaluation in interpreting the results. The selection of best analytical approach and statistical function is not an easy task but requires careful consideration and depends on specific issues/needs of agri-food industries. Although the discussed statistical models have high classification power, still the implementation and their applications for on-site and real-time monitoring are under progress. This chapter describes the significance of spectral data and processing of THz signals emphasizing the future research needs for the development of rapid, simple, convenient, practical, and user-friendly chemometrics integration with spectroscopy and imaging systems. Addressing this research need would help solve many problems of food fraud in agri-food industry and ensure the delivery of safe foods to consumers.

Chapter 5

Molecular Characterization of Biomaterials

5.1 Introduction

The range of the electromagnetic spectrum between the millimeter waves and infrared rays has a broad range of applications in different industrial sectors. The characteristic properties including high directivity, efficient penetration, low energy, and huge capacity for communication made the terahertz (THz) waves applicable in different practices and implementation for industrial use (Wang et al. 2022c). The THz spectroscopy and imaging systems have emerged as the robust analytical tool for qualitative and quantitative measurements. The physical properties of THz waves render a broad range of characterization and sensing applications in many fields such as biomedical engineering, pharmaceutical industry, defense and security, agriculture and food, art conservation, material science, polymers, and packaging (Feng and Otani 2021; Bandyopadhyay and Sengupta 2022). Among the existing spectroscopic techniques, the THz systems are gaining more popularity mainly because of their noninvasive approach to testing samples. With a higher sensitivity of THz for intermolecular interaction rather than molecular composition, THz time-domain spectroscopy (THz-TDS) imaging has the advantage to be used for distinguishing materials with similar chemical compositions having different crystalline structures (Pawar et al. 2013). The THz emissions of 100 GHz to 10 THz frequency range with corresponding 4.14 meV photonic energy fall under the wavelength range of 30 µm to 3 mm (Wan et al. 2020). Such a characteristic frequency range that is much lower than X-rays make THz waves categorized as

DOI: 10.1201/9781003197010-5

nonionizing energy. Hence, THz waves have gained tremendous attention, especially in the non-destructive material characterization of solids like paper, polymer, ceramics, wood, and fabrics. Since THz waves are highly absorbed by the intermolecular bonds like N-H bonds in proteins and H-H bonds in water molecules, THz waves are more useful in the chemical characterization of biological materials at molecular levels (Sun et al. 2021b). Recently, there have been several reports available on the applications of THz waves in biomedical imaging like screening and diagnostic of ailments (Peng et al. 2021), monitoring, and detection of increased water content in blood flow as a result of infectious disease (Sun et al. 2017; Banerjee et al. 2020). However, the applications related to agricultural and food products are still scarce and have a long way to go. Therefore, a better understanding of the current and emerging practices of THz applications in the agri-food sector is required.

Considering this, the chapters of this book have been organized in detailing the concepts of THz spectroscopy and imaging followed by its applications in molecular characterization of biomaterials, use of THz technology as an inspection and identification tool, assessment of plant growth and physiology, and quality monitoring and control. Great efforts were made in organizing the widespread applications of THz systems in the agri-food sector which certainly provides valuable insights for emerging future research in this area. Until today, the growth and applications of THz systems are influenced by the challenges associated with the efficient generation of THz radiation. At this time, significance of THz waves is much appreciated among research communities. This slowly led to the progress in the advancements in THz technologies in generation, modulation, and detection. This upsurge development motivates research in exploring the THz applications dealing with spectroscopy, imaging, communication, and sensing in the agri-food sector. This chapter summarizes the diverse industrial applications of THz, the scope of THz systems in material characterization, and details the qualitative and quantitative measurements of agri-food samples using THz waves.

5.2 Diverse Applications of THz Systems

5.2.1 Security Monitoring

The security applications of THz systems include detection and inspection of baggage for hidden or concealed objects like explosive substances or illicit drugs (Malhotra and Singh 2021). The crystal molecules exert a specific emission under the THz regime that makes it possible to identify and detect harmful or unwanted substances using THz imaging. Due to the powerful penetration of THz waves through packaging materials (Shchepetilnikov et al. 2020), the THz waves are more efficiently used for unambiguous and localized identification of harmful substances within envelopes, parcels, or suitcases (Li et al. 2021c; Sypek and Starobrat 2021). On the other hand, metallic objects like knives and guns are visible and obvious through their solid shape which makes them easier to identify based on a pattern recognition algorithm (Ostrowski et al. 2021). However, it should be taken into consideration that the metallic packages are hardly transparent to THz waves and are often reflective. Hence, X-ray scanners are still a commonly preferred technique over THz scanners for detecting such objects (Guillet et al. 2014). The THz systems could provide supplementary information about the sample in the case of low-density materials and applications related to chemical separation. More interestingly, THz systems can be conveniently used for the detection of liquid explosives as they exhibit different dielectric responses under the THz range which makes them very distinct (Baxter et al. 2021). A study was reported on the comparison of passive imagers for the detection of concealed objects working under THz (1.2 mm; 0.25 THz) and mid-infrared range (3–6 m; 50–100 THz) in different types of fabrics (Kowalski 2019). Results showed that the THz camera could detect the hidden objects because of its nonzero transmission properties through fabrics while the transmission rate through fabrics in thermal cameras is negligible. Thus, this study confirmed that the temperature change in the human body and the concealed objects significantly affect the detection capability of imagers under both the THz and the infrared ranges. Therefore, THz imaging can be conveniently used for the real-time processing of samples. Similar reports are available on exploring the potential of THz systems to detect and identify illicit drugs (Lu et al. 2006; Pan et al. 2010; Tahhan and Aljobouri 2020) and hidden objects (Kowalski et al. 2013; Li et al. 2021e; Mansourzadeh et al. 2021). The THz systems are not only useful in the detection of explosive materials but also to determine the aging of the explosives (Yan and Shi 2022).

5.2.2 Packaging Industry

Most of the applications related to quality assessment of packaging materials require online or on-site monitoring of materials. For which THz systems are more convenient and useful in quality monitoring during their production process. There are several published studies that reported on the measurement of both moisture content and thickness of materials like papers and polymers using THz imaging (Molloy and Naftaly 2014; Ibrahim et al. 2021; Hansen et al. 2022; Naftaly et al. 2022). Mousavi et al. (2009) reported on the comparative assessment of the THz system with other sensors in differentiating two different paper samples. The results of THz systems were on par with other sensing systems yielding similar accuracy. Applications of THz have been reported in the polymer production line include inline monitoring of the polymer compounding process (Zhang et al. 2021c), controlling the quality of the plastic weld joints (Taiber et al. 2021), modulation of conductive properties (Shi et al. 2020), determination of moisture content and identification of fiber orientation defects (Kashima et al. 2020), and measurement of glass transition temperature (Wietzke et al. 2009b). A study reported on the use of two different optical sensing techniques (optical coherence tomography and

THz-TDS) for the quality assessment of paper (Hansen et al. 2022). This study highlights the potential of THz systems in the measurement of thickness, controlling of surface finish, and detection of production defects like voids, cuts, and oil contamination of paper. In another study, THz spectroscopy under transmission mode was used for the quantitative determination of the amount of water in paper (Banerjee et al. 2008). Results proved both phase and amplitude measurements allowed to accurately measure the water content of paper. From the results it can be inferred that THz spectral system could be used as a rapid diagnostic imaging tool for characterization of paper during on-site fabrication process. THz transmission imaging system was employed as a non-destructive alternative to delineating paper and paper products (currency/banknotes) (Ren et al. 2020).

5.2.3 Pharmaceutical Industry

Various applications of THz imaging and their potential significance in the pharmaceutical industry are detailed in Chapter 3 of this book. The molecular characterization has a remarkable application in the pharmaceutical industry (Zeitler 2016). Various polymeric forms could be detected using the THz system as different isomers have different crystalline states with uniquely distinct fingerprint spectra (Guillet et al. 2014). This characteristic of isomers made THz imaging a promising technique to apply for pharmaceutical solids as otherwise; it would suffer from limitations with chirality (Han et al. 2012). The THz imaging procedures are well established for quality control and monitoring of the tablet coatings (Novikova et al. 2018), detection of cracks and delamination (Niwa et al. 2013), and analysis of disintegration profile of core active ingredients (Bawuah et al. 2021). The potential of THz spectroscopy for the inspection of active pharmaceutical ingredients in smart films and tablets was demonstrated (Ornik et al. 2020). The results obtained from the THz system were compared with conventional techniques such as X-ray diffraction (XRD) and differential scanning calorimetry (DSC). The DSC measurements failed to yield a reliable result in detecting the crystalline forms of the active pharmaceutical ingredient while the THz spectroscopy and XRD provided promising results. Thus, the THz spectroscopy being a non-destructive technique is useful in determining the crystalline states of the active pharmaceutical ingredients in paper-based smart films/tablets.

5.2.4 Archaeological Applications

The main concern in archaeology is the dating of the samples through noninvasive procedures for which THz imaging system is a promising solution (Guillet et al. 2014). Gaining knowledge of ancient materials and artifacts is essential in art conservation and preservation of cultural heritage objects. The THz science provides enormous opportunities in the development of non-destructive methods of sampling as well as shows greater flexibility to be integrated with novel diagnostic techniques (Zhang et al. 2017a). However, these applications often demand portable on-site handheld devices for the investigation of samples. Studies were reported on the successful application of THz for the investigation of mural paintings (Inuzuka et al. 2017), applications in revealing the presence of graphite drawing under layers of paints and plasters (Chaban et al. 2021), detection of hidden paint layers (Tu et al. 2021), and quantification of transmittance of several pigments (Catapano et al. 2017). All the above reports demonstrate the potential of THz systems to be used as a diagnostic tool in art and archaeological science. As expected, the THz spectroscopy on integration with THz imaging has greater potential to be used for contact-free testing of materials. Recently the tomographic methods based on THz imaging have gained a lot of research scope by adding a third dimension in the analysis of samples. The combined techniques of THz-TDS and photogrammetric reconstruction were used for analyzing ancient pottery (Mikerov et al. 2020). Further, this integrated approach is proved to be highly efficient in determining the refractive index of samples with complex geometry. In another study, a robotic-based THz imaging was employed for investigating sub-surface structures of Egyptian mummies (Stubling et al. 2019). Results from THz imaging were far superior in terms of depth resolution than conventional computed tomography (CT) scan and further the THz results were comparable to the micro-CT results.

5.2.5 Food Industry

Rising concerns related to food safety put forth the need for the development of novel, rapid, and effortless detection techniques. The identification of harmful foreign objects in food is highly important in delivering safe food. Recently more attention is paid to exploring the applications of THz in food quality monitoring (Khushbu et al. 2021). The THz systems are quite useful in the detection of both metallic and nonmetallic contaminants. The study on moisture quantification for the assessment of shelf life of fresh-cut fruits and vegetable industry is an excellent industrial application (Zhang et al. 2021b). In beverage industry, the THz measurements are quite useful for qualitative assessment of the cork enclosure in the alcoholic beverages bottling line (Hor et al. 2008). The enhanced THz scattering results in contrast images of cork surface and interior that can be used to identify the defects or voids in the cork. In a more recent study, the THz-TDS was used for the discrimination of tea stalks from finished tea products (Sun et al. 2022). It was shown that the combination of THz-TDS along with adaptive iteratively reweighted penalized least-squares (airPLS) and k nearest neighbor (KNN) algorithms yielded promising results in qualitative discrimination of stalk foreign bodies with a 95.7% recognition rate (Figure 5.1). A similar study was reported on exploring the THz systems to detect foreign material in foods with complex composition

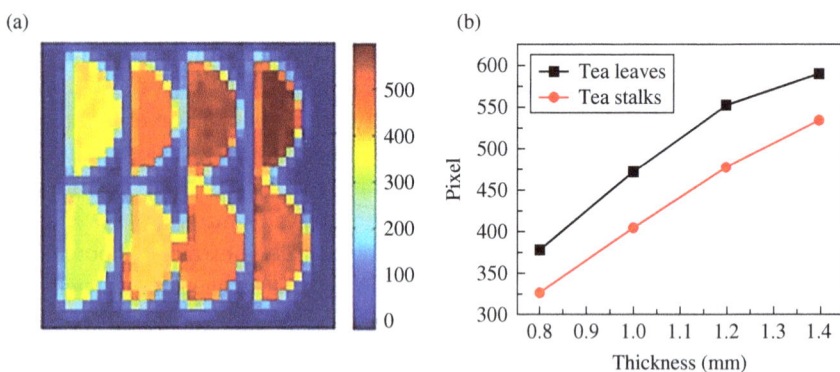

Figure 5.1 THz-TDS imaging for discrimination of tea stalks from finished tea products: (a) tea leaves and tea stalks pressing under transmission mode (images from left to right are pellets with 0.8 ± 0.1 mm, 1 ± 0.1 mm, 1.2 ± 0.1 mm, and 1.4 ± 0.1 mm, respectively. Images on top row and bottom row are pellets of pure tea leaves and pure tea stalks, respectively) and (b) relation curve between average pixel value and thickness of tea leaves and tea stalks. (Source: Sun et al. 2022.)

(Wang et al. 2019a). In their study, the presence of metallic contaminants in sausages was readily located using continuous-wave THz imaging (Figure 5.2). More applications related to THz systems as inspection and identification tool, agricultural applications, and for quality monitoring and control are detailed in the Chapters 6, 7, and 8 of this book.

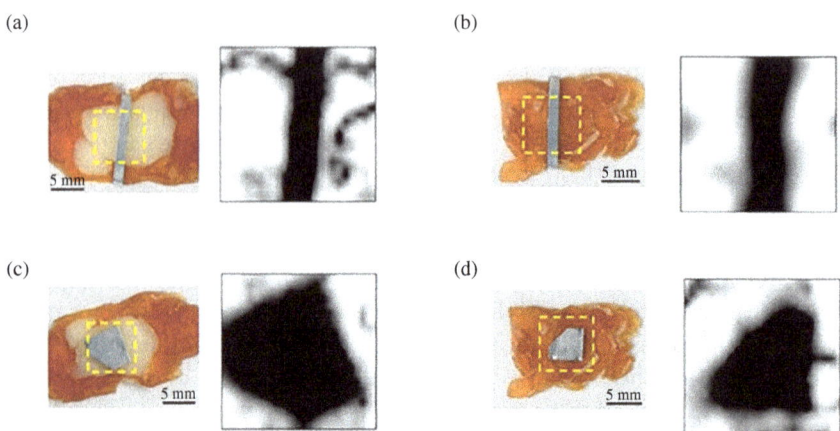

Figure 5.2 Detection of metallic contaminants in sausages using THz imaging: (a) metal strip in fat, (b) metal strip in lean meat, (c) metal polygon in fat, and (d) metal polygon in lean meat. (Source: Wang et al. 2019a.)

5.3 Scope of THz Spectroscopy in Material Characterization

The overlap of the energies of the hydrogen bond (<20 kJ/mol) with the photonic energy of the THz radiation (0.04–2.39 kJ/mol) allows the possibility of spectral investigation in the organic solids as well as liquids (Zeitler 2016). In the case of small molecule organic crystals, there exist the photonic vibrations along with other low-energy intramolecular motions such as torsional and bending at THz frequencies (Kim et al. 2015). This capability to probe molecular interactions of crystals has a great scope in material characterization and identification in agri-food industries. On the other hand, amorphous materials have no unique spectral features observed at THz frequencies. The electromagnetic radiation loss in glasses at a high frequency of spectral bandwidth occurs mainly because of the vibrational density of the state (Taraskin et al. 2006). Only a limited amount of information could be extracted from this universal frequency dependence of spectral range. The THz spectral features can be extracted not only from the solid samples but also from the liquid samples due to associated intermolecular interactions. Although the relaxation process of the liquids occurs at entirely different time scales than glasses, structurally liquids are quite similar to glasses (Zeitler 2016). This is mainly because of the absorbed radiation at THz frequencies during dipole relaxation process occurs on femtosecond to picosecond time scales. In addition to the dielectric relaxation, the intermolecular stretching aids in the THz absorption by hydrogen bonding liquids (Yada et al. 2009). This section describes diverse THz spectral applications in the characterization of two different states of solid materials (crystalline and amorphous solids).

5.3.1 Characterization of Crystalline Materials

5.3.1.1 Identification of Polymorphism

Polymorphism and the associated variations in crystallinity of materials exhibit varied physiochemical properties. The knowledge about these varied material properties is essential as it has a great impact on manufacturing, commercial, therapeutic, and legally permissible limits as in the case of pharmaceutical solids. The THz-TDS can be promisingly employed for the identification of complex polymeric systems, especially in the investigation of the polymorphism phenomenon and the transformation of crystals (Bawuah and Zeitler 2021). Li et al. (2017) explored the potential of THz-TDS in identifying the stereo-complex state of the polymeric crystal. Sensitive response of crystal structure of polylactide was observed where the α-, α′-forms, and stereo-complex crystals exert the absorption peaks at 2.01, 1.82, and 2.09 THz, respectively corresponding to lattice vibrations. Results inferred no identification of chirality among poly (D-lactide) and Poly (L-lactide) (Figure 5.3). Yet the stereo-complex polylactide showed an extra distinct absorption peak at 1.43

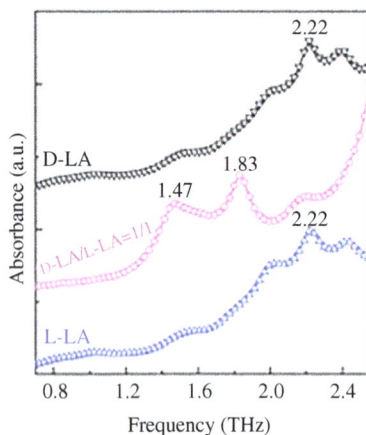

Figure 5.3 **THz spectra of poly(D-lactide) monomer (represented as D-LA), poly(L-lactide) monomer (represented as L-LA), and racemic compound of D-LA and L-LA. (Source: Li et al. 2017.)**

THz compared to homopolylactide. Further, this study showed that the transformation of polylactide from α' to α using THz-TDS was quite slow at 120°C.

The THz pulsed spectroscopy is capable of probing vibrations of long-range crystalline lattice, low-energy torsional vibrations, and hydrogen bonding vibrations. This characteristic advantage made THz pulsed system an ideal tool for the investigation of molecular polymorphism and crystallinity. Thus, the THz pulsed spectroscopy has been successively employed to quantify the solid-state properties, polymorphism levels, and crystallinity in four different drugs (carbamazepine, fenoprofen calcium, indomethacin, and enalapril) (Strachan et al. 2005). In another study, the THz-TDS in transmission mode was applied to analyze five different forms of modifications of furosemide at room temperature at 0.3–1.6 THz (Ge et al. 2009). Different forms of furosemide exhibit different absorption spectra that clearly showed the sensitivity of crystal structures. Further, the results of THz-TDS have been compared with X-ray powder diffractometry to validate and confirm the different forms of modifications of furosemide. Thus, THz-TDS has been proven as a potential analytical technique to investigate different forms of polymorphism in pharmaceutical solids.

5.3.1.2 Crystalline Phase Transitions

The THz-TDS is ideally used for the investigation of complex phase transition behavior in pharmaceutical drugs due to its feature of coherent detection with a rapid spectral acquisition rate and insensitivity to sample temperature (Sibik and Zeitler 2016). Based on the solid-state transformations, the polymeric phase transitions could be resolved through the application of heat to the sample resulting in the emergence of a new polymorph. The associated process of molecular transformation can be assessed by analyzing the kinetics of such processes (Zeitler 2016). Based on this approach, it is possible to analyze the dehydration behavior of the hydrate forms of samples as water molecules move out with increasing temperature due to heating. The detection of the structural collapse of crystals during dehydration is quite possible with high sensitivity of THz-TDS measurements. This deformation results in amorphous phase following the crystallization of the anhydrous form of samples. The unique aspect of THz-TDS over other vibrational spectroscopy techniques is the associated mechanism of dehydration behavior of hydrates (Bian et al. 2021). This includes the rotational transitions of water in vapor phase that exerts unique absorption spectra at THz frequency. These spectral lines are distinguishably characterized as narrow and intense, making them easily identifiable. It is feasible to determine the solid-state transitions and removal of moisture during the *in-situ* analysis of the drying from the sample's surface using THz-TDS. These sequences may occur either simultaneously or could be viewed in two separate distinct steps (Zeitler et al. 2007b). Understanding the chemical nature of the sodium chloride (NaCl) and its hydrate (NaCl.2H$_2$O) from NaCl aqueous solution is vital to studying the complex process of freeze-drying in foods and pharmaceuticals. A study was reported on the detection of NaCl particles in ice as precursors of NaCl.2H$_2$O during freezing of NaCl solution using THz spectroscopy (Ajito et al. 2018). Thus, this study opens a new application of THz systems to monitor the different forms of hydrates in complex food models. The THz-TDS is also adopted to study the solid-state synthesis of co-crystals through size reduction in addition to temperature-induced phase transitions. The THz spectroscopy was used to monitor the dynamic process of mechanochemical construction of two components co-crystal phenazine and mesaconic acid through grinding. The resultant THz spectra showed a clear distinct absorption peak of co-crystal at 1.2 THz (Nguyen et al. 2007).

5.3.1.3 Determination of Glass Transition Point

The THz spectroscopy can be used as an ideal tool to measure the glass transition point of the polymers with high accuracy. The transition from the glassy to rubbery state in the amorphous phase of polymers occurs at a certain range of temperature rather than at a definite temperature. It has been proposed that the segmental motions along the polymeric chains are frozen below the glass transition temperature (Wietzke et al. 2010). This is due to the insufficient volume between the neighboring molecules. It was reported that this kind of transition also has an impact on the refractive index of the polymers at THz frequencies with respect to temperature. To verify this proposed approach, Wietzke et al. (2009b) proposed a work on the applicability and suitability of the THz-TDS in determining the glass transition temperature of the semi-crystalline polyoxymethylene and the results were validated by the conventional DSC approach. Results showed an excellent agreement between the destructive method of DSC and the non-destructive approach of THz spectroscopy (Wietzke et al. 2009b). Further, this study provides valuable insights on the applicability of THz spectroscopy for β and γ transitions which helps in better understanding of glassy states of polymers. The dielectric parameters of different polymers including the refractive index and the coefficient of absorption can be conveniently measured using THz-TDS with high precision (Wietzke et al. 2011). Further, it has been proven that the temperature-dependent refractometric data allows to measure the glass transition temperature and the extrapolation over a broad temperature range below and above the glass transition temperature of the certain polymers.

5.3.2 Characterization of Amorphous Materials

5.3.2.1 Onset of Crystallization

The vibrational density of crystalline states in samples yields stronger absorption and could be used as a sensitive probing for crystallization rather than the spectral baseline. Several studies and researches are progressing in exploiting this sensitivity to better understand the molecular variations associated with onset of crystallization (Engelbrecht et al. 2019; Li et al. 2022; Santitewagun et al. 2022). The hydration and crystallization of the amorphous lactose were studied using THz-TDS (McIntosh et al. 2013). The absence of long-range order in the amorphous state implies lactose to exhibit a different THz spectrum (Figure 5.4). The associated difference allows one to study the transition of amorphous lactose at high humidity into monohydrate forms. An *in-situ* variable temperature THz pulsed spectroscopy was used to study the crystallization of amorphous carbamazepine (Zeitler et al. 2007c). Although there were no distinct spectral features with the

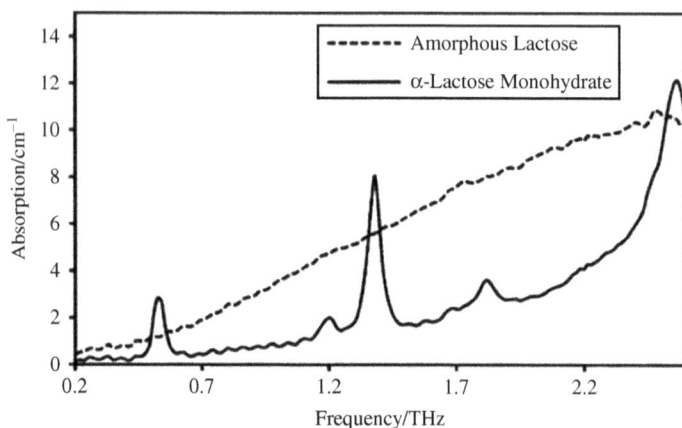

Figure 5.4 THz spectra of freeze-dried amorphous lactose and α-lactose mono-hydrate. (Source: McIntosh et al. 2013.)

disordered materials in a glassy state, the study demonstrated a subtle change in THz spectra with increasing temperature. Thus, it can be possible to study the relaxation and crystallization of materials under varying temperatures using THz pulsed spectroscopy.

5.4 Qualitative and Quantitative Measurements in the Agri-Food Sector

Several unique features of the THz radiation over the other electromagnetic spectra are transient, coherent, transmission, broadband, low energy, fingerprint spectrum, and water absorption property (Qin et al. 2013). All these properties make THz waves a suitable candidate for the non-destructive quality evaluation of agri-food products. The broad range of applications of THz radiation can be categorized into four major classes: spectroscopy, imaging, sensing, and communication which are greatly applied in quality control, inspection, and monitoring in agri-food industries (Figure 5.5). The diverse range of applications of THz waves in agri-food industries dealing with the characterization of macro- and microelements of food is summarized in the following sections. Also, a summary of the THz spectroscopy and imaging applications for the characterization of samples in the agri-food sector is presented in Table 5.1.

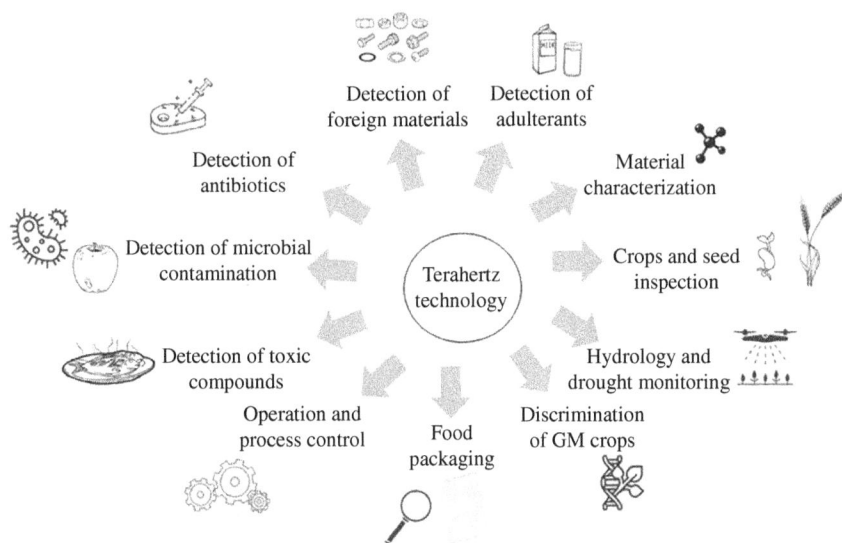

Figure 5.5 Applications of THz technology in agri-food industry.

Table 5.1 Summary of Recent Studies on Application of THz Spectroscopy and Imaging

Analyte	Technique	Process Parameter	Highlights	References
Amino acid mixture in cereal (L-Glutamic acid, L-Glutamine, and L-Tyrosine)	THz-TDS in transmission mode	400–4000 cm^{-1}	THz absorption spectra and concentration profiles of three amino acids components were resolved by MCR-ALS	Zhang et al. (2017d)
Protein conformation – Bovine serum albumin	THz-TDS	0.05–2 THz	THz-TDS plus t-SNE-XGBoost proved to be an effective non-destructive and label-free method for evaluation of protein conformation	Cao et al. (2020)
L-ascorbic acid	FTIR-ATR-THz	20–450 cm^{-1}	Effect of sample temperatures (22°C, 31°C, and 40°C) on model performance of L-ascorbic acid was analyzed	Yulia et al. (2014)
Melamine	THz-TDS	Maximum cut-off frequency of 2.7 THz	Strong absorption peaks were observed at 2, 2.27, and 2.61 THz with redshifts compared to theoretical results	Hwang et al. (2015)
Wheat maltose	THz-TDS in transmission mode	0.1–3.5 THz	Proposed fusion modeling algorithm is effective in quantification of maltose content in wheat samples	Jiang et al. (2020)
Dye – Auramine O	THz-TRS	0.2–1.6 THz	Combined THz spectroscopy and 2DCOS-PLSR is an excellent quantitative analysis method for auramine detection	Zhang et al. (2017b)

Table 5.1 (Continued) Summary of Recent Studies on Application of THz Spectroscopy and Imaging

Starch	Fourier transform THz spectrometry	Frequency range 3–13.5 THz with 0.12 THz resolution	Intensity peak obtained at 9 THz using Fourier transform THz spectroscopy was well correlated with the starch crystallinity from X-ray diffraction	Nakajima et al. (2021)
Moisture mapping	THz time-domain imaging system	Frequency range 0.5–1.3 THz with interval of 0.0031 THz	A dual autoencoder generative adversarial nets guided THz spectral dehulling model was proposed to map the moisture content in the sunflower seeds	Lei et al. (2022)
Ternary saccharide isomer mixtures [D-(–)fructose, D-(+) galactose anhydrous, and D-(+)mannose]	THz-TDS	The absorption bands of the ternary sample located at 1.69–1.75 THz, 1.98 THz, 2.14–2.15 THz, and 2.34 THz	THz-TDS could be used for the analysis of the ternary saccharide isomer mixtures with potential for the nutritional analysis of complex mixtures	Du et al. (2019)
Bisphenol A	THz-TDS	0.1–3 THz	THz-TDS combined with SVR predicts the concentrations of bisphenol A with R^2 0.98	Sun et al. (2021c)
Dipeptides	THz chiroptical spectroscopy	1.2–2 THz	The sensitivity of chiral phonons to smaller structural changes is used to identify physical and chemical differences in an identical formulation of dipeptides in health supplements	Choi et al. (2022)

(Continued)

Table 5.1 (*Continued*) Summary of Recent Studies on Application of THz Spectroscopy and Imaging

Bovine serum albumin proteolysis	THz reflective time-domain polarization spectroscopy (RTDPS) system	The effective frequency spectrum ranges from 0.05 to 2.5 THz, frequency domain resolution is 6.25 GHz, and the SNR ratio is 10^5	THz polarization characteristic spectra of the bovine serum albumin solution have significant differences under different amounts of papain proteolysis	Zhang et al. (2022b)
Sugars (D-glucose, sucrose, and cellulose)	THz-TDS	0.5–2.5 THz	Strongly localized and enhanced THz transmission by nano-antennas effectively increased the molecular absorption and assisted in detection of sugars	Lee et al. (2015)
Saccharides (D-glucose, α-lactose hydrate, and β-maltose hydrate) in solid and liquid states	THz-TDS	0.1–2.5 THz	Each saccharide had a unique spectral characteristic with higher correlation between the THz absorption spectra of the same substance in both the solid and liquid states	Huang et al. (2022a)

Table 5.1 (Continued) Summary of Recent Studies on Application of THz Spectroscopy and Imaging

Osmolytes	THz-TDS	0.3–2.5 THz	A good correlation was observed between the mobility change in terms of water rotational dynamics and the denaturation temperature of ribonuclease A among the studied 15 osmolytes. Results confirmed that the molecular dynamics of water around the protein is a key factor for their denaturation	Hishida et al. (2022)
Melatonin and Circadin	THz-TDS	1.5–4.5 THz	Results showed two characteristic spectral features with a predominant feature at 3.21 THz and a weaker one at 4.2 THz	Puc et al. (2018)
Glucose anhydrate and monohydrate mixtures	THz pulsed spectroscopy (TPS) THz-TDS	0.8–2.2 THz	THz signatures of anhydrate located at 1.44 THz and monohydrate located at 1.82 THz. Results indicate THz spectroscopy has potential applications in industrial monitoring of glucose powders	Yan et al. (2021)
Medicinal herb pollen Typhae – Auramine O	THz-TDS	0.2–1.6 THz	THz-TDS combined with VIP-PLSR has a potential for the rapid and non-destructive prediction of harmful additives residue	Zhang and Li (2018)
Azorubin	THz-TDS	0.2–4.5 THz	THz-TDS can be used as potential tool in the detection of colorant in foods	Leulescu et al. (2021)

5.4.1 Carbohydrates and Sugars

As stated earlier, the ATR-THz spectroscopy combined with chemometrics is used as a quantification tool for analyzing the concentrations of glucose in an aqueous state. Making use of the distinct spectral features of the molecules, the THz-based Fourier transform spectrometer is applied to evaluate the absorption spectra of glucose solutions. Chemically, D- and L-glucose are enantiomers that are difficult to differentiate using the conventional spectroscopy methods. Zheng et al. (2012) applied THz-TDS for the identification of molecules of glucose and fructose. At room temperature, both glucose and fructose have distinct THz spectra. In this study, the Hartree-Fock (HF) function and density functional theory (DFT) were adopted to determine the vibrational frequencies and the ground state structures of glucose and fructose. Results inferred that the differences in THz spectra of the isomers were mainly because of the respective alignment and vibrational patterns of the intermolecular hydrogen bonds and intramolecular covalent bonds. The THz-TDS is conveniently used for differentiating the optical isomers based on their frequency range. In a recent study, Huang et al. (2020a) used THz spectroscopy for qualitative and quantitative identification of different brands of trehalose. The characteristic THz peak was majorly affected by the concentration of trehalose and the optical detection concentration was in the range of 25–55%. Results of THz spectroscopy were validated with the HPLC spectra for the same samples. This study inferred that THz spectroscopy is a suitable and reliable method for quantification of trehalose with variations of less than 5% from HPLC results.

The adulteration of fructose syrup in acacia honey was detected using a THz-TDS operated in ATR mode in combination with chemometric methods. These studies showed that the PLS algorithm was more reliable in detecting fructose concentration in the honey while the PLS-DA algorithm helped in the differentiation of the sources of honey-based on floral origin. Techniques like THz pulsed spectroscopy and THz-TDS are used for the quality monitoring of the crystallization of table sugar and their coatings on the confectioneries (May and Taday 2013). The phase transition of sucrose from aqueous to glassy state was monitored using THz-TDS and the THz imaging was employed for the quality evaluation of the proper coating of sugar in the final product. Results showed that the absorption spectrum of the sugar solution was decreased with time and a series of THz photon bands were observed in the spectra. The resulted spectral band of the sugar was due to the collective excitation of atoms at their crystalline state. The amount of hydration of saccharides can be assessed using THz-TDS. The applications of anhydrous and monohydrate saccharides in food systems are different as anhydrous forms of sugar are used to produce candies and monohydrated glucose is used for direct consumption for instant energy. Both monohydrated and anhydrous glucose exhibit distinct THz spectra at room temperature. The dehydration kinetics of the monohydrated glucose was assessed based on a solid-state theory using THz-TDS. Results showed two more spectral features in addition to the basic spectral bands as reported by

Zheng et al. (2012). The spectral differences were mainly based on the crystalline unit cell. The distinction in the THz spectra was due to the molecular interactions of anhydrous glucose and intermolecular interactions of glucose-glucose and glucose-water molecules in case of monohydrated glucose.

It is possible to determine the molecular stability in solution based on amplitude and phase stabilization using THz technology. Tajima et al. (2016) used a CW double beam THz system with a photonic phase modulator for improving the phase stability. The developed system was used for sensing glucose hydration for understanding the molecular interactions of glucose and water. Under different concentrations of glucose, the Fabry-Perot effect was observed that reveals the transmission attributes of the glucose solutions. This physical event gradually decreased the measurement accuracy of dielectric constants. However, the use of appropriate calibration procedures eventually improved the accuracy of the measurements. Thus, the signal stability was greatly improved, and a THz liquid sensor was successfully used for the characterization of liquid samples.

Interestingly, the determination of hydration shells of the polymers such as inulin, agave fructans, and maltodextrin using THz spectroscopy was compared with the calorimetry methods (Morales-Hernandez et al. 2019). The aqueous polymer solutions showed an inverse increase in absorption coefficient and hydration number with reducing polymer concentrations. The obtained results agreed with the chaotropic model for the hydration shell of polymers. The calculated hydration numbers using THz spectral system were slightly greater than values obtained from the calorimetric analysis. The THz-based real-time monitoring system was used for monitoring the fermentation broth during ethanol production. The mixtures of fermentation broth (sugars and ethanol) were selective to THz frequencies that are based on FP effects. Results showed that the sugar and ethanol mixtures possess the highest THz reflection sensitivity. The effect of the different types of sugars on yeast growth can be analyzed using THz technology based on frequency-averaged reflection measurements (Fawole et al. 2015). Among the tested samples, artificial sweeteners resulted in maximizing the growth of yeast and gas production. Similar approaches using THz-TDS were used for the investigation of dielectric properties of polycrystalline sugars (Sun and Zou 2016), and evaluation of the hydration state of saccharides (Shiraga et al. 2013). Thus, THz technology was successfully used as characterization tool for analyzing different types of sugars and carbohydrates.

5.4.2 Proteins and Amino Acids

The THz spectroscopy has been widely applied for quantification of proteins and amino acids, intermolecular interactions of proteins, and analysis of conformational changes in the protein structures. The THz spectral system was utilized for determining the correlations of the amino acids and their mixtures such as glycine (Gly), serine (Ser), Gly/Ser, and dipeptides of glycine-serine, and serine-glycine (Bian et al. 2020). The THz molecular vibrational intensity of Ser was higher than

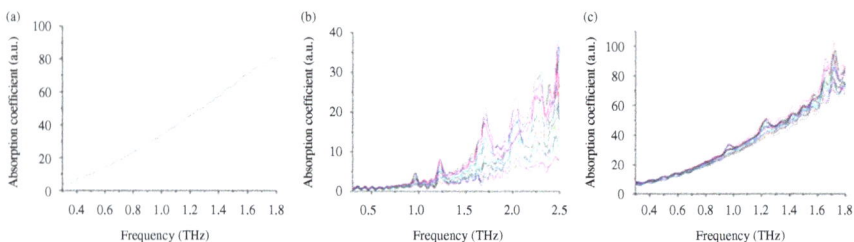

Figure 5.6 THz spectra of amino acids mixture in cereals: (a) foxtail millet powder, (b) amino acids mixtures at different concentrations in polyethylene matrix, and (c) amino acids mixture at different concentrations in foxtail millet matrix. (Source: Zhang et al. 2017d.)

Gly because of the presence of the hydroxyl methyl group on the side chain of the Ser. Also, the spectral results of amino acid mixtures imply the linear addition of the individual absorption spectrum. On the other hand, the THz spectra of dipeptides were distinct from other spectra due to the varied collective vibrations of the terminal groups of dipeptides. Comparatively, the dipeptides exhibit varied and distinct spectra that greatly differ from that of amino acid mixtures. Zhang et al. (2017d) worked on the quantification of the amino acids and chemical structures of ternary mixtures using THz-TDS. Each of the amino acids (L-glutamic acid, L-tyrosine, and L-glutamine) possessed a distinct absorption spectrum in the frequency range of 0.3 and 1.8 THz (Figure 5.6). The results were analyzed using the chemometric statistical models such as PLS and SVM. Also, the spectral data were processed using multiplicative scatter correction (MSC), wave transformation, first derivation, and Savitzky-Golay smoothing (SGS). Among the used processing methods, SVM chemometric method with MSC preprocessing resulted in the highest, stable, and accurate prediction. The resolution of the obtained spectra of amino acids was increased through multivariate curve resolution alternating least square (MCR-ALS) algorithm. Thus, the profiles of amino acids in the cereal media were similar to that of the spectrum of standards (pure amino acids). A similar approach was adopted for quantitative analysis of the ternary amino acids from foxtail millet (Lu et al. 2018). The spectral information of the targeted amino acids was obtained using the Tchebichef image moment method and the respective spectral features were detected based on fingerprint THz spectra. In another study, the concentrations of the dietary supplement of amino acids were quantified using THz-TDS (Ueno et al. 2011). Also, the effectiveness of THz imaging was reported by analyzing two different commercial amino acid tablets. Similar spectral patterns were evident for the wrapped as well as unwrapped samples. Both the experimental and calculated THz spectrum agreed with each other.

THz-TDS as a novel and label-free technology is used to determine the critical micelle concentrations (CMCs) of the surfactants (Yan et al. 2016). The refractive index and the absorption coefficients of different concentrations of ionic surfactant (sodium dodecyl sulfate) and nonionic surfactant (nona oxyethylene monododecyl ether) were measured at a frequency ranging from 0.2 to 1.8 THz. A distinct clear intersection of the THz absorption coefficient was observed with the formation of the micelle. This was due to changes that occurred in the component ratio among the surfactant, bulk water, and hydration water. The corresponding concentration associated with this phenomenon was measured as the CMC of the surfactant. The protein-based hydrophilic and hydrophobic solutes were characterized by the combined methods of THz spectroscopy with molecular simulations (Niehues et al. 2011). A frequency of less than 2.7 THz is sufficient for measuring the optical behavior of peptides and amino acids. A good relation was observed between polarity and solute's hydrophobicity and the solvent's THz absorption. The nature of the hydrophilicity and hydrophobicity of the model particles on the water dynamics was studied using molecular dynamic simulations. Results inferred a remarkable increase in the vibrational density of the oxygen molecules present in solvating water and the absorption of hydrophilic solutes than that of hydrophobic solutes. Recently studies on the applicability of antifreeze glycopeptides due to their diverse range of properties such as regulation of nucleation and control of growth of ice crystals are getting research interest. Due to these properties, antifreeze glycopeptides are greatly used in food applications in the production of ice creams and frozen fruit desserts. Hence, the structural characterization of antifreeze glycopeptides is significant. The relation between the chemical state and the hydration behavior of the antifreeze glycopeptides was reported by Urbanczyk et al. (2017). This study helps in better understanding the interactions among water-protein molecules and the effects of hydration of peptides on protein stability.

5.4.3 Lipids and Fatty Acids

The effect of heat treatments on the physicochemical changes of the cooking oil can be analyzed using THz-TDS (Dinovitser et al. 2017). This study analyzed the sample thickness over the dynamic range. In a window of about 80 ps (picoseconds), a peak was observed at 5 ps for the reference THz pulse. Similar FP spectra and refractive indices were obtained for all types of oil samples at considered thickness. Based on the observations, the best measurements were obtained at low THz frequencies for the thick samples while the thin samples yielded optimal measurements at high frequencies. The oil samples showed similar spectra at frequencies greater than 1 THz. However, the absorption coefficient of the cooking oils had slight variations when the oil samples were heated well above their smoke point. In contrast, different results were obtained by Zhan et al. (2016) who inferred a significant difference in the absorption spectra of fresh and used (gutter) cooking oils. In another report, the dielectric properties of vegetable oils, butter, and lard were examined using THz-TDS (Moller et al. 2010). The resulted absorption spectrum of each sample had unique characteristics for imaginary and real parts of the dielectric function. The THz-TDS can also be used to find the correlation

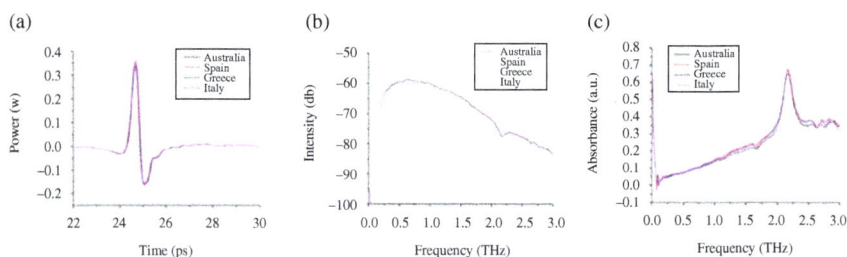

Figure 5.7 THz-TDS spectra of virgin olive oils: (a) frequency domain spectra, (b) absorbance spectra, and (c) absorbance vs frequency plot. (Source: Liu et al. 2018a.)

between the fatty acid chain length and intermolecular forces (Fan et al. 2019). The absorption THz spectra of crystalline fatty acids showed a systematic redshift with an increase in the carbon chain length at the temperature of 96–293 K in the frequency between 0.5 and 3 THz.

The THz-TDS combined with chemometric methods helps in the differentiation of the oils based on the place of origin. Liu et al. (2018a) examined the applicability of THz systems to characterize olive oils based on their geographical source of origin. The differences that exist in the absorption spectra were linked with the respective fatty acid compositions that aid in the classification of oils (Figure 5.7). Chemometric models such as PCA combined with GA, RF, LS-SVM, and BPNN were employed for the screening and classification of oils. Among the tested models, GA combined with LS-SVM yielded in the best result with higher accuracy. A similar work was conducted by Yin et al. (2016) for the identification of edible oils. The average THz absorption spectra were observed to be in the range of 1.5–3.5 THz for the tested oil samples. The combination of PLS, iPLS, and backward interval algorithms was used to validate the performance of the chemometrics used. Results inferred that the GA-PLS algorithm yielded the highest prediction accuracy followed by the genetic algorithm combined with partial least squares (GA-PLS)-DA algorithm with the smallest RMSE, largest R of prediction, and highest accuracy in the classification of oils.

In another study, the iterative model was applied to enhance the prediction accuracy of the optical parameters of the vegetable oils (Jiusheng 2010). A time delay in THz pulse of the oil sample was observed with respect to the reference. The FP effect was obvious with the display of several supplementary peaks in addition to the main THz spectra. The refractive index of the samples showed an inverse relation with the frequency while the power absorption coefficients were directly related to the THz frequency. The spectrum of oil was attributed to molecular vibrations in the THz region with the intermolecular vibrations spotted at frequency between 1.1 and 1.5 THz. The intermolecular vibrations were mainly due to the bending of the hydrogen bonds. About 20% variations were observed in the measurements of the refractive indices and the power absorption coefficients of the oil samples. This was evident due to the use of conventional data processing methods. Thus, the study demonstrated that the accuracy in the prediction of the optical parameters of the oil samples could be enhanced by an iterative algorithm.

Application of THz spectrum for estimation of trans fatty acids (TFA) content in cooked soybean oil was reported by Lian et al. (2019). About 480 oil samples were analyzed in transmission mode at 0.2–1.5 THz. Classification algorithms such as PLS and sub-PLS models were used to estimate the TFA. Results showed that the sub-PLS algorithm showed better prediction than the PLS. Jiang et al. (2011) reported a study on the use of THz spectroscopy for analyzing the chemical structures of saturated and unsaturated fatty acids. Spectral results showed a sharp peak for saturated fatty acids while two distinct peaks were obtained for the unsaturated fatty acids. These spectral peaks represent the carboxylic groups of the fatty acids. This study confirmed that the THz absorbance of the fatty acids depends on concentration. The THz-TDS is also helpful in the characterization of the isomers of the fatty acids. Kang et al. (2011) proposed a study to determine the optical parameters of a trans isomer of oleic acid, i.e., elaidic acid. The associated intra- and intermolecular vibrations resulted in the unique absorption peaks.

5.4.4 Micronutrients

The low-frequency vibrations of vitamins were measured using THz spectroscopy. The peak intensity and THz spectral position were temperature dependent. The vibrational modes of the compounds were measured based on quantum chemical calculations. The THz-TDS was used for quantification as well as detection of intracellular metabolites (vitamins). Suhandy et al. (2011) reported a study on FTIR-ATR THz spectroscopy to quantify the vitamin C in aqueous solutions. The iPLS regression was adapted to select the variables and spectral regions for obtaining a calibration model. Results showed that iPLS model with 5-PLS factors yielded lower standard error in the prediction with better performance than full-spectrum PLS algorithm. As a result, the concentration of L-ascorbic acid in 33 solutions of vitamin C in the range of 3%–21% was successfully quantified using ATR-THz spectroscopy. In another study, L-ascorbic acid, and vitamin B1 were detected using FTIR and THz-TDS spectroscopy (Li et al. 2015a). Results of both the spectroscopic methods were consistent apart from a new peak appearing in the low-frequency region in the FTIR spectrum. The THz absorption spectrum was also used for the detection of L-histidine (0.77 THz) and α-lactose (0.53 THz) in food supplements (Wang et al. 2020c). Among the tested models, linear least square regression (LLSR) showed better performance than PLS regression (PLSR) with R values of 0.9899 and 0.9910 for L-histidine and α-lactose, respectively (Figure 5.8). Further, the estimated accuracies for L-histidine and α-lactose were comparable to that of ion chromatography and HPLC.

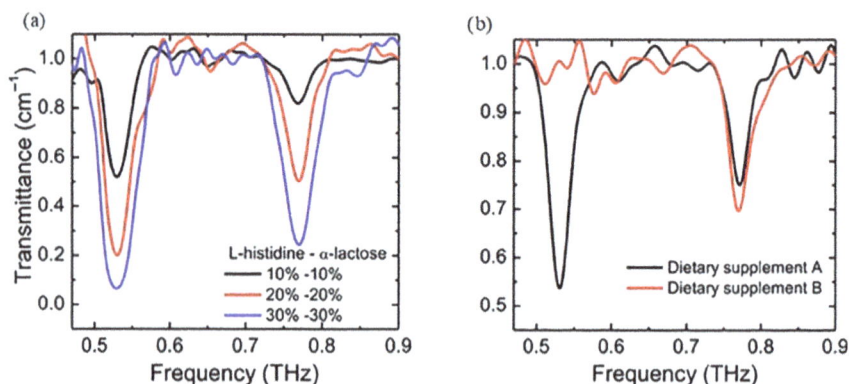

Figure 5.8 THz-TDS spectra of dietary supplements: (a) THz spectra of mixtures of L-histidine and α-lactose in corn starch sample under transmission mode and (b) commercial dietary supplements after removal of baselines. (Source: Wang et al. 2020c.)

5.4.5 *Moisture Content*

All the food components possess different THz absorption coefficients. It was reported that the absorption coefficient of lipids was 20 times, starch 50 times, and proteins 100 times smaller than that of water (Qin et al. 2013). Thus, water is one of the food components that intensely absorbs THz radiation. This characteristic property of water toward THz waves makes THz-TDS a promising nondestructive dynamic tool for moisture measurements of agri-food products. The conventional methods of moisture measurement techniques such as distillation, oven dehydration, and IR spectroscopy methods are laborious and consume more time. Considering these limitations, Chua et al. (2005) published a report on the determination of moisture content of food grains using THz spectroscopy. The crushed wheat grains were sieved to achieve a uniform particle size for homogeneity in the measurements. Resultant apparent spectrum was analyzed by detecting the spectra of wet samples from that of dried samples showing a linear relationship that implies the 18% moisture content samples had the highest absorption. In another study, a micro-structured PCA was employed to determine the moisture content of coffee powder (Yasui and Araki 2005). The exposure of samples to forced humidity for about 3.5 h resulted in significant changes in the transmitted spectra of coffee samples. In another study, THz spectroscopy was used as a noninvasive analytical aid to measure the moisture content of wafers (Parasoglou et al. 2009, 2010). A linear relationship was observed between the moisture content and the amplitude of the THz peak for the dehydrated wafer samples. Results showed that the relative water absorbance was higher at a frequency below 0.6 THz with a small amount of scattering effects of pores. On the other hand, the scattering effects of pores of wafers were quite dominant at frequencies above 0.6 THz with a weakened THz signal.

Glycine solutions are commonly used as an additive in energy beverages and other protein supplements. In an attempt to differentiate the aqueous glycine solution and glycine powder based on their water content, Ogawa et al. (2009) proposed a study on the analysis of THz absorption spectra. Results showed a distinct unique absorption peak for dissolved glycine that was different from that of the glycine powder. Based on the vibrational frequencies of the molecules, THz spectroscopy is not only used to measure water content but also the status of the water in food samples. The refractive index, as well as the absorption coefficient of the condition of water in the crystalline and noncrystalline systems, have been measured based on the plane-wave density functional method. Results of the samples of alcoholic beverages agreed with the reference samples used. Thus, the measurement accuracy greatly depends on the sample's stability at the tested temperature. The nature of the hydrated ions in aqueous salt solutions has been analyzed using THz spectroscopy (Funkner et al. 2012). It was found that the absorption THz spectrum of salt solution is linear and dependent on their concentration. Thus, the THz vibrational spectrum of hydrated ions is helpful in the determination of the dynamics and the structures of ions that are surrounded by water molecules.

The THz-TDS systems are also used as a humidity sensor in food systems. The variations in the hydrophobic and hydrophilic properties due to the humidity change are the basic principle for the use of the THz system as a sensing and monitoring tool in polymeric films. It was reported that an increase in the film thickness due to coating increased the refractive index with a shift in the resonance frequency to lower levels (Shin et al. 2017). Finite-difference time-domain (FDTD) simulation study on guided-mode resonance (GMR) sensors resulted in a good estimation of water absorbance by polyvinyl alcohol-coated sensor. This study proved that the PVA-coated GME sensor can be efficiently used as an inexpensive tool for RH sensing. Results inferred that the moisture content of the PVA film increased to 6% with an increase in the RH to 70% at 20°C.

5.5 Technological Advancements and Futuristic Applications

Compared to other spectroscopic techniques, THz-TDS is more conveniently applied in a vast array of qualitative and quantitative assessments. With recent technological innovations in THz systems, a wide range of optical detection schemes and configurations are developed (Vitiello et al. 2021). This advancement in THz science has led to novel designs of working systems with greater flexibility that broadens its applications in diverse industrial sectors. The time-resolved nature of the system provides unique information on samples that are relatively inaccessible with continuous-wave systems. This feature helps in realizing the applications in a real-time scenario, especially in material characterization, analysis, and quality control (Arteaga et al. 2021). The advantage of using both spectroscopy and imaging techniques together has extended the range of THz radiation applications to meet a wide range of analytical challenges. For instance, the THz spectroscopic systems are well used in the detection of concealed objects and explosives based on their spectral features (Praveena et al. 2022). The analytics of tablets coating and profiling of active ingredients emerged as an ideal realistic method suitable for the pharmaceutical industry (Patil et al. 2022). Similarly, a variety of biological tissues such as vital organs, dermal tissues, and dental parts are promisingly assessed for tumor growth and development using THz waves (Gong et al. 2020). The characterization of polymeric compounds has been illustrated and turned out to be a potential method for analyzing additives during online measurements in the packaging industry (Zhong and Nsengiyumva 2022). Thus, THz science provides a remarkable solution to the modern analytical problems ensuring safety, quality monitoring, and inspection.

Despite these versatile applications, the THz systems suffer from certain hurdles, especially when dealing with the analysis of solid samples. The THz investigation is relatively restricted to dry samples which could be positively overcome with the adoption of innovative methods developed using higher radiant power than the currently available systems (Zhang et al. 2021b). Water is an essential component that exerts intense broad absorption across the THz frequency range. The extent of hydration in dried samples significantly influences the resulting spectral features. Recently the wet powder spectroscopy method was proposed to analyze the variation in biological samples (Cherkasova et al. 2020). For instance, to know about the chemical state of water transition from free to bound forms and vice-versa. In this study, the THz transmission of dry glucose was pressed into pellets and subsequently, the pellets with added water were measured. Results inferred that the THz absorption of bound water falls under a magnitude less than free water. The acquired spectral response described different states from wet powder to the dilute solution. These observations on characteristics of bound water would allow to gain deeper knowledge on determination and prediction of THz response of biochemical characterization. The effects of reflection and scattering may often incur artifacts in acquired THz images which could be eliminated by adoption of appropriate sample preparation techniques and signal preprocessing methods.

In a more recent study, the applicability of the THz-TDS was successfully used in resolving the molecular states of water and retention properties of the Nafion membrane (Alves-Lima et al. 2022). The developed THz-based characterization method demonstrated to have broad industrial applicability and the obtained results were supported by the conventional gravimetric analysis. This study provides useful future insights, especially for the optimization of membranes developed by different thermal treatments. Undoubtedly, THz characterization can be used as a potential technique for rapid testing of membrane performance and allows a better understanding of the material properties in optimizing performance stability.

Over the years, the design of the conventional THz spectroscopic and imaging systems has been subjected to constant investigation for improving its working performance. The advancements in delay line technology provide promising results in improving the scanning rate of imaging systems. In a recent study, the design and fabrication of a 2D PCA array were reported along with its characterization ability to detect both the amplitude and phase of the THz pulse (Henri et al. 2021). The developed PCA array was made from LT-GaAs comprising eight channels with 64 pixels. This unique approach utilizes the spatial light modulator (SLM) to modulate and focus the infrared beam toward PCA array pixels. The use of eight-channel parallel data acquisition assisted in the high acquisition of THz images. Further, the acquisition speed could be enhanced by integrating with high-speed SLMs. These SLMs are commercially available with integrated fast delay line (oscillating and rotary) techniques. Thus, this study highlights the practical importance of the proposed PCA array in the emergence of the real-time next-generation THz imaging systems. In another report, high-speed THz imaging was proposed using a two-way raster scanning without the dwell time being evaluated. The proposed THz-TDS pulsed imaging system is comprised of fiber-coupled PCA and a motorized X-Y linear stage (Han and Kang 2019). The results obtained from the proposed method yield a scanning speed of 328% faster than the conventional scanning. Also, the modal assurance criterion value used to quantify the scanning accuracy was higher than 0.994 without any additional installation of equipment.

A well-designed sub-wavelength structure at the micro- and nanoscale would result in higher output power generation and the emission of broadband THz pulses, which is quite useful for the efficient performance of imaging systems (Malhotra et al. 2018). Also, the use of metamaterial chips and nanomaterial contrast agents helps in overcoming the concerns with the diffraction limit. The random step frequency approach can be applied to assess the approximate range of the 3D surface layer thereby avoiding the range ambiguity that occurs due to the reduction in the sampling frequency. This kind of research advances in the design of the THz antenna for imaging would open a new dimension of characterization, sensing, monitoring, and communication applications under the THz regime. The THz systems have evolved to be developed with reduced apparatus size with the developments and use of femtosecond laser technology. These systems are capable of generation of near-infrared pulses with shorter duration providing increased spectral bandwidth. The transmission efficiency of THz pulses is improved using novel antireflective coatings suitable for THz frequencies that eventually allow for increased throughput and improved signal-to-noise ratio (SNR) (Sun et al. 2021a). All these research investigations are still in progress that would certainly result in the emergence of more powerful analytical solutions using THz waves. Thus, the THz regime of the electromagnetic spectrum provides unique information about the variety of samples of scientific and engineering interest.

5.6 Summary

The THz science has received a higher research potential in recent years and has been widely applied in diverse industrial sectors for process control, monitoring, sensing, and communication applications. The diverse applications of THz in security monitoring, art conservation, packaging industry, pharmaceutical industry, food industry, and biomedical and biosensing applications have been summarized. Also, the scope of THz systems in material characterization and the significance of qualitative and quantitative measurements of agri-food samples are detailed. Different types and configurations of THz spectroscopy and imaging systems are available to be used for selective end-use applications and specific needs. Further many active signs of progress are witnessed in this area in developing improved designs for achieving higher performance efficiency to match up the industrial needs and real-time applications. The technological advancements would bring up a new opportunity in integration with other novel technology like 3D printing. The use of the 3D printer for developing prototype parts of THz systems is forecasted to provide versatility and higher degrees of freedom in the design of THz system configurations. Certainly, this would extend the potential applications of THz in many areas of applied and basic research. Considerations like specific sample preparation techniques, selection of appropriate preprocessing methods and chemometrics, careful data processing methods, standardization, and interpretation of results make THz analysis to be less tedious thereby surpassing the limitation of reduced analytical sensitivity for high moisture samples. Most of the onsite THz applications require handheld portable devices. Currently, the lack of small, portable, economical THz systems limits its applicability. The technological research progress and futuristic applications of THz spectroscopy and imaging applications related to sensing of agri-food samples are briefly described. Indeed, the active research needs of industrial markets and the designs of intelligent THz chips and portable devices would promisingly transform THz science as a reforming technology providing solutions for modern analytics in the future.

Chapter 6

THz Technology

An Inspection and Identification Tool

6.1 Introduction

Food safety has a direct implication on human health and wellness. In today's world, food safety and quality assurance have gained global attention. Different forms of food in our daily diet include whole foods, processed foods, minimally processed foods, beverages, and fresh-cut fruits and vegetables. Regardless of the extent of processing, food safety remains the primary factor that has been considered for the delivery of safe foods with appropriate nutritional composition (Abbas et al. 2022). Hence food safety must be ensured at all the levels of the food supply and distribution chain. Over the years, several food quality safety tests have been used for assessing the quality and safety of the foods. The common conventional methods used for food quality authentication are well-established physicochemical tests, sensory analyses, and microbial analyses (Khaled et al. 2021). However, the conventional techniques based on manual assessment methods have often led to human errors and can be biased, time-consuming, laborious, and unreliable. This highlights the need for the development of rapid noninvasive analytical methods for the current food safety problems faced by the food industry. From the food safety point, any kind of contamination, adulteration, infestation, or pathogenic infection not only decreases the nutritive value but could also cause serious illness (Ali et al. 2022). Indeed, it also significantly affects the economic value of the produce. Hence, many government organizations and regulatory bodies are continuously working in enforcing stringent food regulatory standards ensuring the safe delivery of products.

DOI: 10.1201/9781003197010-6

Considering all the above aspects, various kinds of spectroscopy and imaging methods have evolved over the past two decades. Apart from the traditional optical spectroscopy, other spectroscopic techniques such as infrared (IR) spectroscopy, nuclear magnetic resonance (NMR) spectroscopy, ultraviolet-visible (UV-Vis) spectroscopy, X-ray spectroscopy, and Raman spectroscopy are finding their applications in food quality assessment (Vadivambal and Jayas 2016; Wang et al. 2022a). In line with these spectroscopic techniques, THz spectroscopy is an innovative emerging analytical tool for food safety measurements. In contrast to other spectroscopic techniques, THz spectroscopy has been advantageously used because of its longer spectral band. The THz regime of the electromagnetic spectrum is in between the infrared and microwave regions that are characterized by low photonic energy (4 meV for 1 THz). According to the United States Federal Communications Commission (US-FCC), any radiation with photonic energy higher than 10 eV is considered as ionizing radiation. Hence, THz waves are nonionizing radiation (Ren et al. 2019a). Due to its biomolecule-friendly nature, the THz waves have no negative impacts on photoionization when applied to foods.

Among the various existing forms of THz systems (as discussed earlier in Chapters 2 and 3), the THz time-domain spectroscopy (THz-TDS) is the most used system for material characterization and assessment. The amplitude and phase information of the samples are obtained by measuring the electric field intensity of the THz time-domain pulse in THz-TDS (Gowen et al. 2012). The THz-TDS can be operated in either transmission, reflection, or the attenuated total reflection (ATR) modes. Among these, the transmission mode is the most used method. While the reflection mode is common for measuring the spectrum of liquid samples. On the other hand, the ATR mode can be employed for obtaining a higher optical coefficient. The selection of the operating mode of the THz system depends on the nature of the sample and the purpose of measurements (Wang et al. 2022c). With the development of modern THz sources and multi-performance THz detectors, THz technology has a great scope and potential applications in the agri-food sector (Figure 6.1). In this context, various applications of the THz in the physical and chemical characterization of materials are elaborately discussed in Chapter 5. This chapter covers the THz agri-food applications where THz systems can be used as a potential inspection and identification tool.

6.2 THz System for Safety and Quality Monitoring

Recently, there is a rising expectation from the food manufacturers to ensure standard quality of the processed foods as per regulatory standards. As stated earlier, most of the available food testing methods are destructive and require long testing time. And the current limitation with the available food inspection methods is their inefficiency to detect the foreign bodies within the packed foods. Considering this, the THz system is a potential tool for detecting the contaminants inside the

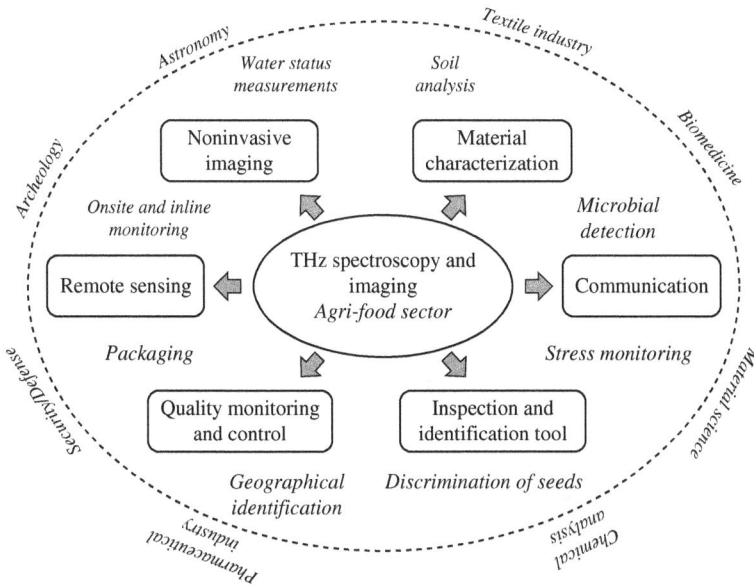

Figure 6.1   Scope and applications of THz in diverse areas of research.

packaging (Hu et al. 2021b). Broadly the food safety hazards can be categorized as physical, chemical, and biological. Examples of physical hazards are nails, screws, nuts, bolts, staples, glass pieces, metal particles, hair, and so on. The pesticides, herbicides, lubricants, heavy metals, detergents, and washing residues belong to chemical hazards. While the microbial contamination of foods is referred to as a biological hazard. Biological hazards breakouts are mainly because of the inadequate sanitation practices in operational plant and cross contamination during processing, handling, storage, and transportation (Singh et al. 2019). This section focuses on the prospects of THz spectroscopy and imaging systems in agri-food sector focusing the use of THz as inspection and identification tool (Table 6.1). For ease of readability, the THz applications are grouped as detection tool of antibiotics, detection of foreign bodies, detection of toxic compounds, and detection of adulterants. The discussions are made based on this grouping in the following sections.

6.2.1 Detection of Antibiotics

The traces of antibiotics in food products pose a serious threat to human health. The antibiotics can be accidentally entered into food streams, especially through animal feed. With the advancements in THz technology, it is possible to detect trace amounts of antibiotics in processed foods. Massaouti et al. (2013) reported a work on the detection of antibiotics and insecticides in honey samples at a very low concentration of up to 0.01 (w/w). The THz absorption peaks of the antibiotics

Table 6.1 **Various Applications of THz Systems as a Tool for Inspection and Material Identification**

Category	Analyte	Process Parameter	Highlights	References
Detection of antibiotics	Fluoroquinolones in fish meal feed	Spectral range 0–2.5 THz; THz frequency resolution 6 GHz	Using THz spectral imaging, the norfloxacin and enrofloxacin concentrations in fish meal feed were precisely visualized	Bai et al. (2022)
	Antibiotics (sixteen types of antibiotics)	Frequency range of 0.2–1.5 THz	All THz spectrum for the sixteen antibiotics were distinct. The use of t-SNE-PSO-SVM model resulted in the highest average accuracy of prediction of 99.91%	Guo et al. (2020b)
	Antibiotics in food and feed matrices	Frequency range of 0.1 and 2 THz	Eight of the eleven antibiotics showed specific fingerprints. The main spectral features of doxycycline and sulfapyridine were still detectable when they were mixed with three food matrices (cereal-based chicken feed, milk powder, and egg powder)	Redo-Sanchez et al. (2011)
	Tetracycline and its degradation products (epitetracycline hydrochloride, anhydrotetracycline hydrochloride, and epianhydrotetracycline hydrochloride – EATCH)	Frequency range of 0.2–1.8 THz; temperature between 4.5 K and 300 K	Results showed that tetracyclines exhibited numerous distinct spectral features in frequency-dependent absorption spectra, which demonstrated the qualitative capacity of THz-TDS	Xie et al. (2019)

Table 6.1 (Continued)   Various Applications of THz Systems as a Tool for Inspection and Material Identification

Detection of foreign bodies	Polymers materials (such as PVC, PE, and PP), metal gaskets, and insects in milk powder	Frequency range of 0.5–1 THz	Results showed that the order of absorption strength from strong to weak was metal pieces, insect, milk powder, PVC, PP, and PE material. These results indicate a promising scope for detection of foreign bodies in milk powder-based on THz-TDS	Hu et al. (2021b)
	Insect matter in Chinese black tea	Frequency range of 0.2–3 THz	THz spectrum combined with pattern recognition algorithm assists in non-destructive detection of tea inclusions of foreign insect bodies	Xu-dong and Jun-bin (2021)
	Insects	Frequency range of 0.3–1.2 THz	Results showed that food materials and foreign substances are distinguished and detected according to the dielectric traces and independent parameters extracted from the complex dielectric constants	Shin et al. (2018b)

(Continued)

Table 6.1 (*Continued*) Various Applications of THz Systems as a Tool for Inspection and Material Identification

Detection of harmful and toxic compounds	Carbendazim in rice flour at different concentrations of 0, 2, 4, 6, 8, 10, 15, 20, 25, 30, 40, 50, and 100%	Frequency range of 0.4–1.4 THz	This study proposed a new pesticide residue detection method based on a deep learning networks of the Wasserstein generative adversarial network (WGAN) and residual neural network (ResNet)	Yang et al. (2022)
	Aqueous solutions of imidacloprid and tetracycline hydrochloride (TCH) at different concentrations of 0.01, 0.02, 0.05, 0.1, 0.2, and 0.5 ppm	The resonant frequencies were found to be at 0.28, 0.45, 0.9, and 1.17 THz	The designed multi-band metamaterial was highly sensitive to detect traces of pesticide residues (imidacloprid and TCH) in food products	Wang et al. (2022d)
	Pesticides in wheat flour (dicofol, chlorpyrifos, chlorpyrifos-methyl, daminozide, imidacloprid, diethyldithiocarbamate, and dimethyldithiocarbamate)	Frequency range of 0.1–3 THz	All the seven pesticides were identified from its characteristic absorption peaks in wheat flour mixtures and a linear relationship were observed between the absorption coefficient and the weight ratio	Maeng et al. (2014)
	Pesticides in flour substrate	Frequency range of 0.1–1.4 THz	The optimized PLS models for both imidacloprid and carbendazim were obtained with R value of 0.9992 in prediction set	Cao et al. (2018)

Table 6.1 (Continued)   Various Applications of THz Systems as a Tool for Inspection and Material Identification

Pyrethroid pesticides (deltamethrin, fenvalerate, and beta-cypermethrin)	Frequency range of 0.06–3.5 THz	Results showed that there was a good matching effect between the THz experimental spectra and the DFT quantum calculation spectra for determination of concentration of multicomponent pesticides	Qu et al. (2018c)
Pesticides – chlorpyrifos, fipronil, carbofuran, dimethoate, methomyl, and thidiazuron	Frequency range of 0.1–3.5 THz	Spectral results showed that all pesticides had a distinct absorption peak. Combination of THz-TDS and DFT was effective in pesticide fingerprint analysis and the molecular dynamics simulations	Qu et al. (2018b)
Carbendazim pesticide	Frequency range of 0.4–1.4 THz	Results illustrate that metamaterial can detect trace amounts of carbendazim, as small as 5 mg/L, which is about 104 times enhancement compared to the squash method for THz-TDS detection	Qin et al. (2018)
Aflatoxin B1 in soybean oil (0 to 20 μg/kg)	Frequency range of 0.1-5 THz	THz spectroscopy was proved to be feasible to detect Aflatoxin B1 at 1 μg/ kg in soybean oil (over 90% accuracy)	Liu et al. (2019b)

(Continued)

Table 6.1 (*Continued*) Various Applications of THz Systems as a Tool for Inspection and Material Identification

| Detection of adulterants | Rice (high-quality rice mixed with low quality) | Frequency range of 0–6.4 THz | Results indicate that an SVM model employing the absorption spectra with a first derivative pretreatment exhibits the best discrimination ability with a prediction accuracy up to 97.33% | Li et al. (2020d) |
| | Adulterated dairy products – skim milk, low-fat milk, skim milk adulterated with fat powder and low-fat milk adulterated with fat powder | Frequency range of 0.1–1.5 THz | The difference between the spectra of samples can be observed with the fatty acids. SVM-DA models resulted in better prediction | Liu (2017a) |

Table 6.1 (Continued) Various Applications of THz Systems as a Tool for Inspection and Material Identification

Detection of additives	Benzoic acid	Absorption peak observed at 1.95 THz	Design and fabrication of metamaterial structure in micro/nanoscale	Hu et al. (2022)
	Benzoic acid in liquid-based foods	Spectral range 0.1–5 THz; THz frequency resolution 7.6 GHz	The designed metamaterial device was used for the effective detection of benzoic acid samples	Hu et al. (2021a)
	Ten different citrate salt powder samples	Absorption peaks found at the frequency range of 0.5–3.0 THz except the ferric citrate and calcium citrate tetrahydrate	THz spectroscopy was efficiently used to detect the citrate salts and differentiate their hydrates	Shen et al. (2022a)
	Six food additives that includes coumarin, 6-methyl coumarin (6-MC), ethyl vanillin (EVA), cinnamic acid (CINN), methyl vanillin (MVA), and vanillyl alcohol (VOH)	Frequency range of 0.5–2 THz	THz spectroscopy combined with manifold learning and SVM model was used for accurate identification of coumarin-based food additives	Chen et al. (2022)

(Continued)

Table 6.1 (*Continued*) Various Applications of THz Systems as a Tool for Inspection and Material Identification

Lithium citrate		Frequency range of 0.5–3 THz	Lithium citrate tetrahydrate showed an obvious absorption peak at 1.66 THz at room temperature. The THz-TDS can be used to detect crystalline hydrates and to study the dehydration kinetics of crystalline hydrates	Gao et al. (2022)
	Determination of glycerol concentration in aqueous glycerol solution	Frequency range of 0.45–0.85 THz	Metamaterial-based THz spectroscopy combined with statistical modeling with line shape features provides a newer strategy for quantitative sensing of glycerol	Liang et al. (2022)
	Alum in sweet potato starch	Frequency range of 0.2–1.6 THz	Results showed that prediction accuracy can reach 99.8% and 99.9%. Comparing the two predictive models, SVM gave better prediction than PLS	Guan and Chao (2019)

Table 6.1 (Continued) Various Applications of THz Systems as a Tool for Inspection and Material Identification

Additive potassium aluminum sulfate dodecahydrate (PASD) in potato starch	Frequency range of 0.2–2 THz	With increase in PASD, the amplitude of the absorption peak and refractive index decreased gradually, which indicated that THz-TDS technique could be used for identification and detection of PASD in potato starch	Liu and Fan (2020)
Talc powder in flour (pure flour, flour/talc powder – 0.5, 1, 3, and 5%)	Frequency range of 0.2–1.8 THz	Spectral characteristics of five kinds of flour/talc powder mixture samples were assessed using THz-TDS spectroscopy in the range of 0.2–1.6THz	Xiao-li and Jiu-sheng (2011)
Benzoic acid additive in wheat flour	Ultra-wide frequency band up to 7 THz	The absorption coefficient had a significant characteristic peak at 1.94 THz, which varies with the increase of benzoic acid concentrations. LS-SVM model could be regarded as an effective tool for quality control of wheat flour	Sun et al. (2019)

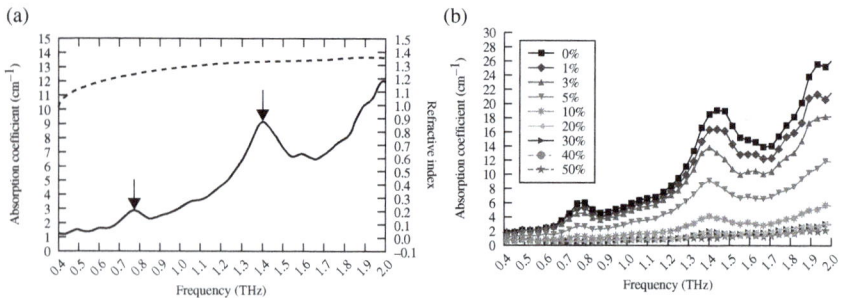

Figure 6.2 THz spectrum of TC-HCl pellet: (a) refractive index (dotted line) and the absorbance (solid line) spectrum at 0.4–2 THz and (b) average absorption spectra at different concentrations of samples. (Source: Qin et al. 2015.)

of the sulfonamide family such as sulfathiazole and sulfapyridine were observed at 0.5–2.6 THz. While amitraz and tetracycline had a featherless spectrum under the same frequency range. With the increase in the THz frequency from 0.5 to 6 THz using a gallium phosphide (GaP) crystal (100 μm), amitraz exhibits a strong resonance peak while the tetracycline showed four distinct spectral peaks. In another study, the THz spectroscopy was combined with the partial least squares regression (PLSR) model to detect and quantify the tetracycline hydrochloride (TC-HCl) in dry samples (Qin et al. 2015). Results showed that an increase in the concentration of TC-HCl significantly increased the absorption coefficient (Figure 6.2). However, this method cannot be adapted to detect TC-HCl in liquid samples. In order to improve the sensitivity of the analytical method and its applicability to liquid samples, metamaterials were used in the THz system (Figure 6.3) (Qin et al. 2016). Thus, the sensitivity of the method was increased by a factor of 10^5 for the detection of TC-HCl at a concentration of 0.1 mg/L.

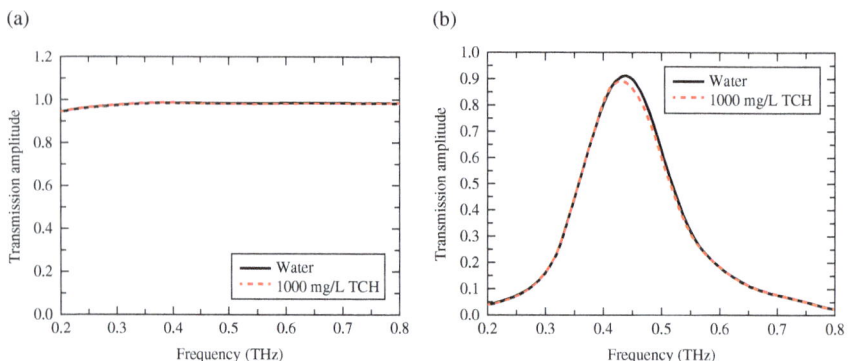

Figure 6.3 The THz amplitude spectra under transmission mode: (a) silicon substrate and (b) metamaterial before and after depositing TCH. (Source: Qin et al. 2016.)

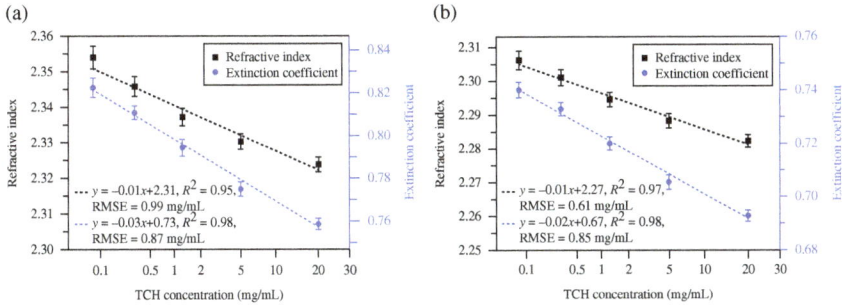

Figure 6.4 Graph showing concentration-dependent complex refractive index at 0.5 THz of tetracycline hydrochloride (TCH): (a) TCH in water and (b) TCH in milk. (Source: Qin et al. 2017c.)

In another study, the same group of researchers used THz-TDS in ATR mode for determining the refractive indices of the TC-HCl solutions (Qin et al. 2017c). It was reported that the refractive index curves of the water and milk were almost similar. The addition of TC-HCl into the milk and water results in a deviation in the curve that lowers the complex refractive indices (Figure 6.4). A similar approach of the THz-TDS with PLSR algorithm was used to quantify chlortetracycline hydrochloride (CCH) and TC-HCl in rice and chicken samples (Wang et al. 2018a). Xie et al. (2015) studied the two-dimensional (2D) version of metamaterial in the THz sensors for improved sensitivity. Due to the higher sensitivity of metamaterials than silicon, the application of meta surface in the THz system was successfully used for the trace detection of kanamycin in liquid samples. A similar approach based on a back-propagation neural network (BPNN) classification model was used to detect three fluoroquinolones in the animal feed samples (Long et al. 2018). The THz spectroscopy was employed for the rapid detection of quinolone antibacterial drug noroxin contamination in poultry products and the developed metamaterial-based THz-TDS method detected noroxin of concentration as low as 0.01 μg/mL (Figure 6.5) (Li et al. 2021b).

6.2.2 Detection of Foreign Bodies

During processing of foods or due to improper cleaning, foreign bodies such as stones, glass pieces, nails, hair, plastics, and metal pieces might contaminate the food. The detection of these materials is vital and crucial before the dispatch of the packaged foods. Conventionally, X-ray imaging is widely used for quality control in food industries (Ok et al. 2014). However, the limitation with X-rays is its ionizing nature and limited detection of low-density contaminants that are optically opaque. Considering these aspects, THz radiation is relatively safe due to its non-ionizing characteristics and good penetration capacity through common packaging materials such as polymers and papers. THz waves were used to identify the

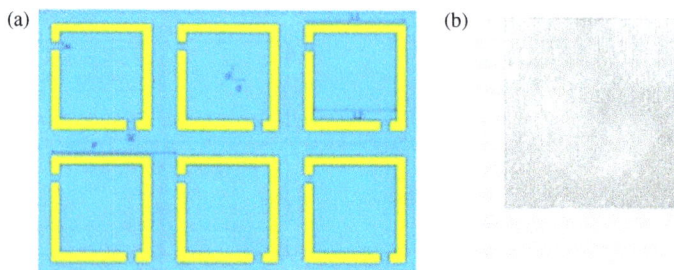

Figure 6.5 Designed metamaterial structure for noroxin detection: (a) schematic diagram of metamaterial structure and (b) photograph of machined metamaterial used for noroxin detection. (Source: Li et al. 2021b.)

insect-infested pecans (Li et al. 2010). The sample composition included nutmeat, inner separators, insect-infested pecans, and shells, each of which possesses different THz absorption. The characteristic absorption of THz radiation by water helps in the differentiation of dead and live insects in the pecans. THz-TDS was employed for detection of very fine contaminants like human hair in food samples (Hiromoto et al. 2013). A polarization-dependent THz-TDS was used to detect a strand of human hair based on the transmission spectra. The positioning of the human hair was similar as that of polarization grating that greatly affects the transmission spectra. Human hairs that are parallel to the THz radiation had lower transmittance than that in perpendicular direction. Also, the calculated standard deviation was small depicting the accuracy of the developed method for detection of fine contaminants.

A sub-THz gyrotron source was used in a CW-THz system for real-time monitoring of foreign objects (Han et al. 2011). A series of gyrotron of two different versions (KG1 and KG2) had been applied for the real-time monitoring of inshell, full, and empty peanuts (Han et al. 2011); and for food inspection with frequency of 0.4 THz (Han 2013), respectively. Sun and Liu (2020a) used THz fingerprint for differentiation of foreign bodies (insects) from green tea. Different types of THz beams used for agri-food applications include Bessel beam, Gaussian beam, and quasi-Bessel beam. A cost-effective scanning method was reported based on the use of 140-GHz Bessel-Gauss beam for the detection of 1 mm line width foreign materials such as paper clips and plastics in food products (Ok et al. 2018). The use of subwavelength transmission mode imaging system seems to be less expensive with a better focusing of beams using a source of 140 GHz (Figure 6.6). Thus, a high-resolution image of 0.52 λ with a 5 λ focal depth was obtained. Ok et al. (2013) reported a comparative study on the use of Gaussian and quasi-Bessel beam-based THz imaging to identify the foreign bodies in noodles. Results showed that the quasi-Bessel beam-based sub-THz imaging resulted in a high-resolution image with sharp peaks that help in the detection of crickets in noodles at about 210 GHz. The shape of the wavefront was greatly influenced by performance of quasi-Bessel

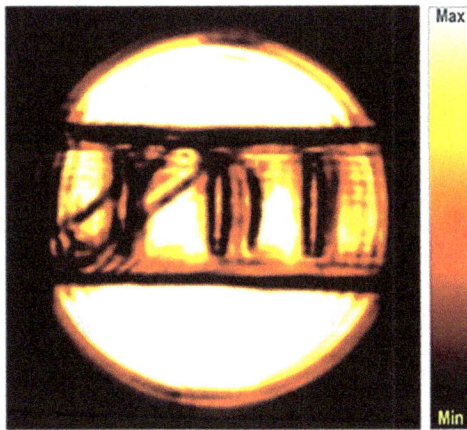

Figure 6.6 THz transmission image of the wrapped chocolate with foreign matters (metal paper clip, mealworm, and dried maggot). (Source: Ok et al. 2018.)

beam and the combination of high-speed quasi-Bessel beam-based THz imaging with broadband detection was highly efficient for the samples with large area. On the other hand, the sub-THz Gaussian beam can be used to detect metal pieces, polymers, and insects in milk powder through focusing by a low f-number lens. With the use of transmission mode, high-resolution image can be obtained for the low moisture content samples. Similar approach has been applied for detection of less-dense objects such as plastics and insects under reflection-mode (Figure 6.7) (Ok et al. 2014).

6.2.3 Detection of Toxic Compounds

Aflatoxins are secondary metabolites produced by fungi that cause severe health implications to humans and ruminants (Singh and Jayas 2011). Conventional methods in practice for the detection of aflatoxins in food and animal feeds are liquid chromatography (LC), high performance liquid chromatography (HPLC), liquid chromatography with tandem mass spectroscopy (LC-MS/MS), and enzyme linked immunosorbent assay (ELISA). Although these sophisticated techniques are highly sensitive and accurate, these methods are destructive and require long time for analysis. Different chemometric techniques like partial least squares (PLS), principal component regression (PCR), support vector machines (SVM), and principal component analysis-support vector machine (PCA-SVM) have been combined with THz spectroscopy for the detection of aflatoxin extracts (Ge et al. 2016). Results showed that the linear regression models, PCR and PLS, were accurate in detecting AFB_1 at concentrations of 1–50 µg/mL. While the non-regression algorithms like SVM and PCA-SVM were accurate to detect AFB_1 at concentrations of 1–50 µg/L (Figure 6.8). The application of THz combined with chemometric

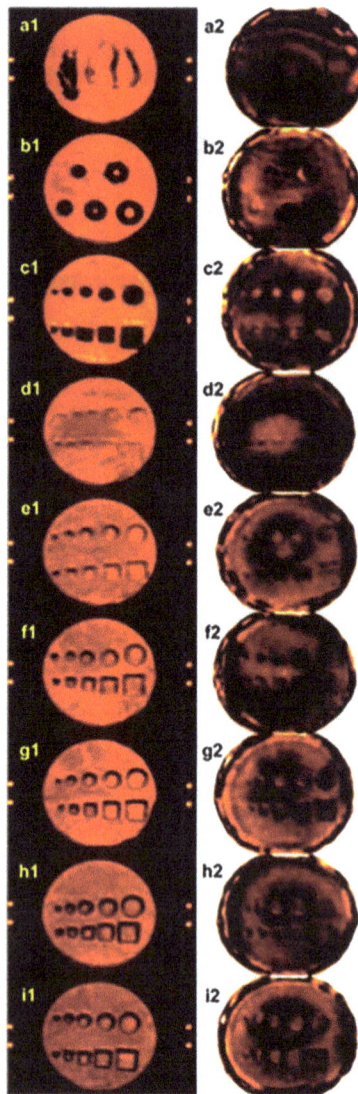

Figure 6.7 THz raster-scanned images of foreign bodies in powdered milk sample: (a) insects, (b) metal pieces, (c) rubber (ethylene propylene diene monomer – EDPM), (d) acrylonitrile-butadiene-styrene copolymer (ABS), (e) polystyrene (PS), (f) polyvinyl chloride (PVC), (g) polycarbonate (PC), (h) polyethylene tere-phthalate (PET), and (i) polymethyl methacrylate (PMMA) (Images on left a1 to i1 are images obtained by transmission mode while images on right a2 to i2 are obtained by reflection mode). (Source: Ok et al. 2014.)

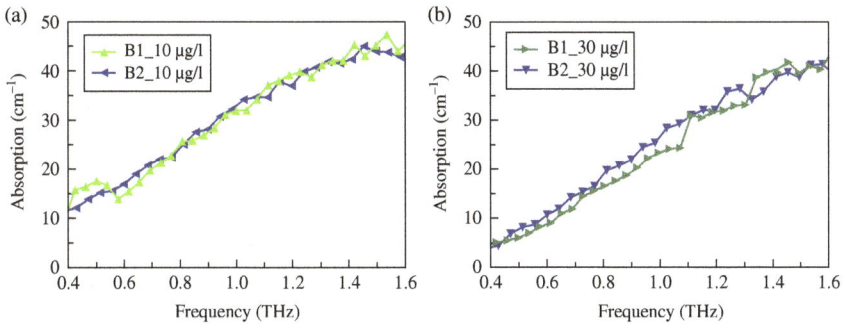

Figure 6.8 The THz absorption coefficient of aflatoxin B1 and aflatoxin B2: (a) 10 µg/L and (b) 30 µg/L. (Source: Ge et al. 2016.)

method on the maize acetonitrile extract assists in the quantification of AFB_1 in the samples in <5 s. Some kinds of preprocessing such as PCA and t-distributed stochastic neighbor embedding (t-SNE) have been used for the conversion of the data set into matrix and to visualize the similarities to quantify AFB_1 in soybean oil (Liu et al. 2019b). Results inferred that the t-SNE pretreatment combined with BPNN classification model yielded a prediction accuracy of about 90% for detecting AFB_1 in soybean oil of 1 µg/kg than other classification algorithms (least square-support vector machine [LS-SVM], random forest [RF], and PLS). However, improvements on the signal processing methods are required for large processing of data. More studies are required in processing of data for enhancing the effectiveness of the THz spectroscopy for rapid inline detection of aflatoxins.

Food additives such as o-phenylphenol and thiabendazole are commonly applied over the surface of fruits such as bananas, grapes, and lemons (Delgado-Blanca et al. 2019). Often, ethylenediaminetetraacetic acid (EDTA) has been added as a preservative to enhance the color and flavor of the canned foods and mayonnaise. The EDTA has been reported to lower blood pressure. In this regard, THz-TDS is helpful to detect the traces of food additives in agri-food products. Yoneyama et al. (2007) used THz waves to detect saccharin, ethylenediaminetetraacetic acid disodium salt (EDTA-2Na), thiabendazole, and o-phenylphenol in liquid and pellet forms. A specially designed membrane was used for holding the solution. This membranous device helped to overcome the problem of distribution deviation as observed with sapphire/silicon plates. Results of absorbance showed that both saccharine and thiabendazole yield three significant peaks in both the liquid and pellet forms. On the other hand, there was only one peak observed for o-phenylphenol in pellet form and two peaks for it in liquid form. While no peaks were observed for EDTA-2Na in solution form, but two significant peaks were evident for EDTA-2Na pellets.

Auramine O (diphenylmethane) (AO), a yellow-colored dye, is commonly incorporated as an adulterant into the food products. It is known to cause bladder

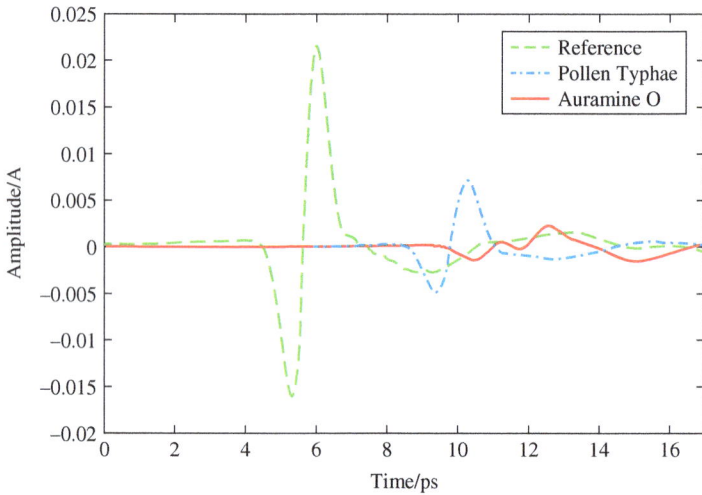

Figure 6.9 The THz-TDS spectra for quantification of Auramine O in pollen Typhae. (Source: Zhang and Li 2018.)

cancer, liver, and DNA damage (Kovacic and Somanathan 2014). THz spectroscopy has been combined with improved PLSR algorithm for detection of AO in the herbs. The adopted PLSR method seems to be fast, reagent free, accurate method for detection of AO. Zhang and Li (2018) reported a study on the implementation of multiple regression analysis on the spectral data with improved analytical performance in the quantification of AO in pollen typhae (Figure 6.9). Addition of benzoic acid (BA) as a preservative in wheat flour may lead to health implications when added in excess concentration. The absorption coefficient of BA was observed at 1.94 THz that increases with increase in the concentrations of BA (Hu et al. 2020). Results showed that the combined THz technology with LS-SVM classification model had better performance in the quantification of BA in wheat flour. Although THz spectroscopy is found to be fast and convenient, more improvements in these systems are required for on-site identification of toxic compounds in food products.

6.2.4 Detection of Adulterants

Food adulteration is one of the serious food safety concerns that lead to severe health implications. Different forms of food adulteration include addition of melamine in milk, addition of glucose and fat powder into dairy products. Chemometric methods along with THz-TDS help in the analysis of spectral data based on pattern recognition. Adulteration of dairy products was determined using combined PCA and SVM-discriminate analysis (SVM-DA) of the THz spectra (Liu 2017a). The intense absorption of spectra reflects the presence of lipids in samples that falls under the range from 0.38 to 1.5 THz. The absorption spectra of skim milk and

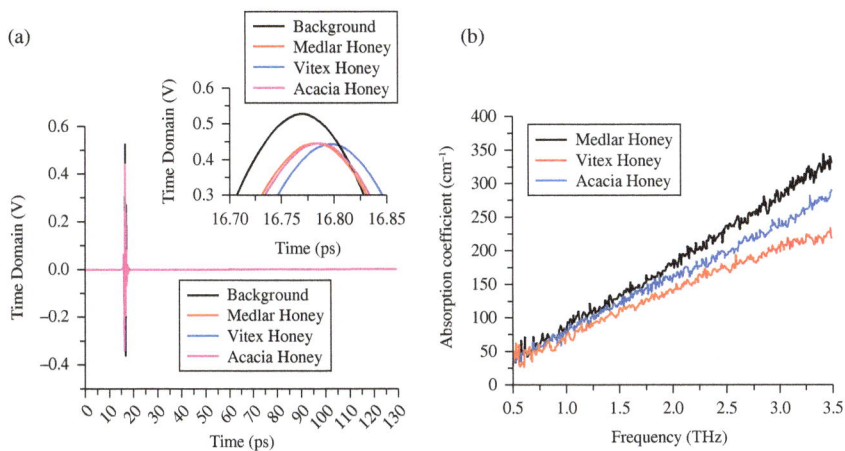

Figure 6.10   The THz-TDS spectra under ATS mode for discrimination of botanical origin of different honey samples: (a) THz spectra of three honey samples and reference and (b) THz absorption spectra of different honey samples. (Source: Liu et al. 2018b.)

low-fat milk samples are distinct that aids in the differentiation of samples. The SVM-DA analysis of adulterated milk showed a 100% specificity and 88.62% sensitivity. Generally, honey is adulterated with high fructose corn syrup. The conventional methods employed to distinguish pure honey are laborious and expensive. A rapid non-destructive quantification approach based on THz-TDS was used to detect the adulteration of Acacia honey with fructose syrup. The spectral raw data were processed using SGS followed by PLS model. Results showed the effectiveness and reliability of the developed approach for the detection of adulteration in honey samples. In another study, different classification methods like PCA, partial least squares-discriminant analysis (PLS-DA), and cluster analysis were adopted to derive information about the botanical origin of honey (Liu et al. 2018b). Among the tested models, PLS-DA was best suited to identify and categorize honey based on the source of floral origin (Figure 6.10). Similar approach has been applied for differentiation of olive oils based on their geographical origin (Liu et al. 2018a). Results of spectral analysis combined with genetic algorithm (GA) and LS-SVM model resulted in higher accuracy (about 96.25%) than other discrimination models used (BPNN, RF, and PCA). Thus, these studies confirmed that THz-TDS combined with chemometrics could be simple and fast technique for on-site detection and identification of adulterants.

Melamine is a nitrogen-rich chemical compound added as an adulterant to increase the protein content into the food products. The tolerable limit of daily intake of melamine is about 0.2 mg/kg of body mass as defined by the World Health Organization (WHO) (Wu and Zhang 2013). Excess intake of melamine than recommended levels cause renal failure and kidney malfunction. The addition of melamine into food products above the recommended levels remains a serious food safety concern. The THz-TDS is used to quantify melamine in infant milk powder (Ung et al. 2009). The absorption spectrum of pure melamine and melamine mixture (melamine, milk powder, and polyethylene) was found to be at 1.99 and 2.29 THz, respectively. With increase in levels of melamine, the absorption peak becomes more distinct and sharper. There exists a linear relationship between the ratio of melamine mixture and their absorption coefficient. Similar studies were reported on melamine in cereal/pulse flour, milk powder, and chocolate powder with or without package materials (Baek et al. 2014) and plastic water cups (Su et al. 2019). Despite the thickness of package, distinct absorption peaks were observed for all food mixtures. Also, fluorescent-based THz detection of melamine was found to be more accurate and selective.

Adulterants such as yellow chalk powder and yellow dyed calcium carbonate have been added in the turmeric powder, making it unsafe for consumption. As an alternative to commonly used acid-HCl method for quantification of adulteration in turmeric, THz spectroscopy was successfully used to detect turmeric powder adulterated with yellow chalk powder (Nallappan et al. 2013). Based on the difference that exists in the spectra of pure and adulterated turmeric, the peaks at 6.3 THz detected turmeric powder adulterated with chalk powder. Sometimes talc powder has been added as an inert carrier and separating agent in bakery products, salt, and dry foods. THz-TDS can be conveniently used to determine the refractive index and the absorption coefficient of talc powder adulterated in flour (Xiao-li and Jiu-sheng 2011).

6.3 Industries in the Development of THz Systems

The fascinating applications of the THz systems in the industrial sectors have drawn the attention of the many researchers and industrial firms around the globe. With developments in advanced optics and photonics, the THz science is witnessing a steadfast growth that symbolizes the gain of trust in this technology for industrial applications. Over the years, there has been active research going on in exploring the potential solutions of THz science for different industrial problems in diverse areas of interest. This includes deployment of THz in security screening, pharma applications, ceramics and material science, automotive, medical field, packaging, agricultural and food applications (Bandyopadhyay and Sengupta 2022). It is evident from the works of literature that lab-scale THz research is actively progressing that provides promising solutions for food quality monitoring. However, the knowledge on the manufacturers of THz systems is required to understand the novel design and emerging working concepts in THz science. This section presents a brief discussion on few of the leading manufacturers of THz systems.

The Terasense Group Inc. (USA) is one of the leading manufacturers of the portable THz imaging systems (cameras, sources, and detectors). As a research-oriented company, the Terasense is continuously working on developing THz products at the cutting edge of scientific breakthrough. It uses the original patent-protected technology in designing new semiconductor detectors with working frequency in sub-THz range from 0.05 to 0.70 THz (TeraSense 2022). The detectors are combined into a compact matrix similar to charge-coupled device (CCD) in a photo camera thereby reducing the system size and cost. These developed detectors are capable of operating at the room temperatures. Also, the company developed novel THz sources that are employed for detection of chemical substances based on THz spectral fingerprints. Considering only agri-food applications, the Terasense has been actively involved in projects dealing with THz instrumentation for the quality monitoring of food grains and oil seeds. Thus, Terasense® technology creates and offers a lot of business opportunities and industrial applications.

The INO is a Canadian leading firm capitalizing on their expertise in areas of optics and photonics. The company offers multiple solutions in imaging systems based on laser and optics technologies (INO 2022). They incorporate the modulation of light to detect, identify, predict, and analyze situations to provide the real-word solutions. They are working on developing THz imaging cameras, THz lenses, and illumination sources. Another firm working on design of modern laser systems is Fyla (Valencia, Spain). It provides different types of pulsed lasers in wide range of spectrum. Some of the Fyla's ultrafast lasers are Iceblink, Zephyr, Rhyo, Cyclone, Cygnus, and Pulsar (Fyla 2022). These are useful for THz generations, spectral imaging, tomography, optical communication, and microprocessing. The MenloSystems (Martinsried, Germany) provides THz-TDS solutions covering traditional pump-probe configurations for spectroscopy and imaging applications, especially for quality control and inspection (MenloSystems 2022). Similar kinds of high-end laser sources applicable for industrial applications are provided by the company Toptica (Munich, Germany). They produce and market a range of products including THz systems and accessories, single mode diode lasers, frequency lasers, picosecond/femtosecond fiber lasers, frequency combs, laser rack systems, tunable diode lasers, multi-laser engines, wavemeters and laser analyzers, optical isolators, and photonicals (Toptica 2022).

Another firm named Alpes Lasers (Switzerland) is the first company that introduced the quantum cascade laser (QCL) on the market and commercialized the first continuous-wave (CW) lasers system (Alpes Lasers 2022). This company offers a broad range of laser sources covering different realms of electromagnetic spectrum. It provides solutions for the applications in packaging, agri-foods, medical/health, environment, ICT, aerospace, defense, and security sectors. The lasers under mid-infrared range are employed for applications related to food safety (pathogen detection, ethylene sensing, monitoring micro-pollutants, and microplastics detection), process optimization (lactose, lipid, and protein analyses in beverages), farming 4.0 (pesticides detection and weed control), and biomarker monitoring for nutrition studies. Thus, these are some of the companies that are actively working on the design of THz systems and offering solutions for industrial applications.

6.4 Summary

Diverse spectroscopy and imaging applications of THz waves as inspection and identification tool are discussed. This chapter highlights versatility of THz systems in assessing different types of solid and liquid agri-food samples. A detailed discussion is provided on the detection of antibiotics, foreign bodies, toxic compounds, and adulterants using THz spectroscopy and imaging systems. The major advantage of THz in food inspection and monitoring is its potential ability to assess even the packaged materials without destroying the structure of package or samples. The works from the literature survey emphasize the integration of THz system with chemometrics for the efficient mining of spectral data. Despite the extensive applications of THz in agri-food sector its broad applications are limited by complex handling of spectral data and high cost of THz equipment. However, the recent developments on THz sources and detectors considerably decrease the cost of instrumentation. The future research must be directed in developing more robust models for qualitative and quantitative measurements based on existing spectral data. This would certainly help in establishing a library based on the available THz fingerprints of spectral data. Further, it has positive effects on decreasing cost and increasing the deployment of onsite and inline THz portable devices for real-time applications. Thus, the THz technology remains as a promising and innovative tool to be used for the noninvasive assessment ensuring food safety and quality assurance.

Chapter 7

THz Technology – Agricultural Applications

7.1 Introduction

The agriculture industry is constantly focusing on the development of rapid, reliable, and non-destructive testing procedures for the quality assessment of agricultural produce. For this purpose, different regimes of electromagnetic spectrums are explored. This has led to the emergence of different types of spectroscopic and imaging systems. Some of the commonly used non-destructive analytical systems are ultraviolet-visible (UV-Vis) spectroscopy, infrared (IR) spectroscopy, Raman spectroscopy, and nuclear magnetic resonance (NMR) spectroscopy (Mei et al. 2021). Recently, the Terahertz (THz) has received the attention of researchers to be used as a potential analytical tool. Much research progress has been well reported on the applicability of THz spectroscopy and imaging as a non-destructive characterization tool in astronomy, aerospace, material science, security screening, art conservation, biomedicine, atmospheric sensing, and communication (Guillet et al. 2014; Patil et al. 2022). However, the THz applications in the agri-food sector are gradually growing in recent years (Wang et al. 2017b; Afsah-Hejri et al. 2019, 2020; Khushbu et al. 2021). The advantage of the THz waves in molecular characterization is their THz spectral fingerprints in analyzing the composition of materials. The dry materials such as paper, fabrics, and polymer packaging materials are transparent to THz radiation that significantly increases the scope of THz in food security screening (Gowen et al. 2012). The nonionizing nature of THz radiation makes it safe to use for the *ex-vivo* analysis of biological foodstuffs (Gong et al. 2020). Despite the potential scope of THz waves, the lack of inefficient THz sources and detectors hinders the extensive exploration of its applicability for industrial applications. Chapter 6 discusses different aspects of THz as inspection tool in food applications. This chapter details the agricultural applications of THz science. It covers the THz applications in the agricultural sector, especially for plant growth monitoring, hydrology and drought/water stress monitoring, crops/seed inspection, and quantitative assessment of spatial leaf constituents and monitoring of water status.

The earliest application of the THz started with the emergence of THz time-domain spectroscopy (THz-TDS). Lately, with the development of THz systems, THz imaging has been found to have great scope in non-destructive analysis. Advantageously, the THz systems of spectroscopy can be potentially coupled with the imaging system. Compared to the conventional imaging system, THz imaging can also be used for 3D tomography (Wang et al. 2022c). Also, THz spectroscopy is very useful in extracting the process parameters like refractive index and absorption coefficient. Thus, spectral data processing and feature extraction form the basic steps in reconstructing the THz images (Guillet et al. 2014).

7.2 Scope of THz in the Agricultural Sector

The THz spectroscopy makes use of electromagnetic radiation of wavelength in the range of 30 μm to 3 mm for the analytical measurements. The THz waves are characterized to be coherent, transient, broadband, high penetration, and low energy due to their spectral position in the electromagnetic spectrum (Patil et al. 2022). The low photonic energy of THz waves conforms to the low energy vibrations or molecular rotations. This makes THz waves advantageously used for the detection of biomolecules such as starch, protein, water, and DNA. The THz spectroscopy has been widely applied for scientific research in pure biology. The applications of THz were eventually broadened to range of different agricultural applications (Figure 7.1). This includes monitoring and quantification of starch in plants, discrimination of transgenic seeds, detection and characterization of plant growth regulators (PGRs), detection of pesticide residues, heavy metal analysis in soil, monitoring of water status in leaves, and drought/water stress monitoring. When combined with appropriate chemometric methods, the THz spectroscopy yields promising results with better classification performance (Liu 2017a; Cao et al. 2018; Liu et al. 2019b). Different spectral techniques acquire data at different sets of wavelengths thereby increasing the complexity of the spectral data analysis. The hidden sample information from the collected spectral data can be visualized by applying efficient modeling algorithms. The THz spectral data obtained from agricultural samples are quite complex. The post-processing of the spectral data must be considered to be equally important as that of the data acquisition for agrifood samples. Hence, the advanced deep learning algorithms are being applied to THz spectral data for effective characterization and classification of materials (Murate et al. 2021).

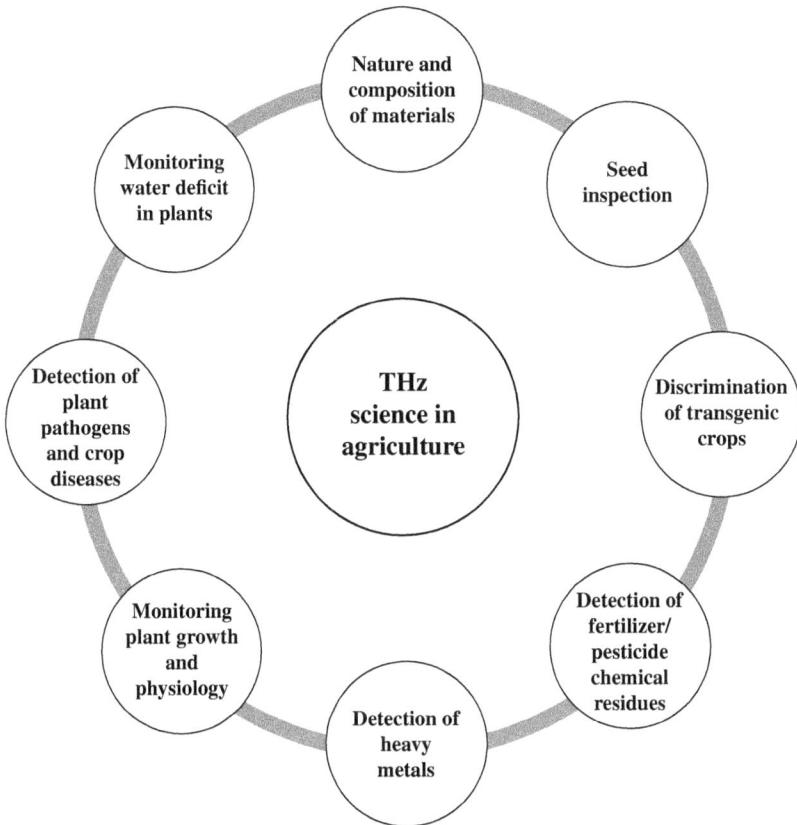

Figure 7.1   Diverse applications of THz science in agricultural sector.

The quantification of macronutrients usually follows a series of extraction and purification steps. For instance, the hydrolysis method is commonly used for the quantification of starch in foods (Mitchell 1990). During which the samples are first extracted using alcohol followed by hydrolyzation to sugars with acid or enzymes. The converted sugars are quantified to determine the starch content. Although this method provides accurate results, it requires strenuous pretreatments. Addressing this technical difficulty, spectroscopic analysis not only provides rapid measurements but also helps to know about the molecular state and chemical structures (Su et al. 2017). Hence, spectroscopic techniques have recently received a greater interest in qualitative and quantitative measurements of metabolites in fruits and vegetables. In contrast to reflecting intramolecular vibrations as in the case of IR and Raman spectroscopy, the THz spectroscopy provides information on intermolecular vibrations (Nakajima et al. 2019). Hence, the THz spectral data are more closely associated with molecular structure and are very useful in the material characterization of agricultural commodities.

The detection, monitoring, and discrimination of the portion of GM organisms in crops and their derived products are most crucial from a biosafety and trade perspectives. Conventionally several methods such as polymerase chain reaction, mass spectrometry, electrophoresis, and chromatography techniques are used. Though these methods are specific, sensitive, and precise, they suffer from the limitations of the high-cost analysis, laborious procedure, time-consuming, and destructive sampling procedures (Liu et al. 2016b). Hence, the conventional methods are not suitable for the online detection of transgenic seeds. Along with other non-destructive techniques (IR spectroscopy, near-IR spectroscopy, and multispectral imaging), THz spectroscopy yields rich information in the detection and distinguishing of transgenic seeds from non-transgenic counterparts.

Similarly, the practice of spraying pesticides on agricultural and horticultural commodities is increased nowadays for promoting the yield. However, it has also increased food safety issues due to remaining of pesticide residues in the finished products. The chemicals 2,6-Dichlorobenzonitrile (2,6-D), 6-benzylaminopurine (6-BA), and imidacloprid are commonly used as insecticides, pesticides, and PGRs. These residues in processed food products such as wheat, rice, maize, and dried fruits have potential to cause health problems in humans. The combined technique of THz spectroscopy along with appropriate machine learning algorithm emerged as a promising method in the detection of the multicomponent pesticide residues. (Ma et al. 2022).

Another food safety concern that receives greater attention in recent days is heavy metal contamination of agricultural fields. The excessive use of pesticides and fertilizers causes heavy metal pollution in the soil. The heavy metal contaminants in the soil such as zinc, lead, cadmium, chromium, arsenic, and copper migrate through the soil-crop cycle (Jing et al. 2018). This ultimately poses an increased threat to human health and life. The direct methods of physical and chemical tests used to monitor heavy metal detection are highly sensitive and accurate. However, due to their laborious test procedures, THz spectroscopy is tested for rapid detection of heavy metal ions in soil and food commodities such as oil (Lu et al. 2022).

The PGRs are the natural hormones that are used for regulating the growth and physiology of crops. Nowadays artificially produced synthetic plant growth hormones are widely applied for increasing the yield and improving the quality of fruits (Suman et al. 2017). Excessive application of PGRs on crops will result in increased residues that are toxic to humans. Efforts are being enforced by the food safety regulatory bodies in establishing the permissible limit of usage. Over the past decades, conventional analytical techniques like capillary electrophoresis, high-performance liquid chromatography, thin-layer chromatography, immune affinity assay, and enzyme-linked immunosorbent assays are used for the detection of PGRs (Du et al. 2021b). However, these methods require complex time-consuming pretreatments. This highlights the need for rapid, reliable, and sensitive analytical techniques in detecting these residues. Thus, the THz-TDS is identified to be an efficient tool in the assessment of residues of plant growth hormone in agricultural commodities.

The ongoing climatic change reflects a remarkable effect on the physiology of crops and agricultural plants. The appropriate phenotypic information of the plants greatly helps in the early detection of disease and improvement of yield (Lew et al. 2020). The advancements in information and technology in today's modern world led to the development of methods for real-time monitoring, detection, and identification of disease outbreaks in plants. As envisaged, the conventional laboratory procedures are laborious and time-consuming. This motivates the need for the development of viable methods to characterize the morphological features of leaves at the cellular level. Techniques based on fluorescence imaging, magnetic resonance imaging, hyperspectral imaging, near-IR spectroscopy, and multiband spectroscopy have been identified to be used for obtaining information about plant phenotypes (Kannan et al. 2022). Although these are advanced techniques utilizing noninvasive procedures, each of them has its limitations. For instance, the techniques based on IR radiation lack sensitivity in tracking the spatial variability in leaf constituents. The magnetic resonance imaging has been considered to be inefficient and inadequate for a long-term research on plant physiology in spite of its higher resolution (Zahid et al. 2022). Considering these aspects, the THz radiation comes to the limelight of plant science research. The interest of the scientific community has recently focused on exploring the THz regime for monitoring plant physiology. The THz-TDS has been proven to be a reliable and effective non-destructive method in the assessment of leaf constituents.

The knowledge of the ratio of the bound and free water content of plant cells has scientific value and practical applications in assessing stress levels and analyzing the stress resistance. The proportion of the free and bound water in cells varies based on the physiological and biochemical metabolism of plants. Various factors that influence its proportion are genetic phenotype, plant stress, and environmental change. The high bound water to free water ratio is beneficial for plants to withstand a drought environment (Ma et al. 2008). Thus, the assessment of the bound and free water contents of leaves is essential in screening the drought resistance and optimization of irrigation patterns. The standard testing methods available for quantification of bound and free water detection are based on weighting and refractometry (Park and Bell 2004). These methods demand prior knowledge and operational skills from workers. Also, monitoring of dynamic water change within the cells is not possible. This put forth the need for the development of label-free technology for water status monitoring. To date, IR and NMR spectroscopy is often used for water status measurements. However, these non-destructive analytical techniques are highly expensive. Hence, research efforts are being made in exploring new methods with simpler analytical procedures. Recently, the THz spectroscopy has emerged as a complementary technique to IR spectroscopy (Zang et al. 2021). The low-frequency vibration of hydrogen bonds under the THz regime has greater prospects in efficiently detecting the biological water. Thus, the THz system has a great range of potential applications in agricultural sector. The various research works on the THz applications in agricultural sector are explained in the next section.

7.3 THz Applications in the Agricultural Sector

The previous section of this chapter explains the scope of THz in the agricultural field and this section provides a detailed summary of various applications of THz waves in the agricultural field (Table 7.1).

7.3.1 Nature and Composition of Materials

Material characterization using THz is one of the promising applications in the agri-food sector. The THz spectroscopy was applied for the starch quantification in germinating mung bean seedlings (Nakajima et al. 2019). Since the absorption spectra of seedlings during the period of germination continue to change, the evaluation of these absorption peaks aids in the quantification of starch. In this study, the mung bean plants of 1–7 days after their germination were examined under THz of frequency range 3–13.5 THz and compared with the standard starch reagent. The absorption peaks of 1-day seedlings and the standard starch had a similar THz spectrum. The detection of a peak at a particular THz frequency was used for monitoring the starch during the germination of mung beans and the peak intensity to quantify starch. Also, the obtained results for mung beans were evaluated by comparing them with the native potato starch. Results showed a high-peak intensity for 1-day-old mung bean seedlings which was mainly due to differences in the structure of the starch molecule. Although the developed method for quantification of starch in germinating seedlings is accurate, the method is destructive and requires pretreatment procedures. Hence, more studies must be carried out in this direction for confirming the applicability of this analytical method. A similar study was reported to estimate protein content of soybeans using THz spectroscopy. The THz-TDS in combination with the dimensional reduction algorithms (such as principal component analysis (PCA), linear discriminant analysis, and generalized discriminant analysis) was demonstrated as an efficient noninvasive tool for the quantification of protein content in soybeans (Wei et al. 2021). The degree of grain budding was assessed using the THz imaging by estimating the maltose concentration in food grains (Jiang et al. 2018).

A similar approach was adopted for the quantification of wheat maltose based on THz spectroscopy and imaging (Jiang et al. 2020). The boosting-based multivariate data fusion and structural risk minimization theory were employed to optimize the model parameters of LS-SVMs. This combined approach resulted in better classification performance and proved to be an efficient approach for the quantification of wheat maltose. In another study, THz imaging with chemometrics (LS-SVM, PLSR, and BPNN) was also used to observe changes within a single wheat kernel at different time periods of germination (Jiang et al. 2016). The first five components from PCA were used as input for LS-SVM, PLSR, and BPNN models to classify seven different germination time periods between 0 and 48 h. Results showed that the integration of THz imaging with chemometrics could be

Table 7.1 Applications of THz in the Agricultural Sector

Agronomic Applications	Analyte	Technique	Process Parameter	Highlights	References
Seed Inspection	Plumpness for intact sunflower seeds	THz transmittance imaging	Frequency range of 0.5–2.0 THz	Compared with the shell, the absorption coefficients and time domain signals of the kernel showed a significant difference in 0.5–2.0 THz. The developed model has an R-value of 0.91	Sun and Liu (2020b)
	Identification of wheat quality	THz-TDS	Frequency range of 0.2–1.6 THz	THz technology combined with PCA-SVM is efficient with accuracy of 95% in the determination of normal, worm-eaten, moldy and sprouting wheat	Ge et al. (2014)
	Characterization of wheat varieties	THz-TDS	Frequency range of 0.2–2 THz	Results demonstrate that THz spectroscopy combined with multivariate analysis can provide rapid discrimination of wheat varieties with prediction accuracy of 0.992	Ge et al. (2015)

(Continued)

Table 7.1 (Continued) Applications of THz in the Agricultural Sector

Determination of geographical origin	Wheat samples from different growing areas (240 samples from 5 provinces)	THz-TDS	Frequency range of 0.1–1.4 THz	Results demonstrated an accurate qualitative analysis of producing area of wheat samples using THz-TDS combined with chemometrics which is useful in security detection and origin tracing	(Shen et al. 2022b)
	Coffee beans (96 samples from three geographical origins)	THz-TDS	Frequency range of 0.5–1.9 THz	THz spectroscopy combined with CNN machine learning method is an efficient approach for classifying geographical origins of coffee beans	(Yang et al. 2021a)
Discrimination of genetically modified (GM) seeds	Transgenic cottonseeds	THz-TDS	Frequency range of 0.1–3.5 THz	Kernel Entropy Composition Analysis (KECA) clustering algorithm was efficiently used for the optimal kernel parameter selection	(Yi et al. 2022)
	GM cottonseeds	THz-TDS	Frequency range of 0.2–1.2 THz	The proposed method for identification of four kinds of GM cottonseeds had a prediction accuracy of 99%	Qin et al. (2017a)

Table 7.1 (Continued)   Applications of THz in the Agricultural Sector

	Transgenic corn oil	THz-TDS	Frequency range of 0–1.5 THz	PLS-DA model with terahertz spectroscopy data is better in the validation set; this model accurately identified transgenic corn oil with accuracy of 98.7%	Liu (2017b)
	GM soybeans	THz-TDS	Frequency range of 0.1–2.5 THz	Results showed that after the iPLS and mean center pretreatment technology, the Grid search-SVM identification model had the best identification with a total accuracy of 98.3%	Wei et al. (2020)
Plant physiology	Plant growth regulators (2,4-D; CPPU; IAA)	THz-TDS	–	Characteristic THz absorption and anomalous dispersion of 2,4-D were found at 1.35, 1.57, and 2.67 THz, those of CPPU were found at 1.77 and 2.44 THz, and the absorption peak of IAA was located at 2.5 THz	Qu et al. (2018a)

(Continued)

Table 7.1 (*Continued*)   Applications of THz in the Agricultural Sector

Discrimination of moldy wheat	THz-TDS	Frequency range of 0.2–1.6 THz	The PCA-SVM method achieved a prediction accuracy of over 95% and was implemented at every pixel in the images to visually classify the moldy wheat	Jiang et al. (2015)
Chlorophyll monitoring	THz-TDS	Frequency range of 0.43 THz chlorophyll-a, 0.73 THz chlorophyll-b, water content 0.557–1.097 THz	With distinct spectra, THz-TDS allows us to monitor chlorophyll and water content of plants	Wagner et al. (2011)
Common fungal plant diseases (apple ring rot, cucumber powdery mildew, and grape gray mold blight)	THz-TDS in transmission mode	Frequency range of 0.1–2 THz	SVM modeling results of the three pathogens at the frequency of 1.376 THz were satisfactory, with a high (93.8%) comprehensive evaluation index F1-score, and a clearly identifiable visualization effect	Li et al. (2020b)

Table 7.1 (*Continued*) Applications of THz in the Agricultural Sector

Water monitoring	Water content in winter wheat leaf	THz-TDS combined with FFT method	Spectra analyzed at 0.3 THz	The predicted correlation coefficient and the root mean square error of the optimal model established by linear regression were 0.812% and 4.4%, respectively	Li et al. (2018)
	Grapevine water status	THz-TDS in reflection mode	Frequency range of 0.14–0.22 THz	Trunk THz time-domain reflection signal proved to be sensitive to changes in plant water availability, as its pattern follows the trend of soil water content and trunk growth variations	Santesteban et al. (2015)
	Water content	THz-TDS	Frequency range of 0.15–0.3 THz	THz measurements can provide a valuable tool for ecological and plant physiological studies, especially in monitoring water stress	Born et al. (2014)

(Continued)

Table 7.1 (*Continued*)   Applications of THz in the Agricultural Sector

Water status of leaves (*Bougainvillea spectabilis*)	THz-TDS	Frequency range of 0.1–3.5 THz	Particle swarm optimization algorithm was employed for a one-off quantitative assay of spatial variability distribution of the leaf compositions based on an extended Landau–Lifshitz–Looyenga model	Zang et al. (2019)
Leaf water dynamics (*Arabidopsis thaliana*)	THz-TDS in transmission mode	Frequency range of 0.1–3 THz	Application of THz-TDS helped in analysis of the water content dynamics in cauline Arabidopsis leaves when plants are subjected to low water availability to follow the real-time water loss kinetics	Castro-Camus et al. (2013)

effective method to distinguish wheat grains at early stages of germination with 92.85%, 93.57%, and 90.71% prediction accuracy for PLSR, LS-SVM, and BPNN models, respectively.

In another study, the fermentation levels of cocoa beans were assessed using a combined THz-hyperspectral imaging technique (Nguyen et al. 2022). The characteristic flavor and taste of processed chocolates are closely associated with the fermentation levels of cocoa beans. The level of fermentation depends on several factors such as drying method, drying temperature, fermentation time, and geographical region. Monitoring of this biochemical process is very important for the grading of cocoa beans. This study provides a newer approach for the characterization of fermentation levels in cocoa beans. A distinctive difference was observed between the fermented and unfermented samples and the proposed imaging technique gave 95% prediction accuracy with testing of 44 samples (Figure 7.2). Similarly, the spoilage of salmon was investigated using THz-TDS (transmission and reflection modes) and electrochemical impedance spectroscopy (EIS) (Nan et al. 2022). The fresh salmon samples were continuously monitored for 20 h until their spoilage (Figure 7.3). Results helped in visualizing four stages of spoilage of samples in both techniques (THz-TDS and EIS). Though the EIS technique is more sensitive, the THz-TDS under reflection mode provided promising results in monitoring salmon spoilage non-destructively.

7.3.2 Seed Quality Evaluation

The capability of THz waves to get transmitted through a material that blocks light makes THz imaging to evolve as a novel imaging tool. The applicability of THz waves results in capturing high-resolution image than the microwaves. The distinct feature of THz waves to apply for agri-food commodities is their sensitivity toward water that helps in quality evaluation of the agricultural products (Tonouchi 2007). It is important to assess the quality of seeds of cereals and oil seeds as the next seasonal production depends on seed vigor. The tetrazolium tests and germination tests are the common testing methods used to determine the seed viability. Although these tests are reliable and accurate, they are destructive (sample), quite time-consuming, and labor-intensive. Hence, the seed producers and scientists are looking for rapid accurate and reliable method for seed identification. The THz spectroscopy and imaging systems can be promisingly used for seed evaluation, especially for discrimination of less viable seeds from healthy ones (Gua et al. 2013). Taking advantage of THz imaging under reflection mode, three different wheat seeds (healthy seed, wrinkled seed, and necrosis seed) were assessed for seed viability. Results of THz images showed the areas of stronger reflection at the location of embryos. Except the area of embryo, the healthy wheat seed had a uniformly distributed inner structure. With the germination status, the inner structure of the seeds (healthy and wrinkled) varied. The THz image of the necrosis seed had no embryo area and showed a clear damage of the inner structure. This kind of

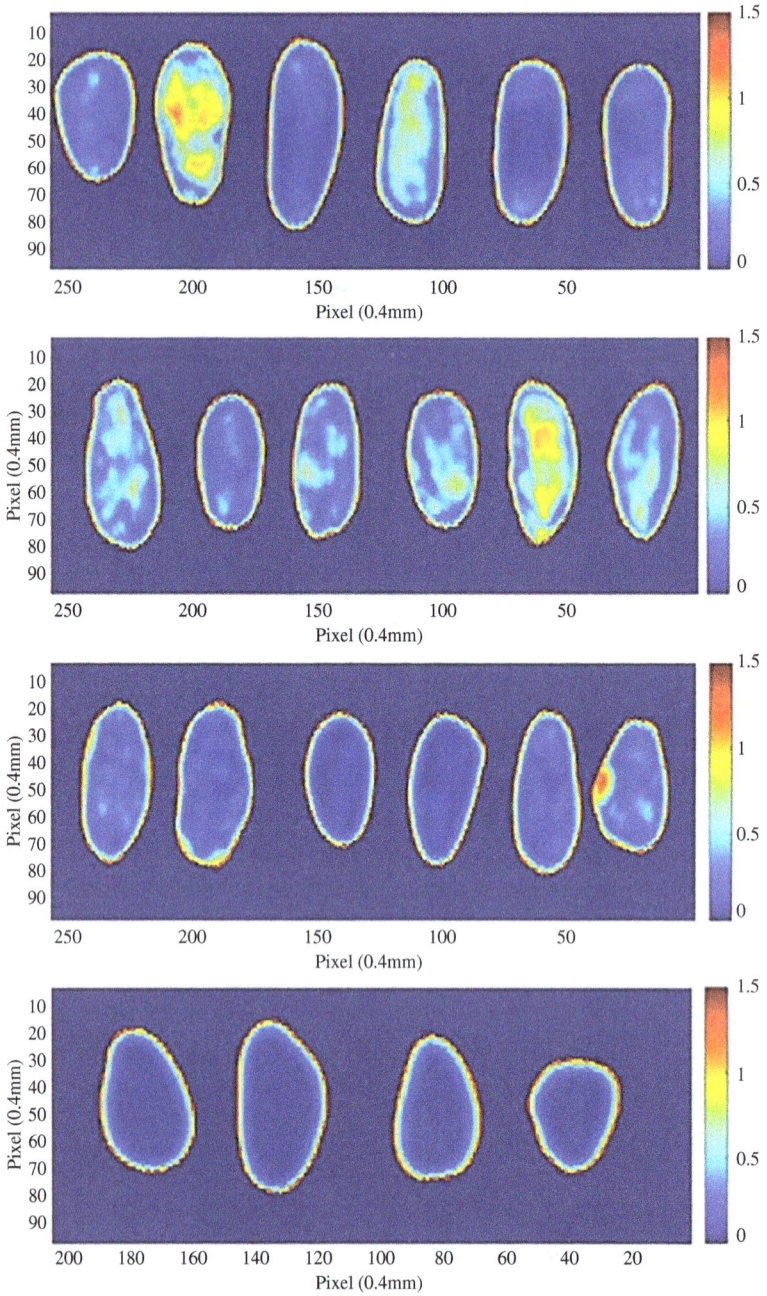

Figure 7.2 THz images showing time-delay difference between the sample and reference for 11 pairs of samples (The value 0.4 mm on the axes represents the spatial scanning step during scanning process). (Source: Nguyen et al. 2022.)

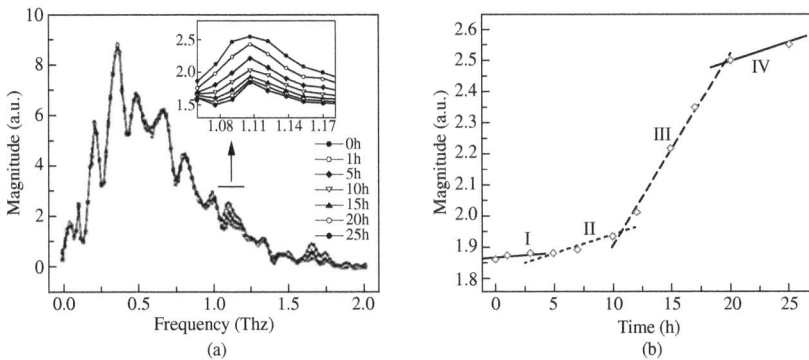

Figure 7.3 **THz spectrum of salmon fish meat: (a) frequency domain under transmission mode at different storage times and (b) rate of change of magnitude of THz over time. (Source: Nan et al. 2022.)**

discrimination of seeds based on their viability is impossible with the naked eye. Thus, THz imaging is proved to be an efficient noninvasive technique to readily distinguish the seeds. In another study, the THz-TDS was used for distinguishing the sound sugar beet seeds from the defective ones (Gente et al. 2016). On comparing the THz images, the sound seeds with the germ showed a higher water content than the defective seeds without germ. As a result, higher THz transmission was observed through the defective seeds than through the sound seeds. When the measured spectral data were converted to frequency domain for the quantitative assessment, there was a clear distinction in phase information between the two classes of seeds.

7.3.3 Discrimination of Transgenic Crops and Seeds

Nowadays, consumers are more cautious about the transgenic foods as the consumption of transgenic food products is perceived to be posing health risks by some consumers; however, perceived health risks have not yet been proven scientifically. In addition, some countries place restrictions on importing transgenic seeds and food products containing transgenic ingredients. Chen et al. (2016) reported a study to distinguish genetically modified (GM) sugar beets using combined chemometrics and THz-TDS. About 84 beet samples (36 GM samples and 48 non-GM samples) were analyzed using three chemometric methods: the DA, PCA, and discriminant partial least squares (DPLS) at 0.2–1.2 THz. Among these tested methods, DPLS yielded 100% accurate discrimination results among GM and non-GM samples. In another study, the THz spectroscopy was used for discrimination of transgenic cottonseeds (Li and Shen 2020). Results showed that there was a small fluctuation in the obtained THz band of cottonseeds and the refractive index of the transgenic seeds was less than that of non-transgenic seeds. Also, the spectral data showed that the transgenic DNA had higher absorption index than the non-transgenic DNA. THz technique is also used for rapid detection and identification of transgenic food products. In a recent study by Liu et al. (2018a), the PCA was combined with the RF, BPNN, LS-SVM, and GA for classification of olive oils. Results showed that LS-SVM and GA resulted in an excellent classification. Further, the THz spectroscopy was combined with few other sets of chemometrics such as PLS, successive projected arithmetic (SPA), and weighted linear discriminant analysis (WLDA) to distinguish transgenic camellia oil (Liu et al. 2019a). The highest classification accuracy was obtained for SPA-WLDA model with 96.5% discrimination rate. These studies confirmed the effectiveness of THz-combined chemometric approach to discriminate oils from transgenic seeds. Similar studies were reported on the potential of THz imaging for the discrimination of transgenic seeds such as rice (Liu et al. 2016b; Hu et al. 2017; Yi et al. 2022), maize (Lian et al. 2017), and processed products like corn oil (Liu 2017b), olive oil (Liu et al. 2018a), cottonseed oil (Liu et al. 2017a), and camellia oil (Liu et al. 2019a).

7.3.4 Seed Inspection and Identification

The yield and productivity of crops are significantly affected by outbreak of plant diseases. The prevention and the control of plant diseases is a must to produce a good quality crop. Although chemical control of plant diseases is effective, excessive use of chemicals would significantly degrade environment in the long run. Therefore, seed breeders and scientists are continuously working to develop disease-resistant cultivars as an environmentally friendly and effective strategy (Nelson et al. 2018). The development and deployment of resistant cultivars demands for the proper identification and functional studies on resistant genes. More recently, the applicability of THz imaging and near-IR hyperspectral imaging was compared for identifying the bacterial blight disease-resistant rice seeds (Zhang et al. 2020a). The 2D spectral images and 1D spectral information were processed using deep learning, machine learning, random forest (RF), convolutional neural network (CNN), SVM, and PLS-DA. Results showed that the highest classification performance was obtained for a CNN discriminate model. The t-SNE method was used for visualizing the processing of CNN data. Further, the study inferred that combined THz imaging with chemometrics was better fast-detecting tool to identify rice seeds that were resistant to bacterial blight than near-IR hyperspectral imaging. The THz spectroscopy integrated with the machine learning was used for the identification of ten different types of soybean seeds (Luo et al. 2019). Results showed that integrated learning method DT_Adaboost (DT_A) combined with SGS and kernel PCA yielded better results with the average accuracy of about 88.3% for detection of variety of soybean seed. A raster-scan THz imaging approach was used for classification of frosted and unfrosted barley spikes (Lee et al. 2020). Both transmission and reflection modes were employed at 0.275 THz for detection of frost damage in barley. The result of this study could help in the deployment of noninvasive inspection of early frost damage in agricultural crops.

7.3.5 Detection of Pesticide Residues

The remains of pesticides in agricultural commodities pose a health concern. THz spectroscopy can be conveniently used for the screening, quantification, and graphical analysis of the pesticide residues. More interest is being paid to detection of pesticides in agri-food products using THz spectroscopy. A novel highly sensitive analytical method was proposed for the detection of pesticide residue methomyl (Lee et al. 2016). This study employed a nano-scale metamaterial-based THz-TDS system for improving the sensitivity of pesticide detection. As a result, the detection sensitivity of 1 ppb was achieved even in solution state of pesticide sample. The reflected THz signal from the nano metamaterial was captured for simple rapid detection of pesticide residues directly on surface of apple without any prior treatment. Like hyperspectral imaging, the THz spectral imaging suffers with limitation of low sensitivity for residue detection. However, the sensitivity of detection could be enhanced by use of metamaterials in THz system. The large size of metamaterials used in THz system has the advantage of ease of manufacturing than metamaterials used in other optical wavebands.

A highly accurate quantitative assessment method was proposed for detecting the pesticide residues in cereal flour. In this study, a BPNN was applied to enhance classification accuracy of samples (Ma et al. 2022). The use of appropriate spectral denoising method and baseline correction significantly helped in improving the spectral quality for achieving highly accurate results for detection of pesticide mixture even at low concentrations (Figure 7.4). Further, the accuracy of the BPNN was enhanced by integrating with GA algorithm. Thus, this study confirmed use of THz spectroscopy along with appropriate machine learning techniques would yield better results in detection of multicomponent pesticide residues at low concentrations.

It is well-known that THz spectroscopy has great application prospects for detection of pesticide residues. However, the tablet method of THz detection has issues of low sensitivity in the detection. Though the inclusion of metamaterials in THz systems improved detection efficiency, it increases complexity in manufacturing systems. Considering these aspects, a novel method for pesticide detection was proposed to detect copper sulfate ($CuSO_4$) with metal grating approach (Tong et al. 2021). The laser micro-machining technology was used to fabricate the metal grating with circular hole array. In contrast to UV lithography, laser micro-machining technology remained to be simpler method with less processing cost. Results showed a clear redshift in the THz transmission peak frequency with change in concentrations of $CuSO_4$ solutions. A minimum concentration of 1 mg/L can be detected with the proposed method. With the use of metal grating, the sensitivity of the THz system was increased by 1000 folds. The feasibility of practical

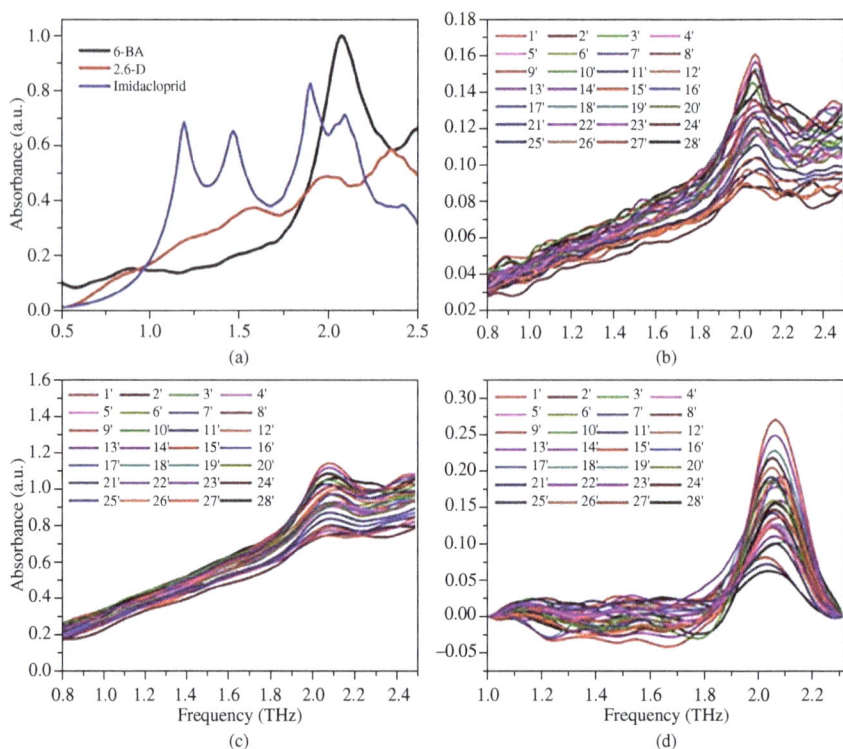

Figure 7.4 THz absorption spectra of three pesticides and ternary pesticide mixtures: (a) absorbance spectra of 6-BA, 2,6-D, and imidacloprid, (b) pesticide mixtures at different concentrations, (c) absorbance spectra of pesticide mixtures using wavelet denoising, and (d) absorbance spectra of pesticide mixtures after wavelet denoising and baseline correction process. (Source: Ma et al. 2022.)

application of this method was assessed by detecting $CuSO_4$ in apples and grapes. A relative error of <5.8% was achieved with application on fruits. Thus, this study confirmed the use of metal grating and THz-TDS can be used for highly sensitive pesticide detection under THz regime.

7.3.6 Detection of Heavy Metals

The soil and water contamination with heavy metals cause significant health risks to consumers. The heavy metal traces from soil and water enter the food commodities through soil-food chain. Various attempts were made for effortless detection of metal contaminants using noninvasive approaches based on spectroscopic techniques. In this regard, THz transmission spectroscopy was used for the detection of heavy metals like copper, mercury, and cadmium in soil (Lu et al. 2022). About

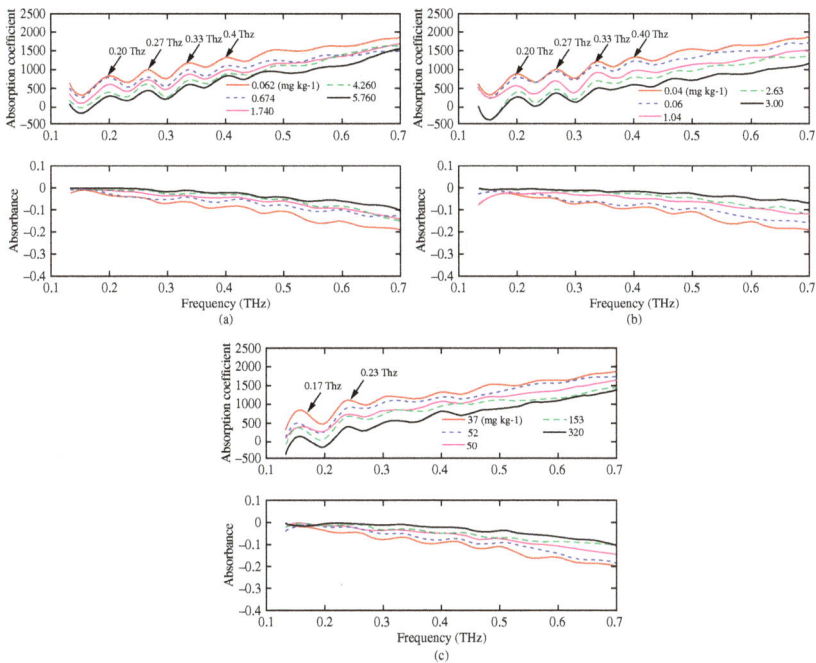

Figure 7.5 THz absorption spectra of heavy metal contaminated soil samples: (a) samples containing Hg, (b) samples containing Cd, and (c) samples containing Cu. (Source: Lu et al. 2022.)

13 soil samples with heavy metals were prepared in planting pots to plant cabbage. The root soil samples were collected at an interval of 7 days for analysis. Results showed increased absorption parameters with time at 0.05–0.7 THz (Figure 7.5). Thus, this study confirms the potential scope of THz in combination with proper machine learning models (probabilistic neural network, RF, backpropagation neural network, and extreme learning machine) could be used for soil analysis to detect heavy metal contaminants. In another study, the THz-TDS was combined with the multivariate statistical technique to identify the lead contaminants (300, 600, 900, and 1200 mg/kg) in soil tablets prepared at different pH (5.5, 7, and 8.5) levels (Li et al. 2021a). Both the bound (Fe-Mn oxide lead) and exchangeable forms of lead had great influence on THz spectral curve that promotes and inhibits the THz absorption, respectively. This study presented a theoretical basis for the rapid detection of lead content in soil using THz technology. In another study, the heavy metal ions such as Pb^{2+}, Zn^{2+}, Cr^{3+}, and Cu^{2+} of different concentrations (50, 300, and 700 ppm) were detected using THz-TDS in soil samples (Li et al. 2011). Results showed that these heavy metal ions had different absorption characteristics in frequency range of 0.1–1.1 THz. The results demonstrated the use of THz technology to detect traces of heavy metal ions in agricultural fields.

Similar approach was adopted for identification and detection of heavy metal pollution in water (Shao et al. 2022). Microalgae were used as the medium in this study and the selected prediction model (partial least squares) showed a best detection time of 6 and 18 h for Pb^{2+} and Ni^{2+}, respectively. From the PCA of microalgae (starch, proteins, lipids, and β-carotene), β-carotene was found to be the most affected substances. Compared to the traditional metal ion detection, the results of THz spectral analysis showed reduced sample volume from 50 to 10 mL, reduced detection time from 20 h to 6 h, and an improved accuracy from 10 ng/mL to 1 ng/mL (Figure 7.6). This study proved that THz-TDS could be potentially used for biological monitoring of heavy metal tainting in water. In another study, the THz

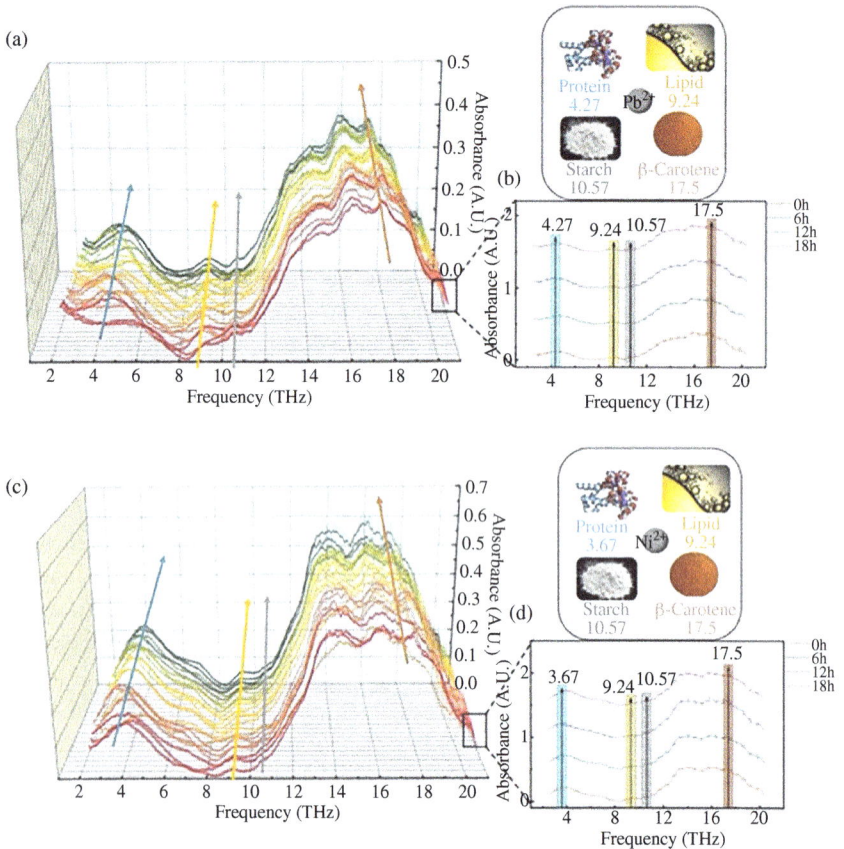

Figure 7.6   THz spectra of microalgae under stress at different concentrations of heavy metal ions at different time intervals (0, 6, 12, and 18 h): (a) seven different concentrations of Pb^{2+}, (b) 5 × 10³ ng/mL Pb^{2+} concentrations, (c) seven different concentrations of Ni^{2+}, and (d) 3 × 10³ ng/mL Ni^{2+} concentrations. (Source: Shao et al. 2022.)

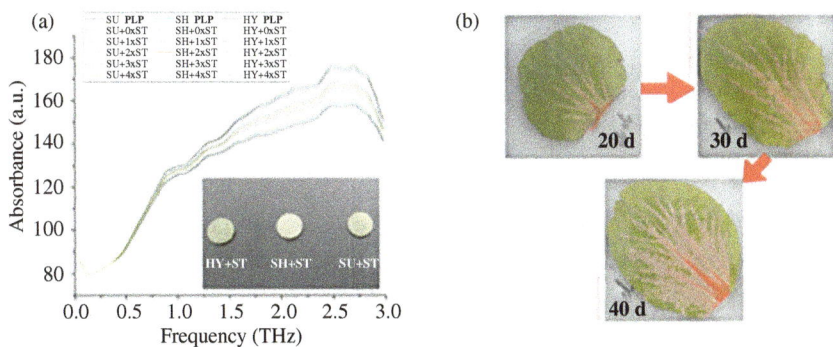

Figure 7.7 **THz analysis of Cd(PC₂)₂ complex of Pak Choi leaves: (a) THz spectra of simulated leaf samples with different concentrations of Cd(PC₂)₂ of three cultivars of Hangyoudong (HY), Shanghaiqing (SH), and Suzhouqing (SU), and (b) distributions of Cd(PC₂)₂ in CdCl2-stressed Suzhouqing pak choi leaves of 20, 30, and 40 days old. (Source: Zhao et al. 2021.)**

spectroscopy along with density functional theory, chemometrics, and dichroism was applied to characterize the cadmium-phytochelatin2 complex in plants (Zhao et al. 2021). The phytochelatin2 chelates the ion Cd^{2+} to form a complex $Cd(PC_2)_2$. The suppression of THz vibrations for PC_2 was observed at 1.03 and 1.71 THz. Further, PC_2 could be used as a potential aid in the detection of Cd in pak choi leaves with the 1.151 ppm limit of detection (Figure 7.7). This study highlights the application of THz for structural analysis, characterization, and quantification of the heavy metal PCs complexes in plants for assessing their phytoextraction performance.

7.3.7 Plant Physiology and Growth

The PGRs are chemical substances synthesized through either microbial fermentation or artificial synthesis. Since the physiological functions of PGRs are same as that of the natural plant hormones, the PGRs regulate crop development and its growth (Basra 2000). With increasing use of PGRs for achieving higher yields, excessive residues of PGRs result in the agricultural soils and plants. This causes environmental pollution and poses potential food safety issues. The rapid detection of PGRs residues in agricultural commodities will help in preventing health issues. A study was conducted on the detection and identification of three different PGRs namely paclobutrazol (PBZ), 6-Benzylaminopurine (6-BA), and maleic hydrazide (MH) using THz-TDS (Qu et al. 2018d). The characteristic THz features were observed at 0.06–4 THz. The spectral noise was removed by applying wavelet threshold denoising method to improve the SNR ratio. The molecular characterization and theoretical calculations of PGRs were made using a density functional theory (Orio and Pantazis 2009). The obtained spectral results showed distinct and unique absorption peaks with the characteristic absorption and dispersion at 2.08 and 3 THz, respectively for 6-BA. While the peaks were observed at 0.71, 1.3, 1.88, and 2.67 THz for PBZ and at 2.34 THz for MH. The THz absorption peaks processed by hard wavelet threshold denoising method agreed with the simulated results from the density functional theory. This study confirms the effectiveness of wavelet threshold denoising in THz-TDs for rapid detection of PGRs. Similar approach was used for THz fingerprint detection of 2,4-Dichlorophenoxyacetic acid (2,4-D), indole-3-acetic acid (IAA), and forchlorfenuron (CPPU) (Qu et al. 2018a). Among the tested eight window functions, four-term Blackmann-Harris function gave better spectral results. Based on the Fresnel formula, the THz optical parameters were obtained at 0.2–3 THz for PGR samples. Results showed different characteristic peaks for three different PGRs with 2,4-D at 1.35, 1.57, and 2.67 THz; IAA at 2.5 THz; and CPPU at 1.77 and 2.44 THz. The refractive index was changed at location of each absorption peaks. The study inferred that the vibrations belonging to each THz absorption peak were characterized by internal molecular stretching and external deformation. This study provides information about theoretical calculation using DFT and experimental guidance for the rapid detection of PGRs and their residues in agricultural products.

Another study was reported on THz sensing of four PGRs: glyphosine, daminozide, gibberellic acid, and naphthaleneacetic acid in frequency ranging from 0.3 to 1.8 THz based on the density functional theory (Du et al. 2021b). In another study, the detection sensitivity was enhanced by combining THz spectroscopy and metamaterials. In this study, a rapid analytical platform for PGR detection was developed based on metamaterial resonator setup (Du et al. 2021c). Results showed that metamaterials could successively detect the PGRs (butylhydrazine and N-N diglycine) at low level of 0.05 mg/L. This study confirmed the potential scope of double formant metamaterial resonator in THz sensing for rapid, low-cost, sensitive detection of PGR residues.

7.3.8 Detection of Crop Diseases

The THz-TDS was employed to detect the plant diseases like late blight and fusariosis in cereals and tuber crops (Penkov et al. 2021). The observations yielded promising results in detecting the phytopathogens, demonstrating that THz-TDS can be efficiently used in measuring the degree of damage in plant tissues. A combined approach of Laser-induced breakdown spectroscopy (LIBS) and THz spectroscopy was used for determination of degree of anthracnose disease in *Camellia oleifera* leaves (Bin et al. 2022). Compared to LDA and PLS-DA models, the SVM showed higher accuracy in qualitative assessment of *C. oleifera* leaves. Results showed that the combined spectrums of THz and LIBS can be efficiently used for detection of anthracnose and can be used for the grading of *C. oleifera* leaves. In another study, three common crop pathogens *Botrytis cinerea*, *Physalospora piricola*, and *Erysiphe cichoracearum* were identified using THz technology (Li et al. 2020b). The pathogenic fungi *B. cinerea*, *P. piricola*, and *E. cichoracearum* are responsible for the

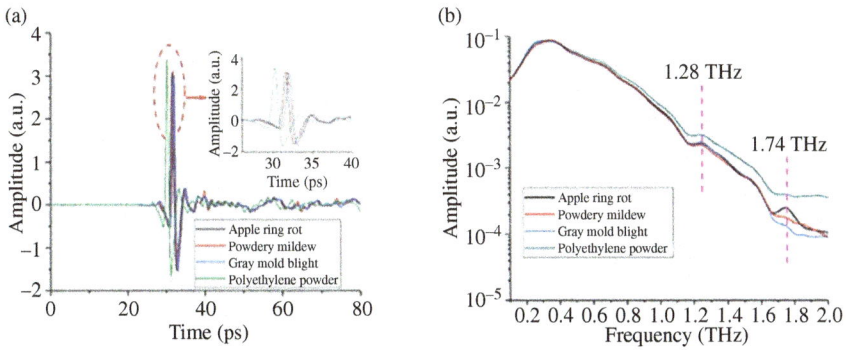

Figure 7.8 Detection of crop pathogens using THz spectroscopy: (a) THz time-domain spectra and (b) THz frequency-domain spectra. (Source: Li et al. 2020b.)

common crop diseases grape gray mold blight, apple ring rot, and cucumber powdery mildew, respectively. The pathogenic samples were prepared as tablets for spectral analysis in the range of 0.1–2 THz. The acquired spectral data were divided using KS algorithm, preprocessed by nonlocal mean filtering followed by SPA algorithm to reduce dimension of data (Figure 7.8). Different classification models such as SVM, KNN, and BPNN were used for identification. The absorption and refraction pattern of the samples in THz region showed a significant difference among different pathogenic samples. The SVM-based classification model provided satisfactory results for all three pathogenic samples at 1.376 THz with evaluation index F1 score as high as 93.8%. This study opens new array of opportunities in the identification of crop pathogens using THz technology. Further these results could be taken as reference for future studies in rapid detection to provide early warning of plant crop diseases. In another study, the fungal infection (*Gnomoniopsis* spp.) in chestnuts was identified using THz imaging. About 50 partially infected chestnuts were analyzed in low THz range imaging system equipped with 0.1 THz source (Girolamo et al. 2021). The correlation between the physical parameters (mass or volume) and light attenuation for sound and infected seeds were significantly different under THz region. The hydrolysis of carbohydrates by the fungal species led to varied water density and structural appearance of chestnuts. This study proved that the light attenuation with mass or volume of chestnut could draw a significant conclusion in identification of the fungi in agricultural commodities such as nuts and fruits.

7.3.9 Hydrology and Drought Stress Monitoring

Climate change is causing global warming and desertification. This leads to depletion of water resources that significantly affect natural ecosystem. This condition gives a warning signal for the efficient use of water bodies that requires intelligent irrigation strategies. Hence, the knowledge about the hydrology is necessary for framing and adopting efficient agricultural practices. This requires *in-situ* monitoring of the time-dependent water distribution in plants for optimal breeding and growth (Jansen et al. 2008). Since THz waves have high water absorption coefficient, THz technology is used for analyzing water distribution and stress monitoring in plants and agricultural crops. Based on the water content, the leaf's refractive index and THz transmission significantly vary that helps in observing the status of water in plants. As a measure to quantify the amount of water in plant leaves, THz-TDS transmission data were used (Gente et al. 2013). The spectral data were processed with iterative algorithm to calculate the relative volumetric fraction of water present in tissue. The obtained spectral results were well correlated with the moisture measurements using gravimetric methods. Thus, THz-TDS could be used as a standard tool for monitoring hydration levels in plants and crops.

In another study, the THz spectra of rapeseed leaves were investigated to assess the leaf water content (Nie et al. 2017). The transmission and absorption THz spectra were measured in the frequency range from 0.3 to 2 THz using THz-TDS. The spectral data were preprocessed by the Savitzsky-Golay method followed by PLS, kernel PLS (KPLS) and Boosting-PLS models for predicting the water content. Results showed that KPLS model best predicted the moisture content of the rapeseed leaves with R values of 0.8508 and 0.8574 based on transmission and absorption spectra, respectively. A similar approach was adopted for determining the water status of leaves of different plants (wintersweet, bamboo, and ginkgo) (Song et al. 2018). Results of THz images based on transmission amplitude showed a greater loss of water from the basal region of the leaves than from the distal region. The results of the estimated water content from the THz spectra agreed with the conventional direct water weight measurements. The spatial and temporal variations of water of damaged ginkgo leaves were investigated to determine the accuracy and sensitivity of the developed method. The flowing of water from basal to distal region during different dehydration periods was consistent with the prediction of string-of-lakes model.

In another study, the mathematical model developed to assess soybean leaf water content was used to examine the effects of different levels of growth media, water stress, and exogenous leaf treatment with abscisic acid (ABA) using THz-TDS (Li et al. 2020c). Results showed that there was a gradual decline in the leaf water content and this decreasing trend was dependent on the moisture-holding capacity of the growth media. Among the tested growth media, soil had the better moisture-holding characteristics in following with seedling matrix and vermiculite matrix. In addition, the stomatal openings in the leaves were significantly decreased at water deficit conditions to preclude excess water loss. However, the ABA treatment resulted in rapid increase in water content of leaf followed by a gradual decrease. This study demonstrated the usefulness of THz radiation combined with modeling approaches for noncontact measurement of water content in plant leaves. All the above studies were based on two-dimensional mapping of water content in plants. With the developments of THz imaging systems, recently a study was reported on three-dimensional (3D) water mapping of succulent leaves (*Agave victoriae-reginae*) (Singh et al. 2020a). The presence of large fractions of fructans was mainly responsible for high water retention of succulent. This study opens a new dimension to the application of THz imaging for 3D mapping of water and helps in the development of high throughput phenotypes with drought tolerance.

7.4 Spectral Assessment in Crop Science

The plant leaf constitutes of solid matter, water, and gas that contributes significantly toward water transport, respiration, and photosynthesis. The abundance and distribution of the solid matter is closely associated with the dynamic metabolic activity of the plants. The lack of adequate concentration and distribution of the leaf constituents shows nutrient deficiency with development of crop diseases (Zang et al. 2019). Hence, qualitative and quantitative assessment of the spatial distribution of leaf constituents would greatly help in better understanding of the plant response to environmental changes. Further quantitative evaluation of leaf constituents helps in establishing a standard procedure for testing samples of different species. This provides a deeper knowledge of the physiological condition of plants, especially transpiration kinetics, water stress management, and accumulation of dry matter content. As against the available standard testing methods such as gravimetry, non-destructive procedures are urgently required for quantification of leaf constituents for continuous monitoring of water status and other plant constituents. In this regard, the THz waves with low-frequency vibration of hydrogen bond have high sensitivity toward water molecules. With greater potential of THz to analyze the spectral information, the monitoring of water status using THz spectroscopy is studied elaborately. The THz band nanosensor network systems were used for high-resolution monitoring of plants (Afsharinejad et al. 2015). The aim of the proposed system is to provide insights on methods of the communication of plants to the scientific community. Despite the challenges associated with high attenuation and limited transmissivity of signals, the proposed THz path-loss model considers THz radiation in a mixed channel of plant leaves and air. This model was based on measured absorption data of individual leaf and future studies should extend the scope of THz nano communication systems to include the attenuation from the stalk, branches, and fruit of the plant. This section provides a detailed discussion on how THz systems are used for qualitative analysis, relative and absolute quantitative analysis of plant constituents, and monitoring of stress levels.

7.4.1 Spatial Distribution of Plant Leaf Constituents

The plant leaves possess heterogenous structures with spatially varied distributions of solid, liquid, and gaseous matter. The associated material composition and distribution correlate with leaf vigor and phylogenic traits. Hence, the assessment of spatial composition of leaf constituents provides knowledge about evolutionary history of plants. The THz spectral imaging technique was employed for quantification of solid content, gas, and water status in leaf (Zang et al. 2019). A particle swarm optimization algorithm based on extended Landau-Lifshitz-Looyenga model was used in the reported method. This algorithm helped in quantitative assay and distribution of constituents of *Bougainvillea spectabilis* leaves. Also, the obtained results of THz imaging showed a good agreement with the gravimetric analysis. Thus, results confirmed that THz-based detection and sensing approach has very good sensitivity to detect even the minute differences in leaf constituents. This highlights the potential of THz to be used as a powerful biosensing tool in monitoring and diagnosis of the crop diseases. The measured THz spectral data under time-domain was used for multilayer structural analysis of sunflower leaves (Abautret et al. 2020b). Results showed that the optical structure of the leaf was observed to be in an eight-layer stack that had a good agreement with the microscopic results. Thus, this study opens newer opportunities in the classification and morphological identification of leaves based on combined reverse engineering technique and THz spectroscopy. In another study, the reflected THz echoes from the sunflower leaf were successfully used for internal assessment of leaf structure such as thickness, complex indices, and geometry (Abautret et al. 2020a).

7.4.2 Monitoring of Stress Levels in Plants

The knowledge of the phenotyping information of plant leaves helps in early detection of crop disease and ultimately aids in yield improvement. The real-time monitoring and evaluation of live plant leaves has received greater research attention in modern precision agriculture and farming. Zahid et al. (2022) reported a study dealing with precise identification and observation of altering physiological changes and behavior of plants at the cellular level. In this study, the plant species (coriander, basil, parsley, coffee, pea-shoot, baby-leaf, and baby-spinach) were dynamically assessed for four consecutive days through integration of machine learning with frequency ranging from 0.75 to 1.1 THz. The measured THz observations were analyzed using support vector machine (SVM), K-nearest neighbor (KNN), and RF algorithms. Among these, the RF showed the highest accuracy of 98.9% followed by KNN (94.6%) and SVM (89.7%) for precise detection and identification of different leaves based on their morphological features. Also, the water-stressed leaves were well classified using RF with the accuracy of 99.4%, demonstrating that THz can be potentially used for successful identification and classification of stressed plant leaves. The feasibility of the advanced THz quantum cascade laser system was assessed for determining the absolute leaf water content (Pagano et al.

2019). Six different leaves, namely, *Corylus avellana, Laurus nobilis, Ostrya carpinifolia, Quercuus ilex, Quercus suber,* and *Vitis vinifera* were considered in the study. This study utilized a simple approach of THz light transmission (2.55 THz) with combined photographic measurement of surface area of leaves for water content measurement. This study also highlighted the potential scope of integration of this technique with gas exchange analyzer and physiological indices such as water use efficiency to assess the plant physiology under both laboratory and field conditions. In future, the study can be extended to estimate the water content of canopy that assists in indirect monitoring of kinetics of leaf invasion by crop pathogens.

7.4.3 Spatiotemporal Variability of Water

The water content of the biological materials is a parameter of high significance providing information about its physiological state. A quantitative method was proposed to analyze the bound and free states of water in plant leaves using THz spectroscopy (Zang et al. 2021). Four different species of plant leaves namely *Bougainvillea spectabilis, Citrus reticulata, Bambusa multiplex,* and *Cinnamomum japonicum* were considered in this study as model samples. The THz spectroscopy integrated with linear absorption model and particle swarm optimization was applied to quantify the free water and bound water in leaves at 0.7 and 0.9 THz, respectively. Results of THz experimental studies had good agreement with traditional measurement techniques (Figure 7.9). Though the chemical states of water in plant leaves were successfully assessed using THz, still more research must be carried out in this direction. Future studies in this area would help in improving the stability and consistency of analysis under diverse extreme environmental conditions. In another study, the THz-TDS was used for testing the water content of soybean leaves (Li et al. 2020c). Based on the obtained spectral results, a

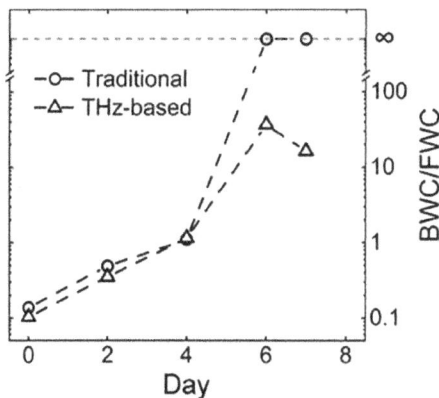

Figure 7.9 Variations in the ratio of bound water content to the free water content of traditional and THz methods. (Source: Zang et al. 2021.)

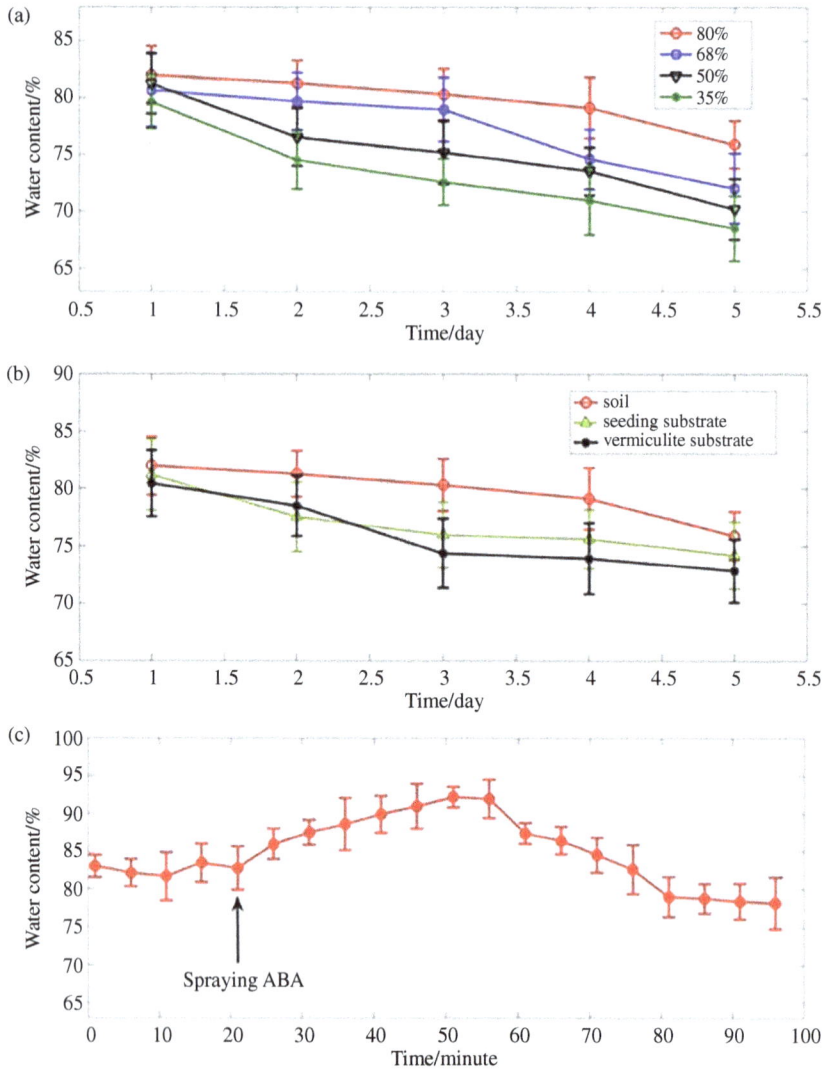

Figure 7.10 THz-TDS for assessment of soybean leaf water content: (a) variations in leaf water content in different growth media at different levels of water deficit, (b) leaf water content of soybean plants grown in varied growth media with time, and (c) effect of ABA application on the soybean leaf water content. (Source: Li et al. 2020c.)

high-precision mathematical model was developed to assess the leaf water content. With the developed model, the effects of water levels on stress (severe deficit – 35%, moderate deficit – 50%, mild deficit – 65%, and normal water supply – 80%), different growth medium (vermiculite, soil, and seedling substrate), and abscisic acid (ABA) leaf treatment were studied (Figure 7.10). With different moisture contents, the leaf water content gets decreased with time at a slower pace under normal supply than that under hydro stress. The observed trend of leaf water content under water stress conditions is highly dependent on the type of growth media. These results were consistent with decrease in stomatal openings to preclude excess water loss. However, the ABA treatment resulted in a significant increase in leaf water content followed by a slower decreasing trend. The variation of stomatal opening as observed under laser scanning confocal microscopy agreed with changes in leaf water content as analyzed macroscopically. These results proved that THz technology in combination with appropriate modeling methods turn out to be an effective noninvasive analytical tool to quantify the leaf water content. In another study, a physical model was proposed predicting the relative water content, water mass per unit area of leaf, and leaf water potential based on THz transmission of leaves (*Arabidopsis thaliana*, *Hedera canariensis*, and *Plantanus racemosa*) (Browne et al. 2020).

7.5 Summary

The THz spectroscopy and imaging are excellent tools for rapid and noninvasive characterization and assessment of agri-food samples. Water is a major constituent of agricultural commodities. The polar nature of water molecules and their characteristic relaxation time ranging from pico- to sub-picosecond remains advantageous feature in diagnosing and mapping biomolecules. Since the energy of THz photons is low, it does not damage the samples. Considering these aspects various novel emerging research works are being carried out in exploring the potential of THz in agricultural sector. This chapter provides a detailed discussion and summary of scope of THz spectroscopy and imaging in agricultural sector. Various applications of THz in analyzing nature and composition of materials, evaluation of seed quality, discrimination of transgenic crops and seeds, seed inspection and identification, detection of pesticide residues, heavy metal residues, plant physiology and growth, crop diseases, hydrology, and drought stress monitoring were discussed in detail. The recent trends of spectral assessment in crop science in tracing the spatial distribution of phytoconstituents, monitoring stress levels in plants, and spatio-temporal variability of water were briefly summarized. On realizing these potential opportunities, the THz science has a great scope in biosensing communications in integration with nanosensors, use of metamaterials, and quantum cascade lasers. Certainly, THz technology has a great potential to modernize analytical techniques for assessing raw and processed agri-food products as well as for enhancing the understanding of interactions of plants with soil microbiome and surrounding environment.

Chapter 8

THz Technology for Quality Monitoring and Control

8.1 Introduction

Most of the foods are perishable due to their high-water activity. Food technologists are continuously working to enhance preservation of foods through various processing techniques for achieving a shelf-stable product. Foodborne illnesses are often caused by food pathogens. Microorganisms such as viruses, bacteria, and fungi are directly linked with the safety and quality of food products (Balali et al. 2020). Food spoilage has become one of the fundamental challenges that raise health concerns. Also, the lack of proper handling of agri-food commodities in the supply and distribution chain at times incurs greater food losses. Hence, it is a must to ensure the safety of food by adopting appropriate detection methods (Liu et al. 2019c). The scientific fraternity has faced major food safety outbreaks due to the dawn of varied strains of pathogens, food allergens from microbial toxins, adulterated foods (Li et al. 2021d), and consumer perception of unknown long-term side effects from genetically modified foods. The common symptoms of foodborne illnesses are diarrhea, vomiting, nausea, dysentery, stomach cramps, and fever. Food poisoning is caused by common pathogens such as *Escherichia coli*, *Campylobacter jejuni*, *Yersinia enterocolitica*, *Listeria monocytogenes*, *Staphylococcus aureus*, *Clostridium* spp., *Salmonella* spp., and *Shigella* spp. (Gourama 2020). The outbreak of such foodborne illnesses all over the world raises the attention of not only the health and medical professionals but also the food scientists to regulate the existing food manufacturing practices. This would significantly help in reducing the outbreaks of foodborne illnesses and

DOI: 10.1201/9781003197010-8

in decreasing food losses due to microbial contamination. As a matter of health concern, it is mandatory to assess and evaluate the food safety and quality in the food processing units and during its distribution through retail to the consumers.

Most of the conventional microbial detection techniques such as microbial enumeration and plating techniques are laborious and time-consuming. Hence, rapid detection and analytical techniques are required to be deployed at food processing sites. The rapid detection approaches are categorized as nucleic acid sequence techniques, immunological techniques, and biosensing techniques (Tahir et al. 2021). Techniques such as polymerase chain reaction (PCR) are based on real-time assessment and multiplex PCR. While loop-mediated isothermal amplification, nucleic acid sequence-based amplification, and DNA microarrays are nucleic acid sequence-based techniques. While mass-based biosensors, optical biosensors, biochemical sensors, and electrochemical biosensors are some of the biosensing approaches used for microbial detection. On the other hand, enzyme-linked immunosorbent assay (ELISA) and lateral flow assays are based on an immunological approach. In contrast to the conventional techniques, the rapid detection methods are efficient, highly sensitive, more specific, and reliable (Law et al. 2015). The conventional detection approaches follow a sampling plan that draws samples at regular interval from liquid samples such as water, processed liquid foods, and wastewater. However, food testing from the processing industry has a complex range of diverse food samples (Mishra et al. 2018). To address this issue, the past decade witnessed molecular methods for microbial detection and assessment. This led to the flexibility of microbial protocols in discovering the vast range of microbial flora in food processing plants. Often the novel assessment tools combine different approaches such as culture-based, biochemical sensing technologies, molecular diagnostic assays, mass spectroscopy, gas chromatography, hyperspectral imaging techniques, Raman spectroscopy, THz spectroscopy, and imaging (Hameed et al. 2018). These are highly advanced and smart techniques that aid in the rapid assessment of microbial analysis. This chapter provides comprehensive knowledge of the applications of THz systems for quality monitoring and control. Various emerging trends in microbial analytical approaches based on THz biosensing are presented. Further, insights on future perspectives and existing limitations in providing solutions for food safety concerns using THz systems are outlined.

8.2 THz Technology for Food Safety

The THz radiation falls between the high-frequency microwaves and long-wavelength far-infrared regions of the electromagnetic spectrum. Due to characteristic spectral features, the THz system has emerged as an efficient tool in material characterization, security screening, remote sensing, and communication (Kundu and Pragti 2022). The THz radiation has predominantly been used as a spectroscopic and imaging tool in basic and applied sciences. Though THz yields a highly resolved image of opaque samples, the nature of samples (like thickness and radiation absorbance) limits its application in biological and chemical studies (Federici et al. 2005). With the recent advancements in THz sources and detectors, THz spectroscopy has become a promising, noninvasive, and label-free approach for biological studies. This section provides a summary of the scope of THz science in biological studies for ensuring food safety and quality monitoring in the agri-food sector. The THz spectroscopy is more advantageous than other conventional microbial testing methods. The culture-oriented standard plate count requires three to five days for the interpretation of microbial count. On the other hand, sophisticated techniques like PCR assays require trained personnel (Ahmadivand et al. 2020). Similarly, fluorescence-based techniques like flow cytometry and epifluorescence microscopy require fluorescent labeling materials for microbial detection. Other optical methods such as light scattering and autofluorescence suffer from less sensitivity in the measurements (Lazcka et al. 2007). Considering these aspects, THz time-domain spectroscopy (TDS) is explored for onsite microbiological investigation in food premises. This has led to a greater research interest in employing THz spectroscopy and imaging for detecting food pathogens. The microbial cell components exert molecular (such as proteins, DNA) motions at a low frequency on excitation with THz radiation (Hameed et al. 2018). This results in a specific spectral feature that assists in the detection of microbes and their metabolites in agri-food products. It was reported that the small variations in the hydration levels between the living state of the bacterial cells of the same species are responsible for different absorption coefficients (Yang et al. 2016a). The corresponding THz spectra allow for the rapid identification and classification of bacterial species.

The transparency of microbes under the THz region is advantageous for their detection; however, the size of microorganisms is smaller than the THz wavelength in the order of 1/100. This size mismatch results in low scattering effects that decrease the sensitivity of measurements (Wang et al. 2018b). Hence, more studies are now being focused on the use of THz antennas and metamaterials for localized generation and enhancement of THz fields for improving sensitivity based on frequency resonance shift. A metallic mesh sensor was used to detect *E. coli* based on the measurement of the dielectric property (Kurita et al. 2014). To capture the *E. coli*, other anti-*E. coli* antibodies were immobilized on the metallic mesh surface. Results inferred a significant correlation between the THz frequency shift and the *E. coli* concentration useful for quantification of microbial cells. However, the resulting higher limit of detection (LOD) of 10^6 CFU/mL remained to be unsatisfactory which put forth the need for further studies in this area. Based on the structural and chemical differences of cells, the gram-negative type of the bacteria was identified using THz in conjunction with appropriate chemometric methods (Berrier et al. 2012). A metamaterial sensor was designed for the rapid, sensitive, and on-site detection of microorganisms. The fabricated sensor had a special design comprised of an array of square rings with a central micro-gap that enabled the detection of microorganisms. Thus, the advanced THz systems with integrated metamaterial and nanosensors provide a promising investigation tool for rapid onsite detection and identification of microbes in the agri-food sector.

8.3 THz Applications in Food Processing Control and Quality Monitoring

8.3.1 Microbial Characterization

Extensive research was reported for the THz detection of cellular components in characterizing the microbial cells and their spores. The spores of *Bacilli* were identified through the detection of bacterial cell component dipicolinic acid (DPA) (Paidhungat et al. 2000). The THz transmission spectrum of DPA exerts a similar feature as that of the spectra of *Bacillus globigii* and *Bacillus thuringiensis* but differs from the *Aspergillus niger* (Fitch et al. 2004). The unique spectral fingerprint of DPA at 1.54 THz coincides with the spectral peak of *Bacillus subtilis* spores at 1.538 THz (Yu et al. 2005). Because of these similar spectral features, it is quite possible to detect the *B. subtilis* spores by detecting the DPA. The calcium complex (Ca-DPA complex) of *Bacillus cereus* is crystalline in nature while the complex is amorphous for *B. subtilis*. Hence, the THz spectral features of DPA and Ca-DPA are useful in the detection of the spectral differences of bacterial components between the *Bacillus* spores (Leuschner and Lillford 2001; Cook et al. 2004). Regardless, the differentiation of *Bacillus* spores based on spectral differences is not always possible as THz signals might be masked by other components. Like the detection of DPA, the DNA of the *B. subtilis* has also been studied under the THz regime. Though there exists a spectral similarity, a distinct correlation was observed between the THz spectra of DNA and the spores (Bykhovski et al. 2005). This was obvious as the microbial cells have more complex components than the spores themselves. The apparent similarity of spectral absorption of *E. coli* DNA and cells showed a significant difference due to the strong THz absorption of hydrogen bonds in DNA molecules. The DNA spectra of *E. coli* and *B. subtilis* exhibit a remarkable difference even though their base-pair compositions are not well distinct (Globus et al. 2012). This considerably helps in the detection and identification of bacterial species based on their response to THz waves. In addition, the THz spectroscopy showed a great potential in the screening of the industrial microbial strains based on their ability to characterize the intracellular metabolites. The intracellular metabolite riboflavin is highly THz absorptive while the cells of *B. subtilis* are less absorptive (Wang et al. 2010a). These distinct THz absorption features of metabolites and cells under the THz range aid in rapid screening of poor riboflavin-producing strains from the high-productivity strains.

Over the years, computational techniques were also employed for better understanding and the prediction of the spectral features of nucleic acids and proteins based on energy minimization, normal mode analysis, and molecular dynamics techniques. A good correlation was observed with the theoretical and experimental results for DNA (Alijabbari et al. 2012), tRNA (Bykhovski et al. 2007), and thioredoxin (Globus et al. 2013) of *E. coli*. The molecular dynamic simulations are useful only with small biomolecules and are not employed to study the entire bacterial cell. This is mainly because of the limited levels of computational abilities (Yang et al. 2016b). However, it could be improved by gaining knowledge of intramolecular relaxation dynamics (Globus et al. 2012). THz spectroscopy is more suitable to monitor the changes in the cellular components due to its unmatchable ability in characterizing the weak interactions of biomolecules. The THz spectroscopy was useful in observing the subtle morphological changes in the micro-vessel endothelial cells. Results showed a decreased THz differential signal after the vascular endothelial growth factor treatment (Liu et al. 2007). This allows for monitoring the changes that occur at the molecular level of cell components. Thus, THz spectroscopy shows a greater scope in the assessment and monitoring of the sterilization effects in biofilms. With this research potential, THz science is envisioned to monitor the growth and morphological changes of microbes over time based on the THz fingerprint spectra. This would open enormous opportunities in areas of biosafety to ensure the effectiveness of the process treatment. For instance, THz is useful in probing the cytoplasm leakage, cell damage, membrane rupture, and DNA damage of the microbial cells after subjecting them to specific sterilization treatment in food premises (Yang et al. 2016b). Likewise, the plasma-treated antibacterial surfaces were characterized using THz spectroscopy (Sulovska and Lehocky 2014). Thus, THz spectroscopy is used as a novel investigating tool in characterizing and analyzing microbial activity in cell engineering and food biotechnology.

8.3.2 *Biochemical Sensing*

Recently, a rapid analytical method based on laser-engraved free-standing THz complementary asymmetric split ring (CASR) metamaterial has been developed for fractionation and identification of organic acids (Zhou et al. 2022). This study used the advanced fabrication method of direct laser writing to fabricate copper foil-based THz devices for distinguishing fruit acids such as citric acid, L-malic acid, and D-tartaric acid. As against the conventional fabrication methods of THz devices such as electron beam lithography, photolithography, and nanoimprinting, the direct laser writing seemed to be flexible, convenient, and low cost. Results showed the designed THz system was able to detect the fruit acid with a concentration of 0.1 mg/L. When compared to conventional acid-base titration, the proposed method is simple and quick. Thus, the proposed method based on a metamaterial-based THz system can be a powerful analytical tool in rapid sensing of *in-situ* analysis of food components.

Another class of microbial products (metabolites) produced by fungal species are mycotoxins (e.g., aflatoxins, ochratoxins, patulin, and fumonisins). The aflatoxins are secondary metabolites produced by species of *Aspergillus parasiticus* and *Aspergillus flavus* (Chakraborty et al. 2021). These aflatoxins are highly toxic, carcinogenic, mutagenic, and teratogenic. Aflatoxins are often found in agricultural products such as maize, peanuts, pistachio nuts, copra, cotton seeds, and even in spices. When the food products contaminated with aflatoxins are ingested it causes severe illness in both human and livestock. Hence, it is most important to screen the agri-food products for the presence of the aflatoxins. The common techniques that are in current practice for the detection of aflatoxins are high-performance liquid chromatography (HPLC), enzyme-linked immunosorbent assay (ELISA), and liquid chromatography coupled to tandem mass spectrometry (LC-MS/MS) (Zhao et al. 2020). In addition to laborious, time-consuming processes, these conventional methods use environmentally unfriendly chemical reagents and are often not suitable for real-time measurements. In contrast to these conventional approaches, the analytical method based on spectroscopy techniques like Fourier transform infrared spectroscopy, near-infrared spectroscopy, and Raman spectroscopy provide details of qualitative and quantitative information about mycotoxins under a single scan with less sample preparation (Lee et al. 2014). In line with these spectroscopic techniques, the THz spectroscopy is a promising analytical aid used for the detection of various ranges of food contaminants, chemical fertilizers, pesticide residues, and antibiotic residues. Considering these applications, the THz-TDS has been applied for the detection of aflatoxin B1 with combined chemometrics in peanut oil (Chen and Lijuan 2014). The peanut oil samples were prepared by artificially contaminating the oil with aflatoxin B1 at different levels of concentrations such as 0, 1, 3, 5, 10, 20, 50, 100, 150, and 200 ppm. Compared to the partial least squares regression (PLSR), the stepwise multiple linear regression model was more stable in the determination of aflatoxin B1. Thus, the results of this study inferred that THz-TDS when combined with an appropriate classification model can be potentially used for the rapid non-destructive analysis of the agri-food products contaminated with aflatoxins.

The metabolic products including proteins, lipids, and carotenoids produced by the microalgae have the utmost industrial significance. This includes the applicability of these metabolites as medicine, biofuels, and nutraceuticals. On realizing the potential of the THz spectroscopy, the THz spectrum of *Haematococcus pluvialis* was analyzed for the detection of metabolites concentration during the algal production process (Shao et al. 2021). This study proposed the simplification of pretreatment procedures of the samples and the use of the THz system aided in enhanced detection speed. The analysis of the algal spectrum showed characteristic peaks at 17.32, 8.69, and 16.22 THz for β-carotene, astaxanthin, and starch, respectively (Figure 8.1). Results proved that THz spectroscopy can be promisingly used for the analytical detection in real-time monitoring of the *H. pluvialis* and its metabolic product astaxanthin.

Likewise, the high-resolution rotational THz spectroscopy has been used for the analysis and fractionation of polar gas molecules, especially for the detection of organic volatile compounds (Galstyan et al. 2021). Generally, the raw and processed foods under improper storage conditions generate a complex mixture of volatiles due to incipient spoilage. The headspace gas mixture can be conveniently assessed using spectroscopic techniques. In a study on food spoilage

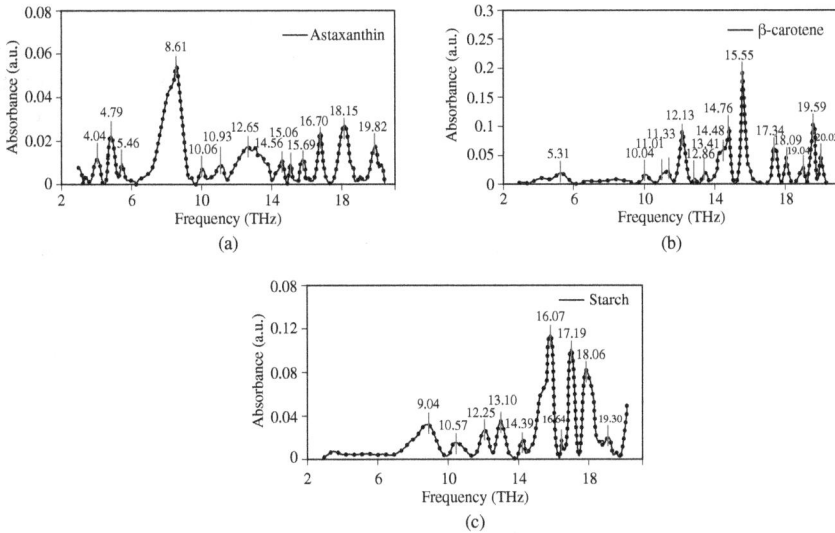

Figure 8.1 **THz absorption spectra of metabolites of *Haematococcus pluvialis*:** (a) astaxanthin, (b) β-carotene, and (c) starch (vertical red lines indicate absorption peaks at different frequencies for astaxanthin, β-carotene and starch, respectively and error bars are labeled on each data). (Source: Shao et al. 2021.)

assessment, the THz spectroscopy was employed for monitoring the generation and emission of hydrogen sulfide (H_2S) in the headspace of packed Atlantic salmon fillets (Hindle et al. 2018). The results obtained from the THz spectral spectroscopy were validated by comparing it with the conventional method of selective ion flow tube mass spectrophotometry (SIFT-MS). This study provided promising results in the monitoring of H_2S production in nitrogen-filled (100%) packs under refrigerated storage. Based on these results, the scope of the THz spectroscopy can also be extended for the detection and quantification of molecular components of interest in other perishable foods as a spoilage assessment tool.

8.3.3 Food Microbial Analysis

The quality control and monitoring of food processes must include the detection and the estimation of microorganisms as one of its main aspects. Most of the microbial testing techniques are based on culturing methods that take a long time. Other sophisticated molecular analytical detection involves complex procedures with the LOD falling between the range of 10 and 100 CFU/mL. Compared to these methods, the application of THz spectroscopy for the detection of microbes is quite simple and fast which has the advantage of see-through ability (Yang et al. 2016b).

Due to the penetration characteristics, THz waves are conveniently used for detecting microorganisms in food products along with packaged materials. Another advantage of THz spectroscopy is the ability of THz radiation to differentiate the live and dead cells. Considering the absorption coefficient, bioparticles such as bacteria, fungi, and yeast cells have low absorption coefficient that allows most of the THz radiation to impregnate through these materials. The macro cellular components such as proteins, lipids, and genetic material of biomolecules possess a distinct absorption coefficient that is reflected in the fingerprint THz radiation spectra (Wilmink and Grundt 2011).

In another study, the spectral characteristics obtained from FTIR spectroscopy in the sub-THz region (10–25 cm^{-1}) were used to quantify the *E. coli* and *B. subtilis* (Globus et al. 2012). The developed method showed improved sensitivity and reliability. The THz-TDS was used as bacterial sensor that worked in evanescent mode and detected the *E. coli* at 0.7 THz in concentration of 10^4–10^9 CFU/mL. The sensitivity of detection of films coated with bacterial cells has been improved with the use of THz plasmonic antennas (Berrier et al. 2012). Based on the structural differences in the cell components and cell wall, the THz transmission varies that aids in the detection of gram type bacteria. Microbial sensors based on metamaterials depend on the resonance effects due to analyte deposition that operates at the THz frequency range enabling the detection of bacteria, fungi, and yeast. The compatibility of the micro-sized gap of THz metamaterials with the size of microbial cells made THz spectroscopy to be ideal for the detection of microorganisms (Park et al. 2014). THz spectroscopy has a low sensitivity that limits its application for detecting microorganisms because of mismatch of bacterial cell size (1–3 μm) and wavelength of T waves (300 μm) at 1 THz (Yang et al. 2016b). This results in lowering the scattering loss during the quantification of bacterial cells. However, this limitation of sensitivity can be improved by using nanoantenna with width of 5 μm slot. Recently, a rapid, label-free identification method was developed using CW-THz for detection of microorganisms at different living states. The developed THz bacterial sensor has an R-value of 0.977 for *E. coli* (Yang et al. 2017).

With the developments of THz components, THz slot antennas were used for visualizing the bacteria, yeast, and mold in the THz region. It was reported that because of the higher dielectric constants yeast cells are more apt for optimizing the THz microbial sensors (Park et al., 2014). As stated earlier, THz spectroscopy has been used for monitoring the fermentation process. In a study conducted by Fawole et al. (2015), the effect of sugar and artificial sweeteners on yeast (*Saccharomyces cerevisiae*) growth was reported. Interestingly, the vigorous bubbling was observed with the solutions of artificial sweeteners and the maximal yeast growth was attained within 24 min. On the other hand, aspartame and sucralose also yielded maximum yeast growth with emission of CO_2 bubbles. Wessel et al. (2013) developed a contactless THz biosensor for monitoring the yeast growth. The developed biosensor works based on the number of yeast cells in a compartment that reflects a change in average permittivity.

8.3.4 Operation and Process Control

Quality is one of the key criteria that decides the consumer acceptability of food products. The use of modern spectroscopy techniques for online/on-site detection and monitoring of operation and process control is urgently needed. THz spectroscopy and imaging are one of the efficient non-destructive analytical techniques that helps in online monitoring of process flow. In THz spectroscopy, the refractive index and the absorption coefficient could be mapped in two-dimensional coordinates that aid in the quality assessment of the food materials. Based on this quantification and resulting THz fingerprint, the presence of insects and impurities in milk powder and sugar has been reported (Shin et al., 2018a). In another study, the quality of the green tea was assessed by acquiring the THz spectrum and measuring refractive index of tea samples in the range of 0.2–1.5 THz. In this study, the acquired data were analyzed using least squares-support vector machine (LS-SVM), back propagation artificial neural network (BPANN), and naive Bayes classification models. Among these, LS-SVM showed the best classification for Chinese green tea (Xi-Ai et al. 2011). The advanced chemometric linear and nonlinear models are useful in distinguishing the state of the food materials. For instance, the THz images of the confectionery creamy samples have been characterized for the added hydrocolloids (Hadjiloucas et al. 2010). Here, the real and imaginary parts of the THz images were linked with the sample's refractive index to assess its state.

A quasi-optical THz imaging system combined with network analyzer was applied for quality control of the agri-food products (Etayo et al. 2011). The quality of leaf sample was assessed for the defects, holes, and foreign objects based on the differences in absorption, reflection, and transmission of THz signals. The developed system resulted in a high-resolution image used for the detection of traces of foreign materials, layers, and thickness control. Another quality defect commonly encountered in dry powder samples is caking effect which significantly affects the flowability. The particle size of the powder samples ranges between 75 and 300 µm and can be characterized using THz imaging at frequency of 0.1–4 THz. The progress of crystallization and the thickness of the sugar coatings have been monitored using THz-TDS operated in ATR mode (May and Taday 2013). Results showed that absorption spectra were dependent on time with the absorption of sugar solution gradually decreasing in 50 h and thereafter the crystallization process starting with the evidence of formation of crystalline material at about 58.5 h. The presence of antibiotics and somatic cells significantly affects the milk quality. Naito et al. (2011) used a modified ATR THz-TDS system with the addition of temperature controller into Fourier transform THz spectrometer for detecting somatic cells in milk samples. Results showed that the use of temperature controlling unit greatly improved the performance efficiency of the system in detecting the somatic cells by reducing the error in temperature changes. In another study, the freshness of meat was evaluated using THz-TDS (Huang et al. 2015). The differences in the time domain and the absorption coefficient of the samples aided in the differentiation of fresh and deteriorated meat samples. Resulting changes in the decomposition of meat tissues and variation in the water content were mainly responsible for the distinct THz spectra.

8.3.5 Food Packaging

Polymeric films used as food packages are inspected for flaws and leaks to prevent quality loss. Detection of channel leakage in food packages remains challenging with the conventional testing methods. THz-TDS has been used for detecting the water-filled and air-filled defects in polymeric packages (Morita et al. 2005). The differences in the optical parameters have been used for the detection of defects in the water-filled channels and air-filled channels. Since the standard polymers are transparent to low-frequency THz waves, the incorporation of additive would reflect a deviation from the standard THz spectra. A study was reported on the detection of additive content in polymeric compounds using THz-TDS operated in transmission mode (Wietzke et al. 2007a). With the extracted spectral data based on the refractive indices of compounds, the presence of magnesium hydroxide in linear low-density polyethylene; calcium carbide and silicon dioxide in polypropylene; and glass fibers in polyamide were detected. In another study, the fiber-coupled THz spectroscopy was used for analysis of molten polymers and compounding process (Jansen et al. 2008). In this study, the effect of temperature and pressure on properties of polypropylene, polypropylene-$CaCO_3$, and PA6-glass-fiber compounds were examined. Results confirmed the detection of volumetric additive content in the polymers based on THz refractive index of the chemical compound. Thus, THz spectroscopy can characterize the additives used in polymers for aiding in the quality control of polymeric compounding processes.

With developments in manufacturing practices and production planning, new novel additives such as silica and nanotubes allow for improvements in material design and quality. A small variation of about 1% in the polymeric mixtures can be easily recognized using THz-TDS (Peters et al. 2013). The inline monitoring of extrusion process of elastomer using THz-TDS greatly helps in visualizing the quality and minimizing the waste. The combination of THz imaging with computational techniques has been reported for the investigation of material dispersion (Kulya et al. 2017). In this study, the THz pulse time-domain holography (PTDH) was used for imaging the concealed objects in materials with complex spatial distribution of optical characteristics of sodium polyacrylate. This study confirms the ability of THz imaging based on PTDH approach to enhance the visualization of concealed objects in a complex dispersion medium. In another study, the THz-TDS was applied for the monitoring of the loading levels of multi-walled carbon nanotube (MWNT) in resin films (Peters et al. 2012). There was a good correlation observed between the measured THz transmission spectra and the MWNT concentrations (0.1–1 wt%). Thus, this study demonstrates the ability of THz-TDS imaging system to localize and visualize the material accumulations in polymeric films.

As a contact-free emerging technique, THz-TDS is being employed for monitoring of phase change and the glass transition in polymers. The amorphous phase change of the material would reflect a change in the temperature-dependent refractive index of the polymer under THz regime (Wietzke et al. 2009b). Based on this principle, the glass transition temperature of semicrystalline polyoxymethylene was determined using THz spectroscopy, and the obtained results were validated using conventional differential scanning calorimetry (DSC) measurements. THz spectroscopy can be used as a complementary technique for visualizing the phase transition of polymers. In a recent study, the developed novel phase change materials based on tetradecanol and lauric acid were characterized for their phase-changing properties (Aytan et al. 2019). Among the tested samples, only L20 sample yields spectral features at 0.85 and 1.1 THz while there was no distinct spectral data observed for the other samples. Other applications of THz spectroscopy in packaging sector are the measurement of water sensitiveness, determination of film or paper thickness, measurements of THz birefringence for monitoring orientation of fibers in reinforced plastics, inspection of adhesive bonds as well as plastic weld joints (Wietzke et al. 2010). With the advantage of excellent dielectric contrast of materials at THz frequencies, the defects that are invisible to ultrasonic measurements and X-rays can be well predicted using the THz systems.

8.4 Perspectives and Limitations

As stated earlier, there is urgent need to come up with a novel rapid detection tool for microbial detection in the agri-food sector to prevent food losses. This need has spurred up a lot of research efforts for non-destructive qualitative and quantitative assessment methods as an alternative to standard plate count procedures. Spectroscopic and spectral imaging are closely related techniques that share similar advantageous features. Compared to the traditional detection methods, the results from spectral assessment of agri-food samples are faster with precise measurements. Hence, the spectroscopy as well as imaging tools have a great scope for the online microbial assessment in food industries. In this regard, the THz science offers enormous range of solutions to food safety and quality assurance problems. The quick assessment of microbial detection remains to be top priority at the manufacturing point to eliminate serious outbreak of health risks from contaminated foods. The combined technique of THz spectroscopy and imaging offers promising solution in the distinct identification of individual microbial strains of same species (Yang et al. 2016b). Also, it remains a better probing tool in assessment of live and dead cells. With the development of novel THz system components, the biosensing applications of THz for detection of toxins and microbial metabolites are quite possible. The THz spectroscopic systems under total reflection mode have been used for *in-situ* real-time monitoring of cytoplasmic leakage in epithelial cells (Liu et al. 2007). All these above-mentioned progresses pave a way for the development of THz spectral database of food microbes for implementation in rapid detection and identification of food pathogens.

Although THz science shows greater scope in food quality monitoring and control, the THz applications in the agri-food sector are at their infant stage. Some challenges such as the limited THz sensitivity with the identification and detection of microbes are mainly because of size mismatch between THz wavelength and the microbial cells. In accordance with the Rayleigh theory, the mismatch of sizes causes low scattering cross-section that decreases the reliability of spectral data from small quantity of samples. However, this limitation can be overcome by the use of advanced THz antennae and metamaterials (Lawler et al. 2020). This would result in the enhancement of sensitivity of detection at the molecular level with the three orders of magnitude using antenna of 5 μm wide slot. Also, the use of two-dimensional metamaterials results in increased enhancement of sensitivity through the generation of strong localized THz near-fields at resonances for bacterial detection. In addition, the strong water absorptive behavior of THz poses another challenge that limits the THz applications in detection of food pathogens. However, the THz spectroscopy is efficiently used for intracellular and extracellular assessment of the water content of bacterial cells (Tros et al. 2017). Certainly, the THz science has gained greater interest in microbial investigation of food samples. Many active research works are under progress in this area focusing on the improvement of THz systems for sensitive detection. Thus, THz spectroscopy is at its infancy that requires great research efforts to exhibit distinct advantage over Raman spectroscopy for microbial detection (Pahlow et al. 2015). This highlights the need for knowledge to bridge up the research gap in implementation of efficient rapid, noninvasive, portable, real-time quality monitoring THz systems for qualitative and quantitative assessment of food pathogens, allergens, microbial toxins, and their metabolites. Hence, the future research should be focused on the design and fabrication of compact biosensors using novel metamaterials and nanoantenna with improved detection sensitivity of microbes. Also, a considerable research attention must be paid toward the selection of appropriate chemometric models for unbiased results with better classification. With these research insights, the THz spectroscopy and imaging technique can be potentially used as a complementary method to other spectroscopic techniques in the microbial detection.

8.5 Summary

The THz science shows higher research potential in detection of food microbes and their metabolites. This chapter provides an overview on diverse scope of THz spectroscopy and imaging systems for the identification and classification of food pathogens. From the reports in the literature, the THz system is envisaged to provide one-step solution as a simple, rapid, cost-effective, noninvasive characterization of microbes and their metabolites. With significant improvements in THz system components, it is quite possible to distinguish virulent strain from the non-virulent ones. Further, the THz systems provide promising results in continuous monitoring of the industrial fermentation process. This chapter provided a detailed discussion on various applications of THz for microbial characterization, biochemical sensing, food microbial analysis, operation and process control, and food packaging. The ongoing research advancements in the development of THz systems with integrated metamaterial would certainly improve sensitivity of microbial detection. The knowledge on THz spectral response of diverse microbes as well THz spectra of microbial cellular components can be grouped together forming a THz spectral library for rapid detection and assessment of the microbial spoilage of foods. By addressing the technical issues of size mismatch and measurement sensitivity, the THz spectroscopy would emerge as a promising probing technique for the microbial evaluation and monitoring the shelf-life of agri-food products.

Chapter 9

Future Prospects, Opportunities, and Challenges

9.1 Background

The agricultural commodities, horticultural crops, and aquaculture products are significant contributors to the economies of most countries around the globe. The proper handling, storage, transportation, trade, and marketing play key roles in improving the agricultural economy and enhance returns to producers and retailers along with consumer satisfaction. The agri-food commodities are heterogenous with inherent variability requiring adjustments in conditions of their processing and storage. Hence, the knowledge on the intrinsic and extrinsic quality of the food materials would aid in establishing good manufacturing practices (Eissa et al. 2022). Consumers are becoming more specific in knowing the processing conditions of their daily foods with growing awareness related to dietary pattern, lifestyle, food-borne illness, and allergies. Therefore, food classification and discrimination of the agri-food commodities are most important both from consumer and industrial viewpoint. Traditionally the food quality assessment is being carried out using visual inspection. This kind of manual grading is time consuming, laborious, and requires skilled personnel (Afsah-Hejri et al. 2020). To overcome these constraints, several laboratory procedures have been developed. Most of the laboratory methods are often destructive. On realizing the need for the rapid, non-destructive assessment of food samples, spectroscopic and imaging methods are introduced. Some of the well-established non-destructive analytical methods of industrial significance are optical imaging, thermal imaging, infrared spectroscopy, nuclear magnetic

DOI: 10.1201/9781003197010-9

resonance imaging, X-ray imaging, and hyperspectral imaging (Vadivambal and Jayas 2016; Su et al. 2017). Each of these spectral imaging techniques has their own advantages and limitations given in Vadivambal and Jayas (2016). The THz spectroscopy and imaging are other non-destructive probing tools for determining the chemical nature and characterization of agri-food samples.

The THz regime of the electromagnetic spectrum falls between the microwave and infrared radiation. The region of the electromagnetic spectrum from 0.1 to 10 THz remained unexplored until mid-1990s (Samanta et al. 2022). Hence, this region of the electromagnetic spectrum has been popularly known as THz gap. This was mainly because of the unavailability of efficient THz sources and detectors, thus making it difficult to build powerful instrumentation systems under the THz spectral range. The successive growth in modern-day THz time-domain spectroscopy (THz-TDS) has been realized in the late 1990s with the development of the ultrafast lasers. Nowadays, Gunn diodes, super lattice electronic devices, and photonic sources like photoconductive dipole antennae are being used (Shalini and Madhan 2022). The THz systems are designed to operate in three different modes (reflection, transmission, and total attenuated reflection) and in either spatial or time domain. The wide spectral window of the THz region complemented with synergistic spectral features of microwave and infrared radiation opens enormous novel scientific opportunities. The major applications of the THz science in the agri-food sector are spectroscopy, imaging, sensing, and communication (Ren et al. 2019a; Feng and Otani 2021). Regarding the material characterization, the THz regime is useful in studying the vibrational, rotational, and torsional modes of biomolecules in the low frequency range. Further, the nonionizing nature (2–40 meV photon energy) of THz waves (Kar 2020) can be applied advantageously to characterize the agri-food samples without damaging the sample. Earlier chapters of this book summarize the broad scope and scientific developments of THz science in the agri-food sector. This chapter provides a summarized discussion on the emerging trends of THz waves, practical implementation, prospects, opportunities, and challenges of THz science in context to the agri-food sector.

9.2 Viewpoints on THz Waves

The applications of THz for analyzing the agri-food samples are at a stage of infancy. The THz waves are promisingly applied to characterize the vibrational, torsional, and rotational modes of the agri-food samples (Baxter and Guglietta 2011). The THz waves show high sensitivity toward water molecules since the THz waves coincide with the vibrations of hydrogen bonds. However, the strong water absorptive nature of THz limits its application in analyzing the perishable foods (Qin et al. 2013). Considering this limitation, analytical approaches based on vibrational spectroscopy remain advantageous in providing point-based spectral information. With the development of THz light sources and detectors, the THz spectral imaging techniques are found to be a powerful analytical tool. It is quite useful in mining the information on intramolecular and intermolecular interactions including hydrogen bonding, hydrophobic interactions, and van der Waals forces in agri-food samples (Liu et al. 2020). In this context, this section describes the various viewpoints of THz waves. Until today, the THz regime is not fully explored to realize the applicability of THz waves as probing tool for material characterization of agri-food samples. The lack of active and passive system components makes THz domain to remain as a 'black box'. This is mainly because the development of the components of THz systems relies on both the approaches of electronics and photonics of electromagnetic spectrum. On one side, the technologies based on electronics are used for THz signal generation and manipulation. While on the other hand, the visible light of the electromagnetic spectrum is used for the optical fiber-based transmission of light. It is well known that the frequency range of THz is very low for optical technology as well as high for electronics technology (Kar 2020). Therefore, the conventional devices used for either the microwaves or optical devices are difficult to be integrated for THz applicability. Hence, the THz region is known to be a quasi-optic domain which is a combination of both electronics and photonics. The active research on THz science has made "THz gap" as no more and it is slowly emerging as frontier research and applications area in biological sciences.

The THz signals are invisible to naked eye and could pass through materials such as fabrics, paper, wood, cardboard, plastics, ceramics, and masonry. This characteristic feature of THz waves found enormous applications in security screening and monitoring (Pawar et al. 2013) and can be applied to assess packaged foods. The THz signals are promisingly applied for identifying the concealed objects. Since the penetration depth of THz waves is comparatively less than the microwaves, the THz signals get reflected from metallic objects. Further higher frequency of THz waves results in better image resolution than microwaves. Therefore, the THz waves have advantages of optical waves (spectroscopic applications of chemical discrimination) as well as radio waves (penetrating power in zero visibility). The generation and detection of THz signals is a real challenge that limits their applications (Ferguson and Zhang 2002). Conventionally used bulk semiconductor devices for microwaves and radio waves cannot be scaled down for direct operation at THz frequency. Considering this difficulty, many works are in progress for the development of efficient THz system components. From the literature, the works on optical-based generation of THz waves are evident (Saeedkia and Safavi-Naeini 2008; Blanchard et al. 2010; Wilke and Sengupta 2017). In optical heterodyning technique, the signals from two different laser sources are tuned to yield desired frequency difference for generation of THz signals using photodiode or photomixers (Liebermeister et al. 2021). This technique would result in THz signal in the microwatt power. Another technique is the quantum cascade laser that results in generation of the THz signal in the milliwatt power with frequency of broader bandwidth (Hayton et al. 2013). The free electron laser is accelerator-based THz generation technique resulting in the broad range of frequency ranges from a few hundred milliwatt to watt power level. Each kind of these THz generation techniques has its own pros and cons in terms of power usage, cost, process complexity, and amount of cryogenic cooling. Similarly, the THz signal detection and reconstruction of THz images is another challenging task. The THz detection devices and sensors based on photodetectors have seen a steadfast progress in the present decade (Yang et al. 2018; Qiu and Huang 2021). Other techniques based on Schottky diodes or quantum limited detectors are some of the recently used devices for THz detection (Huang et al. 2022b; Koul and Kaurav 2022).

9.3 Prospects of THz Spectral Imaging

The THz spectrum has the following unique characteristic features to be advantageously used for the spectral imaging applications in the agri-food sector (Li et al. 2022):

- The wavelength of THz radiation corresponds to photonic energy of 1.2–12.4 meV that is well below the earth's environmental background. The THz radiation with the characteristic low energy aids in the material characterization with photoionization of biological samples.
- The THz waves possess the characteristic properties of radio waves enabling them to penetrate through the packaging materials like paper, polymers, ceramics, and fabrics. This 'see-through' property of THz radiation is employed in identification of foreign materials.

Continuous research efforts are being taken in commercializing the THz technology for spectral and imaging applications. The THz technology was first pioneered in the Bell Laboratories by a group of researchers (Hu and Nuss 1995; Mittleman et al. 1996). The implementation and use of THz realm of electromagnetic spectrum are attributable to intramolecular and intermolecular vibrations. The THz radiation is viewed to possess better resolution at shorter wavelengths than longwave infrared light and higher penetration than short-wave infrared light. Indeed, the frequency of THz radiation does not match the molecular vibrations of specific functional groups. The THz spectra are influenced by the weak covalent bonds, ionic bonds, intramolecular hydrogen bonds, and Van der Waal forces (Ruggiero 2020). Thus, this wealth of unique fingerprint spectral information from THz realm sometimes makes it difficult to interpret. Hence, to better understand the THz spectral data, several analytical methods based on stoichiometric analysis and logical analysis means are adopted. This is evident from the works in literature on the use of chemometrics in analyzing metal complexes in plants (Zhao et al. 2021), discrimination of transgenic crop seeds (Liu et al. 2016a, 2016b; Chen et al. 2016; Yi et al. 2022), authenticity of *Fritillariae Cirrhosae* bulbs (Du et al.

2021a), identification of adulterated food products (Liu 2017a; Li et al. 2020d; Bin et al. 2021), and moisture measurements (Shen et al. 2021; Gong et al. 2022). The greatest advantage of integration of spectral analysis with chemometrics helps in reduction of dimensionality of obtained spectral data. This involves filtration of characteristic variables and optimizes data size for convenience. Further, the application of chemometrics helps in the extraction of pertinent information about the molecular structure of macronutrients like carbohydrates, proteins, and lipids (Li et al. 2022). The subsequent chemometric analysis of spectral data improves the accuracy and efficiency of the scientific results of THz spectral data.

9.4 Constraints and Barriers

The major barriers to the practical adoption of the THz spectroscopy and imaging systems are the high cost of THz sources and detectors. Although significant research progress is being made toward developing low-cost THz system components (Lucyszyn et al. 2013; Probst et al. 2015; Ozturk et al. 2017), it still seems to take quite a long time for rendering economically viable applications. Most of the THz imaging systems possess low acquisition speed. Though high-speed THz systems are possible based on subregion sampling of THz waveforms, these systems are based on specific applications (Gowen et al. 2012). Also, a high signal-to-noise ratio remains a constraint to access certain regions of THz spectrum. However, this limitation can be eliminated by acquisition of multiple scans and averaging. Another limitation is with the detection of moisture content because the THz spectroscopy is not suitable for high moisture samples due to high absorptive power of THz radiation by water. Sometimes even thin samples (1 mm) remain to be opaque to THz above 0.7 THz. In such cases, reflection THz imaging could be used (He et al. 2006). However, the differences in the optical path length cause appearance of standing waves and generate image artifacts. The variation in the particle size of the biological samples is another constraint that influences the refractive index of the agri-food samples. This limits the practical application of THz systems in quality monitoring of fresh produce. The scattering effects also significantly affect the THz absorption measurements of heterogenous samples (Baek et al. 2014). Further, the scientific knowledge about the effect of measurement conditions on accuracy, repeatability, and performance of THz spectra is obscure. Considering the agri-food applications, the process optimization with respect to measurement conditions is most critical. Further the integration of THz system with rugged software for postprocessing of spectral data complicates the prediction procedures. Adoption of complementary spectroscopic techniques in optimizing process to control quality remains as a challenge in implementing standard operating procedures and protocols in agri-food processing facilities. Some of the common challenges encountered with agri-samples are strong THz water absorption, scattering effects due to soil particles or other foreign materials, low sensitivity and limited limit of detection as in case of pesticide detection, and complex post-spectral analysis. These limitations can be eventually addressed with the advanced developments in photonics and electronics. Significant research effort must be made to reduce the cost associated with THz components and sensing devices, improving the limit of detection and accuracy, temperature-dependent sensing, signal-to-noise ratio, software performance, and overall design of instrumentation.

9.5 Recent Trends in THz System Components

The THz spectroscopy and imaging offer various applications related to identification, detection, and discrimination of samples to ensure safety and quality monitoring of agri-food materials. The advancements in material science and nanotechnology foster the development, fabrication, and design of new system components like nanoantenna sensing chips for wireless communication (Singh et al. 2020b; Khodadadi et al. 2022). Other advancements like the developed high quality THz camera are already under deployment (Stantchev et al. 2021; Zatta et al. 2021). Practically, the agricultural and food applications need a low-cost, affordable, real-time monitoring, and noninvasive system. This would certainly be possible with the THz spectroscopy and imaging system. Some of the emerging trends in THz spectral imaging techniques are as follows: exploitation of matured technology through optical conversion of THz signal, fabrication of THz diffractive, reflective, and refractive optics for imaging applications by three-dimensional (3D) printing technology, use of THz plasmonic chip-based spectroscopic components toward system miniaturization, technology integration through merging multiple orthogonal spectroscopic techniques under single detection platform, and exploring near-field imaging mode in THz spectral imaging of biomaterials (Bandyopadhyay and Sengupta 2022).

The development of sources based on the THz parametric generation system was reported to be less influenced by refraction, scattering, and multiple reflections (Murate and Kawase 2018). This is mainly because of narrow linewidth of source as well as the large area of detection. More research should be focused on the theoretical studies on interactions of agri-food samples with THz waves. This would provide deep knowledge to better understand intermolecular interactions under THz region. Special analytical sensing tools are still required to explore the potential applications related to microbial safety of agri-food products. The THz quantum cascade laser are in use for the sensing of gases (Sampaolo et al. 2021) that could be adopted for sensing of headspace gases in food packages. To realize applications of THz for quality monitoring, sensing, and communication, highly sensitive THz sensors, high-power sources, and detectors are required. In addition to research on design of novel THz system components, the development of the exclusive database for different agri-food samples in THz frequency range is critical to foresee the practical implementation of THz spectroscopy and imaging applications in the agri-food sector. Attention must also be paid to standardized establishment of THz measurement systems, components, standards, and electromagnetic compatibility (Tonouchi 2007; Akyildiz et al. 2022). Although it seems to take a long time for the development of practical applications, the current research trends would make it within reach soon. On looking at the historic evolution of the THz systems, the THz technology is forecasted to have a great promising scope in the agri-food sector. Undeniably, more attention must be paid to the coordination of different efforts on the range of applications of THz science in the agri-food sector. This book on "Terahertz Technology: Principles and Applications in the Agri-Food Industry" is an initial attempt in exploring the broad scope of THz agri-food applications and feasibility of implementation THz system as a solution for quality monitoring and screening in the agri-food industry.

9.6 Summary

The THz technology is an interdisciplinary field providing unmatchable research opportunities. The unprecedented feasibilities of THz science promote technological innovation, national security, and economical development. The inherent ability of molecular interaction of THz radiation with biomolecules results in revealing numerous physical and chemical information. Hence, the research progress on THz system components grabbed the attention of scientists and food technologists to be applied for material characterization, sensing, screening, monitoring, and communication in the agri-food sector. Although the basic research and application of THz technologies in the agri-food sector witnessed a gradual growth, the commercial applications are predicted to have a long way to go. Addressing the research gap in development of novel and portable THz system components like sources and detectors will help in implementation of feasible THz spectroscopy and imaging systems in near future. Research efforts should focus to enrich scientific knowledge on underlying biological phenomenon of THz and material interactions at macro and micro scale of agri-food samples. In addition, the existing challenges on different aspects of process variables like effects of temperature, water content, and sample nature on THz spectral data acquisition should be considered to enhance the image resolution and improve performance analytical efficiency of THz system. This book has explored the potential of novel THz regime of electromagnetic spectrum for agri-food applications. The current research trends in THz science would certainly lead to a breakthrough and pave the way for emerging of novel noninvasive tool for practical implementation in the agri-food sector.

References

Abautret, Y., D. Coquillat, M. Zerrad, et al. 2020a. "Optical Thickness of a Plant Leaf Measured with THz Pulse Echoes." In *Terahertz, RF, Millimeter, and Submillimeter-Wave Technology and Applications XIII*, 11279:112790L.

Abautret, Y., D. Coquillat, M. Zerrad, et al. 2020b. "Terahertz Probing of Sunflower Leaf Multilayer Organization." *Optics Express* 28 (23): 35018–37.

Abbas, H., L. Zhao, N. Faiz, H. Ullah, J. Gong, and W. Jiang. 2022. "One Belt One Road Influence on Perishable Food Supply Chain Robustness." *Environment, Development and Sustainability* 24: 9447–63.

Acharyya, A. 2019. "Three-Terminal Graphene Nanoribbon Tunable Avalanche Transit Time Sources for Terahertz Power Generation." *Physica Status Solidi (A)* 216 (18): 1900277.

Acharyya, A., and J.P. Banerjee. 2014. "Prospects of IMPATT Devices Based on Wide Bandgap Semiconductors as Potential Terahertz Sources." *Applied Nanoscience* 4 (1): 1–14.

Adam, A.J.L. 2011. "Review of Near-Field Terahertz Measurement Methods and Their Applications." *Journal of Infrared, Millimeter, and Terahertz Waves* 32 (8): 976–1019.

Adegbenjo, A.O., L. Liu, and M.O. Ngadi. 2020. "Non-Destructive Assessment of Chicken Egg Fertility." *Sensors* 20 (19): 5546.

Afsah-Hejri, L., E. Akbari, A. Toudeshki, T. Homayouni, A. Alizadeh, and R. Ehsani. 2020. "Terahertz Spectroscopy and Imaging: A Review on Agricultural Applications." *Computers and Electronics in Agriculture* 177: 105628.

Afsah-Hejri, L., P. Hajeb, P. Ara, and R.J. Ehsani. 2019. "A Comprehensive Review on Food Applications of Terahertz Spectroscopy and Imaging." *Comprehensive Reviews in Food Science and Food Safety* 18 (5): 1563–621.

Afsharinejad, A., A. Davy, B. Jennings, and C. Brennan. 2015. "Performance Analysis of Plant Monitoring Nanosensor Networks at THz Frequencies." *IEEE Internet of Things Journal* 3 (1): 59–69.

Ahi, K., N. Asadizanjani, S. Shahbazmohamadi, M. Tehranipoor, and M. Anwar. 2015. "Terahertz Characterization of Electronic Components and Comparison of Terahertz Imaging with X-Ray Imaging Techniques." In *Terahertz Physics, Devices, and Systems IX: Advanced Applications in Industry and Defense*, 9483:94830K.

Ahi, K., S. Shahbazmohamadi, and N. Asadizanjani. 2018. "Quality Control and Authentication of Packaged Integrated Circuits Using Enhanced-Spatial-Resolution Terahertz Time-Domain Spectroscopy and Imaging." *Optics and Lasers in Engineering* 104: 274–84.

Ahmadivand, A., B. Gerislioglu, Z. Ramezani, A. Kaushik, P. Manickam, and S.A. Ghoreishi. 2020. "Femtomolar-Level Detection of SARS-CoV-2 Spike Proteins Using Toroidal Plasmonic Metasensors." *ArXiv Preprint ArXiv:2006.08536.*

Ajito, K., Y. Ueno, J. Kim, and T. Sumikama. 2018. "Capturing the Freeze-Drying Dynamics of NaCl Nanoparticles Using THz Spectroscopy." *Journal of the American Chemical Society* 140 (42): 13793–97.

Akyildiz, I.F., C. Han, Z. Hu, S. Nie, and J.M. Jornet. 2022. "Terahertz Band Communication: An Old Problem Revisited and Research Directions for the Next Decade." *IEEE Transactions on Communications* 70 (6): 4250–85.

Albert, S., and F.C.D. Lucia. 2001. "Fast Scan Submillimeter Spectroscopy Technique (FASSST): A New Analytical Tool for the Gas Phase." *CHIMIA International Journal for Chemistry* 55 (1–2): 29–34.

Ali, S., W.A.S. Shah, M.A. Shah, et al. 2022. "Adulteration and Safety Issues in Nutraceuticals and Functional Foods." In *Advances in Nutraceuticals and Functional Foods*, 79–103. London: Apple Academic Press.

Alijabbari, N., Y. Chen, I. Sizov, T. Globus, and B. Gelmont. 2012. "Molecular Dynamics Modeling of the Sub-THz Vibrational Absorption of Thioredoxin from E. Coli." *Journal of Molecular Modeling* 18 (5): 2209–18.

Almeida, M., K.E. Torrance, and A.K. Datta. 2006. "Measurement of Optical Properties of Foods in Near-and Mid-Infrared Radiation." *International Journal of Food Properties* 9 (4): 651–64.

Almond, N.W., X. Qi, R. Degl'Innocenti, et al. 2020. "External Cavity Terahertz Quantum Cascade Laser with a Metamaterial/Graphene Optoelectronic Mirror." *Applied Physics Letters* 117 (4): 41105.

Alpes Lasers. 2022. "Alpes Lasers – Applications." Alpes Lasers, Switzerland. 2022. Accessed 14th May 2022, https://www.alpeslasers.ch/applications/#aggro

Al-Sharabi, M., D. Markl, T. Mudley, et al. 2020. "Simultaneous Investigation of the Liquid Transport and Swelling Performance during Tablet Disintegration." *International Journal of Pharmaceutics* 584: 119380.

Alsharif, M.H., A.H. Kelechi, M.A. Albreem, S.A. Chaudhry, M.S. Zia, and S. Kim. 2020. "Sixth Generation (6G) Wireless Networks: Vision, Research Activities, Challenges and Potential Solutions." *Symmetry* 12 (4): 676.

Alves-Lima, D., J. Song, X. Li, et al. 2020. "Review of Terahertz Pulsed Imaging for Pharmaceutical Film Coating Analysis." *Sensors* 20 (5): 1441.

Alves-Lima, D.F., X. Li, B. Coulson, et al. 2022. "Evaluation of Water States in Thin Proton Exchange Membrane Manufacturing Using Terahertz Time-Domain Spectroscopy." *Journal of Membrane Science* 647: 120329.

Arefi, A., P.A. Moghaddam, K. Mollazade, A. Hassanpour, C. Valero, and A. Gowen. 2015. "Mealiness Detection in Agricultural Crops: Destructive and Nondestructive Tests: A Review." *Comprehensive Reviews in Food Science and Food Safety* 14 (5): 657–80.

Arikawa, T., M. Nagai, and K. Tanaka. 2008. "Characterizing Hydration State in Solution Using Terahertz Time-Domain Attenuated Total Reflection Spectroscopy." *Chemical Physics Letters* 457 (1–3): 12–17.

Arshad, M.S., S. Zafar, B. Yousef, et al. 2021. "A Review of Emerging Technologies Enabling Improved Solid Oral Dosage Form Manufacturing and Processing." *Advanced Drug Delivery Reviews* 178: 113840.

Arteaga, H., N. Leon-Roque, and J. Oblitas. 2021. "The Frequency Range in THz Spectroscopy and Its Relationship to the Water Content in Food: A First Approach." *Scientia Agropecuaria* 12 (4): 625–34.

Auston, D.H., K.P. Cheung, and P.R. Smith. 1984. "Picosecond Photoconducting Hertzian Dipoles." *Applied Physics Letters* 45 (3): 284–86.

Aytan, E., Y.S. Aytekin, O. Esenturk, and M.V. Kahraman. 2019. "Fabrication and Characterization of Photocrosslinked Phase Change Materials by Using Conventional and Terahertz Spectroscopy Techniques." *Journal of Energy Storage* 26: 100989.

Bacon, D.R., T.B. Gill, M. Rosamond, et al. 2020. "Photoconductive Arrays on Insulating Substrates for High-Field Terahertz Generation." *Optics Express* 28 (12): 17219–31.

Badia-Melis, R., P. Mishra, and L. Ruiz-Garcia. 2015. "Food Traceability: New Trends and Recent Advances. A Review." *Food Control* 57: 393–401.

Baek, S.H., H.B. Lim, and H.S. Chun. 2014. "Detection of Melamine in Foods Using Terahertz Time-Domain Spectroscopy." *Journal of Agricultural and Food Chemistry* 62 (24): 5403–07.

Bai, J., Y. Ni, B. Li, H. Wang, J. Zhu, and C. Dong. 2022. "Quantitative Analysis of FQs Antibiotics Content in FMF Using THz Spectral and Imaging Technology." *Spectrochimica Acta Part A: Molecular and Biomolecular Spectroscopy* 264: 120284.

Balali, G.I., D.D. Yar, V.G.A. Dela, and P. Adjei-Kusi. 2020. "Microbial Contamination, an Increasing Threat to the Consumption of Fresh Fruits and Vegetables in Today's World." *International Journal of Microbiology* 2020: 13.

Balasekaran, S., K. Endo, T. Tanabe, and Y. Oyama. 2010. "Patch Antenna Coupled 0.2 THz TUNNETT Oscillators." *Solid-State Electronics* 54 (12): 1578–81.

Bandyopadhyay, A., and A. Sengupta. 2022. "A Review of the Concept, Applications and Implementation Issues of Terahertz Spectral Imaging Technique." *IETE Technical Review*, 39(2): 471–89.

Banerjee, A., C. Chakraborty, and M. Rathi Sr. 2020. "Medical Imaging, Artificial Intelligence, Internet of Things, Wearable Devices in Terahertz Healthcare Technologies." In *Terahertz Biomedical and Healthcare Technologies*, 145–65. Amsterdam, Netherlands: Elsevier.

Banerjee, D., W.V. Spiegel, M.D. Thomson, S. Schabel, and H.G. Roskos. 2008. "Diagnosing Water Content in Paper by Terahertz Radiation." *Optics Express* 16 (12): 9060–66.

Basra, A. 2000. *Plant Growth Regulators in Agriculture and Horticulture: Their Role and Commercial Uses*. New York, NY: CRC Press.

Bawuah, P., T. Ervasti, N. Tan, J.A. Zeitler, J. Ketolainen, and K. Peiponen. 2016. "Noninvasive Porosity Measurement of Biconvex Tablets Using Terahertz Pulses." *International Journal of Pharmaceutics* 509 (1–2): 439–43.

Bawuah, P., D. Markl, D. Farrell, et al. 2020. "Terahertz-Based Porosity Measurement of Pharmaceutical Tablets: A Tutorial." *Journal of Infrared, Millimeter, and Terahertz Waves* 41 (4): 450–69.

Bawuah, P., D. Markl, A. Turner, et al. 2021. "A Fast and Non-Destructive Terahertz Dissolution Assay for Immediate Release Tablets." *Journal of Pharmaceutical Sciences* 110 (5): 2083–92.

Bawuah, P., and J.A. Zeitler. 2021. "Advances in Terahertz Time-Domain Spectroscopy of Pharmaceutical Solids: A Review." *TrAC Trends in Analytical Chemistry* 139: 116272.

Baxter, H.W., A.A. Worrall, J. Pang, R. Chen, and B. Yang. 2021. "Volatile Liquid Detection by Terahertz Technologies." *Frontiers in Physics* 9: 107.

Baxter, J.B., and G.W. Guglietta. 2011. "Terahertz Spectroscopy." *Analytical Chemistry* 83 (12): 4342–68.

Berrier, A., M.C. Schaafsma, G. Nonglaton, J. Bergquist, and J.G. Rivas. 2012. "Selective Detection of Bacterial Layers with Terahertz Plasmonic Antennas." *Biomedical Optics Express* 3 (11): 2937–49.

Berry, C.W., N. Wang, M.R. Hashemi, M. Unlu, and M. Jarrahi. 2013. "Significant Performance Enhancement in Photoconductive Terahertz Optoelectronics by Incorporating Plasmonic Contact Electrodes." *Nature Communications* 4 (1): 1–10.

Bian, Y., X. Zhang, Z. Zhu, X. Wu, X. Li, and B. Yang. 2020. "Investigation of the Correlations between Amino Acids, Amino Acid Mixtures and Dipeptides by Terahertz Spectroscopy." *Journal of Infrared, Millimeter, and Terahertz Waves* 42: 1–12.

Bian, Y., X. Zhang, Z. Zhu, and B. Yang. 2021. "Vibrational Modes Optimization and Terahertz Time-Domain Spectroscopy of L-Lysine and L-Lysine Hydrate." *Journal of Molecular Structure* 1232: 129952.

Biedron, S.G., J.W. Lewellen, S.V. Milton, et al. 2007. "Compact, High-Power Electron Beam Based Terahertz Sources." *Proceedings of the IEEE* 95 (8): 1666–78.

Bin, L., L. Bing, L. Yan-de, and W. Jian. 2021. "Detection of Adulteration of Kudzu Powder by Terahertz Time-Domain Spectroscopy." *Journal of Food Measurement and Characterization* 15 (5): 4380–87.

Bin, L., W. Qiu, Z. Chao-Hui, et al. 2022. "Research on Anthracnose Grade of Camellia Oleifera Based on the Combined LIBS and THz Technology." *Plant Methods* 18 (1): 1–13.

Biswas, A., S. Sinha, A. Acharyya, et al. 2018. "1.0 THz GaN IMPATT Source: Effect of Parasitic Series Resistance." *Journal of Infrared, Millimeter, and Terahertz Waves* 39 (10): 954–74.

Blanchard, F., G. Sharma, L. Razzari, et al. 2010. "Generation of Intense Terahertz Radiation via Optical Methods." *IEEE Journal of Selected Topics in Quantum Electronics* 17 (1): 5–16.

Bogue, R. 2018. "Sensing with Terahertz Radiation: A Review of Recent Progress." *Sensor Review* 38 (2): 216–22. https://doi.org/10.1108/SR-10-2017-0221.

Booske, J.H. 2008. "Plasma Physics and Related Challenges of Millimeter-Wave-to-Terahertz and High Power Microwave Generation." *Physics of Plasmas* 15 (5): 55502.

Born, N., R. Gente, D. Behringer, et al. 2014. "Monitoring the Water Status of Plants Using THz Radiation." In *2014 39th International Conference on Infrared, Millimeter, and Terahertz Waves (IRMMW-THz)*, 1–2.

Bouveyron, C., G. Celeux, T.B. Murphy, and A.E. Raftery. 2019. *Model-Based Clustering and Classification for Data Science: With Applications in R.* Vol. 50. Cambridge: Cambridge University Press.

Brock, D., J.A. Zeitler, A. Funke, K. Knop, and P. Kleinebudde. 2012. "A Comparison of Quality Control Methods for Active Coating Processes." *International Journal of Pharmaceutics* 439 (1–2): 289–95.

Brown, E.R. 2003. "THz Generation by Photomixing in Ultrafast Photoconductors." In *Terahertz Sensing Technology: Volume 1: Electronic Devices and Advanced Systems Technology*, edited by Dwight L Woolard, William R Loerop, and Michael S Shur, 147–95. Singapore: World Scientific Publishing Company. https://doi.org/10.1142/5244.

Brown, E.R., K.A. McIntosh, K.B. Nichols, and C.L. Dennis. 1995. "Photomixing up to 3.8 THz in Low-Temperature-Grown GaAs." *Applied Physics Letters* 66 (3): 285–87.

Browne, M., N.T. Yardimci, C. Scoffoni, M. Jarrahi, and L. Sack. 2020. "Prediction of Leaf Water Potential and Relative Water Content Using Terahertz Radiation Spectroscopy." *Plant Direct* 4 (4): e00197.

Burford, N.M., and M.O. El-Shenawee. 2017. "Review of Terahertz Photoconductive Antenna Technology." *Optical Engineering* 56 (1): 10901.

Bykhovski, A., T. Globus, T. Khromova, B. Gelmont, and D. Woolard. 2007. "An Analysis of the THz Frequency Signatures in the Cellular Components of Biological Agents." *International Journal of High Speed Electronics and Systems* 17 (02): 225–37.

Bykhovski, A., X. Li, T. Globus, et al. 2005. "THz Absorption Signature Detection of Genetic Material of E. Coli and B. Subtilis." In *Chemical and Biological Standoff Detection III*, 5995:198–207.

Cao, B., H. Li, M. Fan, W. Wang, and M. Wang. 2018. "Determination of Pesticides in a Flour Substrate by Chemometric Methods Using Terahertz Spectroscopy." *Analytical Methods* 10 (42): 5097–104.

Cao, C., Z. Zhang, X. Zhao, and T. Zhang. 2020. "Terahertz Spectroscopy and Machine Learning Algorithm for Non-Destructive Evaluation of Protein Conformation." *Optical and Quantum Electronics* 52: 1–18.

Casado-Gavalda, M.P., Y. Dixit, D. Geulen, et al. 2017. "Quantification of Copper Content with Laser Induced Breakdown Spectroscopy as a Potential Indicator of Offal Adulteration in Beef." *Talanta* 169: 123–29.

Castro-Camus, E., M. Palomar, and A.A. Covarrubias. 2013. "Leaf Water Dynamics of Arabidopsis Thaliana Monitored In-Vivo Using Terahertz Time-Domain Spectroscopy." *Scientific Reports* 3 (1): 1–5.

Catapano, I., M. Picollo, and K. Fukunaga. 2017. "TeraHertz Waves and Cultural Heritage: State-of-the-Art and Perspectives." In *Sensing the Past Geotechnologies and the Environment*, edited by N Masini and F Soldovieri, 16: 313–23. Cham: Springer.

Catapano, I., and F. Soldovieri. 2019. "THz Imaging and Data Processing: State of the Art and Perspective." *Innovation in Near-Surface Geophysics* 1: 399–417.

Chaban, A., G.J. Tserevelakis, E. Klironomou, R. Fontana, G. Zacharakis, and J. Striova. 2021. "Revealing Underdrawings in Wall Paintings of Complex Stratigraphy with a Novel Reflectance Photoacoustic Imaging Prototype." *Journal of Imaging* 7 (12): 250.

Chakraborty, S.K., N.K. Mahanti, S.M. Mansuri, M.K. Tripathi, N. Kotwaliwale, and D.S. Jayas. 2021. "Non-Destructive Classification and Prediction of Aflatoxin-B1 Concentration in Maize Kernels Using Vis-NIR (400-1000 Nm) Hyperspectral Imaging." *Journal of Food Science and Technology* 58 (2): 437–50.

Chan, W.L., J. Deibel, and D.M. Mittleman. 2007. "Imaging with Terahertz Radiation." *Reports on Progress in Physics* 70 (8): 1325.

Chantry, G.W. 1971. "Submillimetre Spectroscopy." AGRIS, Food and Agriculture Organization of the United Nations. New York, NY: Academic Press.

Chen, M., and X. Lijuan. 2014. "A Preliminary Study of Aflatoxin B1 Detection in Peanut Oil by Terahertz Time-Domain Spectroscopy." *Transactions of the ASABE* 57 (6): 1793–99.

Chen, Q., S. Jia, J. Qin, Y. Du, and Z. Zhao. 2020. "A Feasible Approach to Detect Pesticides in Food Samples Using THz-FDS and Chemometrics." *Journal of Spectroscopy* 2020: 10.

Chen, Q., C. Zhang, J. Zhao, and Q. Ouyang. 2013. "Recent Advances in Emerging Imaging Techniques for Non-Destructive Detection of Food Quality and Safety." *TrAC Trends in Analytical Chemistry* 52: 261–74.

Chen, S., C. Luo, H. Wang, L. Peng, B. Deng, and Z. Zhuang. 2018. "Three-Dimensional Terahertz Coded-Aperture Imaging in Space Domain." *IEEE Access* 6: 32727–36.

Chen, T., Z. Li, X. Yin, F. Hu, and C. Hu. 2016. "Discrimination of Genetically Modified Sugar Beets Based on Terahertz Spectroscopy." *Spectrochimica Acta Part A: Molecular and Biomolecular Spectroscopy* 153: 586–90.

Chen, T., L. Ma, Z. Tang, and L. Xiao Yu. 2022. "Identification of Coumarin-Based Food Additives Using Terahertz Spectroscopy Combined with Manifold Learning and Improved Support Vector Machine." *Journal of Food Science* 87 (3): 1108–18.

Chen, X., H. Liu, Q. Li, et al. 2015a. "Terahertz Detectors Arrays Based on Orderly Aligned InN Nanowires." *Scientific Reports* 5 (1): 1–11.

Chen, Z., Z. Zhang, R. Zhu, Y. Xiang, Y. Yang, and P.B. Harrington. 2015b. "Application of Terahertz Time-Domain Spectroscopy Combined with Chemometrics to Quantitative Analysis of Imidacloprid in Rice Samples." *Journal of Quantitative Spectroscopy and Radiative Transfer* 167: 1–9.

Cherkasova, O.P., M.M. Nazarov, M. Konnikova, and A.P. Shkurinov. 2020. "THz Spectroscopy of Bound Water in Glucose: Direct Measurements from Crystalline to Dissolved State." *Journal of Infrared, Millimeter, and Terahertz Waves* 41 (9): 1057–68.

Chevalier, P., A. Amirzhan, F. Wang, et al. 2019. "Widely Tunable Compact Terahertz Gas Lasers." *Science* 366 (6467): 856–60.

Cho, Y.T. 2018. "Fabrication of Wire Grid Polarizer for Spectroscopy Application: From Ultraviolet to Terahertz." *Applied Spectroscopy Reviews* 53 (2–4): 224–45.

Choi, W.J., K. Yano, M. Cha, et al. 2022. "Chiral Phonons in Microcrystals and Nanofibrils of Biomolecules." *Nature Photonics* 16: 366–73.

Chua, H.S., J. Obradovic, A.D. Haigh, et al. 2005. "Terahertz Time-Domain Spectroscopy of Crushed Wheat Grain." In *IEEE MTT-S International Microwave Symposium Digest, 2005*, 4.

Ciccoritti, R., M. Paliotta, T. Amoriello, and K. Carbone. 2019. "FT-NIR Spectroscopy and Multivariate Classification Strategies for the Postharvest Quality of Green-Fleshed Kiwifruit Varieties." *Scientia Horticulturae* 257: 108622.

Consolino, L., S. Bartalini, and P.D. Natale. 2017. "Terahertz Frequency Metrology for Spectroscopic Applications: A Review." *Journal of Infrared, Millimeter, and Terahertz Waves* 38 (11): 1289–315.

Cook, D.J., B.K. Decker, G. Dadusc, and M.G. Allen. 2004. "Through-Container THz Sensing: Applications for Biodetection." In *Chemical and Biological Standoff Detection*, 5268:36–42.

Cook, D.J., and R.M. Hochstrasser. 2000. "Intense Terahertz Pulses by Four-Wave Rectification in Air." *Optics Letters* 25 (16): 1210–12.

Cortes, V., A. Rodriguez, J. Blasco, et al. 2017. "Prediction of the Level of Astringency in Persimmon Using Visible and Near-Infrared Spectroscopy." *Journal of Food Engineering* 204: 27–37.

Cristofani, E., F. Friederich, S. Wohnsiedler, et al. 2014. "Nondestructive Testing Potential Evaluation of a Terahertz Frequency-Modulated Continuous-Wave Imager for Composite Materials Inspection." *Optical Engineering* 53 (3): 31211.

Crowe, T.W., T. Globus, D.L. Woolard, and J.L. Hesler. 2004. "Terahertz Sources and Detectors and Their Application to Biological Sensing." *Philosophical Transactions of the Royal Society of London. Series A: Mathematical, Physical and Engineering Sciences* 362 (1815): 365–77.

Cui, H., X. Zhang, J. Su, Y. Yang, Q. Fang, and X. Wei. 2015. "Vibration-Rotation Absorption Spectrum of Water Vapor Molecular in Frequency Selector at 0.5–2.5 THz Range." *Optik* 126 (23): 3533–37.

Cunha, C.L., A.S. Luna, R.C.G. Oliveira, G.M. Xavier, M.L.L. Paredes, and A.R. Torres. 2017. "Predicting the Properties of Biodiesel and Its Blends Using Mid-FT-IR Spectroscopy and First-Order Multivariate Calibration." *Fuel* 204: 185–94.

Curl, R.F., F. Capasso, C. Gmachl, et al. 2010. "Quantum Cascade Lasers in Chemical Physics." *Chemical Physics Letters* 487 (1–3): 1–18.

Damyanov, D., A. Batra, B. Friederich, T. Kaiser, T. Schultze, and J.C. Balzer. 2020. "High-Resolution Long-Range THz Imaging for Tunable Continuous-Wave Systems." *IEEE Access* 8: 151997–2007.

D'Arco, A., M.D. Fabrizio, V. Dolci, M. Petrarca, and S. Lupi. 2020. "THz Pulsed Imaging in Biomedical Applications." *Condensed Matter* 5 (2): 25.

Davies, A.G., and E.H. Linfield. 2007. "Molecular and Organic Interactions." In *Terahertz Frequency Detection and Identification of Materials and Objects*, 91–106. Dordrecht: Springer.

Delfanazari, K., R.A. Klemm, H.J. Joyce, D.A. Ritchie, and K. Kadowaki. 2020. "Integrated, Portable, Tunable, and Coherent Terahertz Sources and Sensitive Detectors Based on Layered Superconductors." *Proceedings of the IEEE* 108 (5): 721–34.

Delgado-Blanca, I., A. Ruiz-Medina, and P. Ortega-Barrales. 2019. "Novel Sequential Separation and Determination of a Quaternary Mixture of Fungicides by Using an Automatic Fluorimetric Optosensor." *Food Additives & Contaminants: Part A* 36 (2): 278–88.

Dey, P., and J.N. Roy. 2021. "Terahertz Spintronics." In *Spintronics: Fundamentals and Applications*, 185–200. Singapore: Springer Singapore. https://doi.org/10.1007/978-981-16-0069-2_8.

Diebold, S., S. Nakai, K. Nishio, et al. 2016. "Modeling and Simulation of Terahertz Resonant Tunneling Diode-Based Circuits." *IEEE Transactions on Terahertz Science and Technology* 6 (5): 716–23.

Dinovitser, A., D.G. Valchev, and D. Abbott. 2017. "Terahertz Time-Domain Spectroscopy of Edible Oils." *Royal Society Open Science* 4 (6): 170275.

Dorney, T.D., R.G. Baraniuk, and D.M. Mittleman. 2001. "Material Parameter Estimation with Terahertz Time-Domain Spectroscopy." *JOSA A* 18 (7): 1562–71.

Du, C., X. Zhang, and Z. Zhang. 2019. "Quantitative Analysis of Ternary Isomer Mixtures of Saccharide by Terahertz Time Domain Spectroscopy Combined with Chemometrics." *Vibrational Spectroscopy* 100: 64–70.

Du, H., W. Chen, Y. Lei, et al. 2021a. "Discrimination of Authenticity of *Fritillariae Cirrhosae* Bulbus Based on Terahertz Spectroscopy and Chemometric Analysis." *Microchemical Journal* 168: 106440.

Du, X., Y. Wang, X. Zhang, et al. 2021b. "A Study of Plant Growth Regulators Detection Based on Terahertz Time-Domain Spectroscopy and Density Functional Theory." *RSC Advances* 11 (46): 28898–907.

Du, X., X. Zhang, Y. Wang, et al. 2021c. "Highly Sensitive Detection of Plant Growth Regulators by Using Terahertz Time-Domain Spectroscopy Combined with Metamaterials." *Optics Express* 29 (22): 36535–45.

Dugan, M.A., J.S. Melinger, and A.C. Albrecht. 1988. "Terahertz Oscillations from Molecular Liquids in CSRS/CARS Spectroscopy with Incoherent Light." *Chemical Physics Letters* 147 (5): 411–19.

Eissa, A., L. Helyes, E. Romano, A. Albandary, and A. Ibrahim. 2022. "Thoughts for Foods: Imaging Technology Opportunities for Monitoring and Measuring Food Quality." In *A Glance at Food Processing Applications*, 1–50, London: IntechOpen.

Ellrich, F., M. Bauer, N. Schreiner, et al. 2020. "Terahertz Quality Inspection for Automotive and Aviation Industries." *Journal of Infrared, Millimeter, and Terahertz Waves* 41 (4): 470–89.

Engel, J., J. Gerretzen, E. Szymanska, et al. 2013. "Breaking with Trends in Pre-Processing?" *TrAC Trends in Analytical Chemistry* 50: 96–106.

Engelbrecht, S., K.H. Tybussek, J. Sampaio, J. Bohmler, B.M. Fischer, and S. Sommer. 2019. "Monitoring the Isothermal Crystallization Kinetics of PET-A Using THz-TDS." *Journal of Infrared, Millimeter, and Terahertz Waves* 40 (3): 306–13.

Ergun, S., and S. Sonmez. 2015. "Terahertz Technology for Military Applications." *Journal of Management and Information Science* 3 (1): 13–16.

Eroglu, A., M. Dogan, O.S. Toker, and M.T. Yilmaz. 2015. "Classification of Kashar Cheeses Based on Their Chemical, Color and Instrumental Textural Characteristics Using Principal Component and Hierarchical Cluster Analysis." *International Journal of Food Properties* 18 (4): 909–21.

Etayo, D., J.C. Iriarte, I. Palacios, et al. 2011. "THz Imaging System for Industrial Quality Control." In *2011 IEEE MTT-S International Microwave Workshop Series on Millimeter Wave Integration Technologies*, 172–75.

Exter, M.V., C. Fattinger, and D. Grischkowsky. 1989. "Terahertz Time-Domain Spectroscopy of Water Vapor." *Optics Letters* 14 (20): 1128–30.

Fan, S., M.T. Ruggiero, Z. Song, Z. Qian, and V.P. Wallace. 2019. "Correlation between Saturated Fatty Acid Chain-Length and Intermolecular Forces Determined with Terahertz Spectroscopy." *Chemical Communications* 55 (25): 3670–73.

Farhat, M., A.M. Amer, V.B. Cunningham, and K.N. Salama. 2020. "Numerical Modeling for Terahertz Testing of Non-Metallic Pipes." *AIP Advances* 10 (9): 95112.

Fawole, O., K. Sinha, and M. Tabib-Azar. 2015. "Monitoring Yeast Activation with Sugar and Zero-Calorie Sweetener Using Terahertz Waves." In *2015 IEEE SENSORS*, 1–4.

Federici, J.F., B. Schulkin, F. Huang, et al. 2005. "THz Imaging and Sensing for Security Applications – Explosives, Weapons and Drugs." *Semiconductor Science and Technology* 20 (7): S266.

Feng, C., and C. Otani. 2021. "Terahertz Spectroscopy Technology as an Innovative Technique for Food: Current State-of-the-Art Research Advances." *Critical Reviews in Food Science and Nutrition* 61 (15): 2523–43.

Feng, C., C. Otani, and Y. Ogawa. 2022. "Innovatively Identifying Naringin and Hesperidin by Using Terahertz Spectroscopy and Evaluating Flavonoids Extracts from Waste Orange Peels by Coupling with Multivariate Analysis." *Food Control* 137: 108897.

Ferguson, B., S. Wang, D. Gray, D. Abbot, and X.C. Zhang. 2002. "T-Ray Computed Tomography." *Optics Letters* 27 (15): 1312–14.

Ferguson, B., and X. Zhang. 2002. "Materials for Terahertz Science and Technology." *Nature Materials* 1 (1): 26–33.

Fitch, M.J., C. Dodson, D.S. Ziomek, and R. Osiander. 2004. "Time-Domain Terahertz Spectroscopy of Bioagent Simulants." In *Chemical and Biological Standoff Detection II*, 5584:16–22.

Fortier, T., and E. Baumann. 2019. "20 Years of Developments in Optical Frequency Comb Technology and Applications." *Communications Physics* 2 (1): 1–16.

Fosodeder, P., S. Hubmer, A. Ploier, R. Ramlau, S.V. Frank, and C. Rankl. 2020. "Terahertz Computer Tomography for Plastics Extrusion (TACTICS)." In *OSA Technical Digest (Optical Society of America)*. Washington, DC: OSA Publishing.

Fujita, K., S. Hayashi, A. Ito, M. Hitaka, and T. Dougakiuchi. 2019. "Sub-Terahertz and Terahertz Generation in Long-Wavelength Quantum Cascade Lasers." *Nanophotonics* 8 (12): 2235–41.

Funkner, S., G. Niehues, D.A. Schmidt, et al. 2012. "Watching the Low-Frequency Motions in Aqueous Salt Solutions: The Terahertz Vibrational Signatures of Hydrated Ions." *Journal of the American Chemical Society* 134 (2): 1030–35.

Fyla. 2022. "Lasers." Fyla Laser, Spain. 2022. Accessed 14th May 2022, https://fyla.com/laser/.

Gallerano, G.P., and S. Biedron. 2004. "Overview of Terahertz Radiation Sources." In *Proceedings of the 2004 FEL Conference*, 1:216–21.

Galstyan, V., A. D'Arco, M.D. Fabrizio, N. Poli, S. Lupi, and E. Comini. 2021. "Detection of Volatile Organic Compounds: From Chemical Gas Sensors to Terahertz Spectroscopy." *Reviews in Analytical Chemistry* 40 (1): 33–57.

Galvao, R.K.H., S. Hadjiloucas, A. Zafiropoulos, G.C. Walker, J.W. Bowen, and R. Dudley. 2007. "Optimization of Apodization Functions in Terahertz Transient Spectrometry." *Optics Letters* 32 (20): 3008–10.

Gao, J., Y. Li, J. Liu, et al. 2022. "Terahertz Spectroscopy Detection of Lithium Citrate Tetrahydrate and Its Dehydration Kinetics." *Spectrochimica Acta Part A: Molecular and Biomolecular Spectroscopy* 266: 120470.

Garbacz, P. 2016. "Terahertz Imaging – Principles, Techniques, Benefits, and Limitations." *Problemy Eksploatacji* 1: 81–92.

Gbur, G., and E. Wolf. 2001. "Relation between Computed Tomography and Diffraction Tomography." *JOSA A* 18 (9): 2132–37.

Ge, H., Y. Jiang, F. Lian, Y. Zhang, and S. Xia. 2015. "Characterization of Wheat Varieties Using Terahertz Time-Domain Spectroscopy." *Sensors* 15 (6): 12560–72.

Ge, H., Y. Jiang, F. Lian, Y. Zhang, and S. Xia. 2016. "Quantitative Determination of Aflatoxin B1 Concentration in Acetonitrile by Chemometric Methods Using Terahertz Spectroscopy." *Food Chemistry* 209: 286–92.

Ge, H., Y. Jiang, Z. Xu, F. Lian, Y. Zhang, and S. Xia. 2014. "Identification of Wheat Quality Using THz Spectrum." *Optics Express* 22 (10): 12533–44.

Ge, M., G. Liu, S. Ma, and W. Wang. 2009. "Polymorphic Forms of Furosemide Characterized by THz Time Domain Spectroscopy." *Bulletin of the Korean Chemical Society* 30 (10): 2265–68.

Gente, R., N. Born, N. Voß, et al. 2013. "Determination of Leaf Water Content from Terahertz Time-Domain Spectroscopic Data." *Journal of Infrared, Millimeter, and Terahertz Waves* 34 (3–4): 316–23.

Gente, R., S.F. Busch, E. Stubling, et al. 2016. "Quality Control of Sugar Beet Seeds with THz Time-Domain Spectroscopy." *IEEE Transactions on Terahertz Science and Technology* 6 (5): 754–56.

Ghann, W., and J. Uddin. 2017. "Terahertz (THz) Spectroscopy: A Cutting Edge Technology." *Terahertz Spectroscopy – A Cutting Edge Technology*, edited by Jamal Uddin, 3–20. London. http://dx.doi.org/10.5772/62805.

Ghivela, G.C., and J. Sengupta. 2021. "Space Charge Studies in Graphene Based Avalanche Transit Time Devices." *Superlattices and Microstructures* 155: 106899.

Gilbert, P. 1972. "Iterative Methods for the Three-Dimensional Reconstruction of an Object from Projections." *Journal of Theoretical Biology* 36 (1): 105–17.

Girolamo, F.V.D., M. Pagano, A. Tredicucci, et al. 2021. "Detection of Fungal Infections in Chestnuts: A Terahertz Imaging-Based Approach." *Food Control* 123: 107700.

Globus, T., T. Dorofeeva, I. Sizov, et al. 2012. "Sub-THz Vibrational Spectroscopy of Bacterial Cells and Molecular Components." *American Journal of Biomedical Engineering* 2 (4): 143–54.

Globus, T., I. Sizov, and B. Gelmont. 2013. "Teraherz Vibrational Spectroscopy of E. Coli and Molecular Constituents: Computational Modeling and Experiment." *Advances in Bioscience and Biotechnology* 4: 11. https://doi.org/10.4236/abb.2013.43A065.

Glyavin, M.Y., A.G. Luchinin, G.S. Nusinovich, et al. 2012. "A 670 GHz Gyrotron with Record Power and Efficiency." *Applied Physics Letters* 101 (15): 153503.

Goncharsky, A.V., and S.Y. Romanov. 2017. "Iterative Methods for Solving Coefficient Inverse Problems of Wave Tomography in Models with Attenuation." *Inverse Problems* 33 (2): 25003.

Gong, Z., D. Deng, X. Sun, J. Liu, and Y. Ouyang. 2022. "Non-Destructive Detection of Moisture Content for Ginkgo Biloba Fruit with Terahertz Spectrum and Image: A Preliminary Study." *Infrared Physics & Technology* 120: 103997.

Gong, A., Y. Qiu, X. Chen, Z. Zhao, L. Xia, and Y. Shao. 2020. "Biomedical Applications of Terahertz Technology." *Applied Spectroscopy Reviews* 55 (5): 418–38.

Gordon, R. 1974. "A Tutorial on ART (Algebraic Reconstruction Techniques)." *IEEE Transactions on Nuclear Science* 21 (3): 78–93.

Gordon, R., R. Bender, and G.T. Herman. 1970. "Algebraic Reconstruction Techniques (ART) for Three-Dimensional Electron Microscopy and X-Ray Photography." *Journal of Theoretical Biology* 29 (3): 471–81.

Gourama, H. 2020. "Foodborne Pathogens." In *Food Safety Engineering*, edited by Ali Demirci, Hao Feng, and Kathiravan Krishnamurthy, 25–49. Cham: Springer International Publishing. https://doi.org/10.1007/978-3-030-42660-6_2.

Gowen, A.A., C. O'Sullivan, and C.P. O'Donnell. 2012. "Terahertz Time Domain Spectroscopy and Imaging: Emerging Techniques for Food Process Monitoring and Quality Control." *Trends in Food Science & Technology* 25 (1): 40–46.

Granato, D., P. Putnik, D.B. Kovacevic, et al. 2018. "Trends in Chemometrics: Food Authentication, Microbiology, and Effects of Processing." *Comprehensive Reviews in Food Science and Food Safety* 17 (3): 663–77.

Grubin, H.L., V.V. Mitin, E. Scholl, and M.P. Shaw. 2013. *The Physics of Instabilities in Solid State Electron Devices*. New York, NY: Springer Science & Business Media.

Gua, T., Z. Ding, D. Zhang, et al. 2013. "Evaluation of Wheat Seeds by Terahertz Imaging." In *2013 6th UK, Europe, China Millimeter Waves and THz Technology Workshop (UCMMT)*, 1–2.

Guan, A., and Y. Chao. 2019. "Quantitative Analysis of Alum Based on Terahertz Time-Domain Spectroscopy Technology and Support Vector Machine." *Optik* 193: 163017.

Guillet, J.P., B. Recur, L. Frederique, et al. 2014. "Review of Terahertz Tomography Techniques." *Journal of Infrared, Millimeter, and Terahertz Waves* 35 (4): 382–411.

Guo, F., C. Pandey, C. Wang, et al. 2020a. "Generation of Highly Efficient Terahertz Radiation in Ferromagnetic Heterostructures and Its Application in Spintronic Terahertz Emission Microscopy (STEM)." *OSA Continuum* 3 (4): 893–902.

Guo, J., D. Hu, Q. Liu, L. Chen, Z. Xiong, and L. Shang. 2020b. "A Reliable Method for Identification of Antibiotics by Terahertz Spectroscopy and SVM." *Journal of Spectroscopy*, 2020. https://doi.org/10.1155/2020/8811467.

Haddad, G.I., J.R. East, and H. Eisele. 2003. "Two-Terminal Active Devices for Terahertz Sources." *International Journal of High Speed Electronics and Systems* 13 (02): 395–427. https://doi.org/10.1142/S0129156403001788.

Hadjiloucas, S., G.C. Walker, and J.W. Bowen. 2010. "Quantifying Consumer Perception of Foodstuffs Using THz Spectrometry." In *35th International Conference on Infrared, Millimeter, and Terahertz Waves*, 1–2. https://doi.org/10.1109/ICIMW.2010.5612357.

Hameed, S., L. Xie, and Y. Ying. 2018. "Conventional and Emerging Detection Techniques for Pathogenic Bacteria in Food Science: A Review." *Trends in Food Science & Technology* 81: 61–73.

Han, D., and L. Kang. 2018. "Nondestructive Evaluation of GFRP Composite Including Multi-Delamination Using THz Spectroscopy and Imaging." *Composite Structures* 185: 161–75.

Han, D., and L. Kang. 2019. "High-Speed THz Imaging Using Two-Way Raster Scanning Method without Dwell Time." *Journal of Mechanical Science and Technology* 33 (3): 1079–86.

Han, P., X. Wang, and Y. Zhang. 2020. "Time-Resolved Terahertz Spectroscopy Studies on 2D Van Der Waals Materials." *Advanced Optical Materials* 8 (3): 1900533.

Han, S. 2013. "Compact Sub-THz Gyrotrons for Real-Time T-Ray Imaging." In *2013 IEEE 14th International Vacuum Electronics Conference (IVEC)*, 1–2.

Han, S., W.K. Park, Y. Ahn, W. Lee, and H.S. Chun. 2012. "Development of a Compact Sub-Terahertz Gyrotron and Its Application to t-Ray Real-Time Imaging for Food Inspection." In *2012 37th International Conference on Infrared, Millimeter, and Terahertz Waves*, 1–2.

Han, S., W.K. Park, and H.S. Chun. 2011. "Development of Sub-THz Gyrotron for Real-Time Food Inspection." In *2011 International Conference on Infrared, Millimeter, and Terahertz Waves*, 1–2.

Hansen, R.E., T. Baek, S.L. Lange, et al. 2022. "Non-Contact Paper Thickness and Quality Monitoring Based on Mid-Infrared Optical Coherence Tomography and THz Time Domain Spectroscopy." *Sensors* 22 (4): 1549.

Hayton, D.J., A. Khudchenko, D.G. Pavelyev, et al. 2013. "Phase Locking of a 3.4 THz Third-Order Distributed Feedback Quantum Cascade Laser Using a Room-Temperature Superlattice Harmonic Mixer." *Applied Physics Letters* 103 (5): 51115.

He, J., T. Dong, B. Chi, S. Wang, X. Wang, and Y. Zhang. 2020. "Meta-Hologram for Three-Dimensional Display in Terahertz Waveband." *Microelectronic Engineering* 220: 111151.

He, J., J. Ye, X. Wang, Q. Kan, and Y. Zhang. 2016. "A Broadband Terahertz Ultrathin Multi-Focus Lens." *Scientific Reports* 6 (1): 1–9.

He, M., A.K. Azad, S. Ye, and W. Zhang. 2006. "Far-Infrared Signature of Animal Tissues Characterized by Terahertz Time-Domain Spectroscopy." *Optics Communications* 259 (1): 389–92.

He, Y., X. Bai, Q. Xiao, F. Liu, L. Zhou, and C. Zhang. 2021. "Detection of Adulteration in Food Based on Nondestructive Analysis Techniques: A Review." *Critical Reviews in Food Science and Nutrition* 61 (14): 2351–71.

Hempel, H., T. Unold, and R. Eichberger. 2017. "Measurement of Charge Carrier Mobilities in Thin Films on Metal Substrates by Reflection Time Resolved Terahertz Spectroscopy." *Optics Express* 25 (15): 17227–36.

Henri, R., K. Nallappan, D.S. Ponomarev, et al. 2021. "Fabrication and Characterization of an 8 × 8 Terahertz Photoconductive Antenna Array for Spatially Resolved Time Domain Spectroscopy and Imaging Applications." *IEEE Access* 9: 117691–702.

Herman, G.T. 1995. "Image Reconstruction from Projections." *Real-Time Imaging* 1 (1): 3–18.

Herman, G.T., and L.B. Meyer. 1993. "Algebraic Reconstruction Techniques Can Be Made Computationally Efficient (Positron Emission Tomography Application)." *IEEE Transactions on Medical Imaging* 12 (3): 600–09.

Hindle, F., L. Kuuliala, M. Mouelhi, et al. 2018. "Monitoring of Food Spoilage by High Resolution THz Analysis." *Analyst* 143 (22): 5536–44.

Hiromoto, N., N. Shiba, and K. Yamamoto. 2013. "Detection of a Human Hair with Polaization-Dependent THz-Time Domain Spectroscopy." In *2013 38th International Conference on Infrared, Millimeter, and Terahertz Waves (IRMMW-THz)*, 1–2.

Hisazumi, J., T. Suzuki, N. Wakiyama, H. Nakagami, and K. Terada. 2012. "Chemical Mapping of Hydration and Dehydration Process of Theophylline in Tablets Using Terahertz Pulsed Imaging." *Chemical and Pharmaceutical Bulletin* 60 (7): 831–36.

Hishida, M., R. Anjum, T. Anada, D. Murakami, and M. Tanaka. 2022. "Effect of Osmolytes on Water Mobility Correlates with Their Stabilizing Effect on Proteins." *The Journal of Physical Chemistry B*. 126 (13), 2466–75. https://doi.org/10.1021/acs.jpcb.1c10634.

Ho, L., R. Muller, K.C. Gordon, et al. 2008. "Applications of Terahertz Pulsed Imaging to Sustained-Release Tablet Film Coating Quality Assessment and Dissolution Performance." *Journal of Controlled Release* 127 (1): 79–87.

Ho, L., R. Muller, K.C. Gordon, et al. 2009. "Terahertz Pulsed Imaging as an Analytical Tool for Sustained-Release Tablet Film Coating." *European Journal of Pharmaceutics and Biopharmaceutics* 71 (1): 117–23.

Hocine, M., N. Balakrishnan, T. Colton, et al. 2014. "*Wiley StatsRef: Statistics Reference Online.*" Hoboken, NJ: John Wiley & Sons, Ltd.

Hoffmann, V., C. Moser, and A. Saak. 2019. "Food Safety in Low and Middle-Income Countries: The Evidence through an Economic Lens." *World Development* 123: 104611.

Hor, Y.L., J.F. Federici, and R.L. Wample. 2008. "Nondestructive Evaluation of Cork Enclosures Using Terahertz/Millimeter Wave Spectroscopy and Imaging." *Applied Optics* 47 (1): 72–78.

Hovenier, J.N., M.C. Diez, T.O. Klaassen, et al. 2000. "The P-Ge Terahertz Laser-Properties under Pulsed-and Mode-Locked Operation." *IEEE Transactions on Microwave Theory and Techniques* 48 (4): 670–76.

Hu, B.B., and M.C. Nuss. 1995. "Imaging with Terahertz Waves." *Optics Letters* 20 (16): 1716–18.

Hu, J., R. Chen, Z. Xu, et al. 2021a. "Research on Enhanced Detection of Benzoic Acid Additives in Liquid Food Based on Terahertz Metamaterial Devices." *Sensors* 21 (9): 3238.

Hu, J., Y. Li, R. Chen, Y. He, and Y. Liu. 2022. "Establishment and Optimization of Temperature Compensation Model for Benzoic Acid Detection Based on Terahertz Metamaterial." *Infrared Physics & Technology* 123: 104101.

Hu, J., Y. Liu, Y. He, X. Sun, and B. Li. 2020. "Optimization of Quantitative Detection Model for Benzoic Acid in Wheat Flour Based on CARS Variable Selection and THz Spectroscopy." *Journal of Food Measurement and Characterization* 14 (5): 2549–58.

Hu, J., Z. Xu, M. Li, Y. He, X. Sun, and Y. Liu. 2021b. "Detection of Foreign-Body in Milk Powder Processing Based on Terahertz Imaging and Spectrum." *Journal of Infrared, Millimeter, and Terahertz Waves* 42 (8): 878–92.

Hu, X., W. Lang, W. Liu, X. Xu, J. Yang, and L. Zheng. 2017. "A Non-Destructive Terahertz Spectroscopy-Based Method for Transgenic Rice Seed Discrimination via Sparse Representation." *Journal of Infrared, Millimeter, and Terahertz Waves* 38 (8): 980–91.

Huang, H., S. Shao, G. Wang, P. Ye, B. Su, and C. Zhang. 2022a. "Terahertz Spectral Properties of Glucose and Two Disaccharides in Solid and Liquid States." *iScience*, 25: 104102.

Huang, H., D. Wang, W. Li, et al. 2017. "Continuous-Wave Terahertz Multi-Plane in-Line Digital Holography." *Optics and Lasers in Engineering* 94: 76–81.

Huang, L., C. Li, B. Li, M. Liu, M. Lian, and S. Yang. 2020a. "Studies on Qualitative and Quantitative Detection of Trehalose Purity by Terahertz Spectroscopy." *Food Science & Nutrition* 8 (4): 1828–36.

Huang, T., L. Yin, Y. Shuang, J. Liu, Y. Tan, and P. Liu. 2019a. "Far-Field Subwavelength Resolution Imaging by Spatial Spectrum Sampling." *Physical Review Applied* 12 (3): 34046.

Huang, X.B., D.B. Hou, P.J. Huang, Y.H. Ma, X. Li, and G.X. Zhang. 2015. "The Meat Freshness Detection Based on Terahertz Wave." In *Selected Papers of the Photoelectronic Technology Committee Conferences Held June–July 2015*, 9795:97953D.

Huang, Y., Y. Shen, and J. Wang. 2022b. "From Terahertz Imaging to Terahertz Wireless Communications." *Engineering,* In Press.

Huang, Y., R. Singh, L. Xie, and Y. Ying. 2020b. "Attenuated Total Reflection for Terahertz Modulation, Sensing, Spectroscopy and Imaging Applications: A Review." *Applied Sciences* 10 (14): 4688.

Huang, Y., Z. Yao, C. He, et al. 2019b. "Terahertz Surface and Interface Emission Spectroscopy for Advanced Materials." *Journal of Physics: Condensed Matter* 31 (15): 153001.

Hussain, N., D. Sun, and H. Pu. 2019. "Classical and Emerging Non-Destructive Technologies for Safety and Quality Evaluation of Cereals: A Review of Recent Applications." *Trends in Food Science & Technology* 91: 598–608.

Hwang, Y.H., Y.H. Noh, D. Seo, et al. 2015. "Analysis of the Terahertz Absorption Spectrum of Melamine." *Bulletin of the Korean Chemical Society* 36 (3): 891–95.

Ibrahim, M.E., D. Headland, W. Withayachumnankul, and C.H. Wang. 2021. "Nondestructive Testing of Defects in Polymer-Matrix Composite Materials for Marine Applications Using Terahertz Waves." *Journal of Nondestructive Evaluation* 40 (2): 1–11.

Idehara, T., S.P. Sabchevski, M. Glyavin, and S. Mitsudo. 2020. "The Gyrotrons as Promising Radiation Sources for THz Sensing and Imaging." *Applied Sciences* 10 (3): 980.

INO. 2022. "Terahertz Imaging." INO, Canada. 2022. Accessed 14th May 2022, https://www.ino.ca/en/solutions/thz/.

Inuzuka, M., Y. Kouzuma, N. Sugioka, K. Fukunaga, and T. Tateishi. 2017. "Investigation of Layer Structure of the Takamatsuzuka Mural Paintings by Terahertz Imaging Technique." *Journal of Infrared, Millimeter, and Terahertz Waves* 38 (4): 380–89.

Jacobs, K.J.P., B.J. Stevens, T. Mukai, D. Ohnishi, and R.A. Hogg. 2015. "Non-Destructive Mapping of Doping and Structural Composition of MOVPE-Grown High Current Density Resonant Tunnelling Diodes through Photoluminescence Spectroscopy." *Journal of Crystal Growth* 418: 102–10.

Jahani, S., S.K. Setarehdan, D.A. Boas, and M.A. Yucel. 2018. "Motion Artifact Detection and Correction in Functional Near-Infrared Spectroscopy: A New Hybrid Method Based on Spline Interpolation Method and Savitzky – Golay Filtering." *Neurophotonics* 5 (1): 15003.

Jansen, C., S. Wietzke, and M. Koch. 2012. "Terahertz Spectroscopy of Polymers." In *Terahertz Spectroscopy and Imaging*, edited by K E Peiponen, A Zeitler, and M Kuwata-Gonokami, Springer Series in Optical Sciences, 171: 327–53, Berlin, Heidelberg. https://doi.org/10.1007/978-3-642-29564-5_13.

Jansen, C., S. Wietzke, M. Scheller, et al. 2008. "Applications for THz Systems: Approaching Markets and Perspectives for an Innovative Technology." *Optik & Photonik* 3 (4): 26–30.

Jepsen, P.U., D.G. Cooke, and M. Koch. 2011. "Terahertz Spectroscopy and Imaging – Modern Techniques and Applications." *Laser & Photonics Reviews* 5 (1): 124–66.

Jepsen, P.U., J.K. Jensen, and U. Moller. 2008. "Characterization of Aqueous Alcohol Solutions in Bottles with THz Reflection Spectroscopy." *Optics Express* 16 (13): 9318–31.

Jha, S.N. 2015. *Rapid Detection of Food Adulterants and Contaminants: Theory and Practice.* San Diego: Academic Press.

Ji, Y., F. Fan, S. Xu, et al. 2019. "Terahertz Dielectric Anisotropy Enhancement in Dual-Frequency Liquid Crystal Induced by Carbon Nanotubes." *Carbon* 152: 865–72.

Jia, D., J. Xu, T. Xin, C. Zhang, and X. Yu. 2019. "Multifocal Terahertz Lenses Realized by Polarization-Insensitive Reflective Metasurfaces." *Applied Physics Letters* 114 (10): 101105.

Jiang, F.L., I. Ikeda, Y. Ogawa, and Y. Endo. 2011. "Terahertz Absorption Spectra of Fatty Acids and Their Analogues." *Journal of Oleo Science* 60 (7): 339–43.

Jiang, Y., H. Ge, F. Lian, Y. Zhang, and S. Xia. 2015. "Discrimination of Moldy Wheat Using Terahertz Imaging Combined with Multivariate Classification." *RSC Advances* 5 (114): 93979–86.

Jiang, Y., H. Ge, F. Lian, Y. Zhang, and S. Xia. 2016. "Early Detection of Germinated Wheat Grains Using Terahertz Image and Chemometrics." *Scientific Reports* 6 (1): 1–9.

Jiang, Y., H. Ge, and Y. Zhang. 2018. "Quantitative Determination of Maltose Concentration in Wheat by Using Terahertz Imaging." *Spectroscopy and Spectral Analysis* 38 (10): 3017.

Jiang, Y., H. Ge, and Y. Zhang. 2019. "Detection of Foreign Bodies in Grain with Terahertz Reflection Imaging." *Optik* 181: 1130–38.

Jiang, Y., H. Ge, and Y. Zhang. 2020. "Quantitative Analysis of Wheat Maltose by Combined Terahertz Spectroscopy and Imaging Based on Boosting Ensemble Learning." *Food Chemistry* 307: 125533.

Jing, F., X. Chen, Z. Yang, and B. Guo. 2018. "Heavy Metals Status, Transport Mechanisms, Sources, and Factors Affecting Their Mobility in Chinese Agricultural Soils." *Environmental Earth Sciences* 77 (3): 1–9.

Jiusheng, L.. 2010. "Optical Parameters of Vegetable Oil Studied by Terahertz Time-Domain Spectroscopy." *Applied Spectroscopy* 64 (2): 231–34.

Jooshesh, A., L. Smith, M. Masnadi-Shirazi, et al. 2014. "Nanoplasmonics Enhanced Terahertz Sources." *Optics Express* 22 (23): 27992–8001.

Jordens, C., and M. Koch. 2008. "Detection of Foreign Bodies in Chocolate with Pulsed Terahertz Spectroscopy." *Optical Engineering* 47 (3): 37003.

Jordens, C., S. Wietzke, M. Scheller, and M. Koch. 2010. "Investigation of the Water Absorption in Polyamide and Wood Plastic Composite by Terahertz Time-Domain Spectroscopy." *Polymer Testing* 29 (2): 209–15.

Kang, X., G. Zhang, X. Chen, P. Huang, D. Hou, and Z. Zhou. 2011. "Terahertz Spectroscopic Investigation of Elaidic Acid." *Spectroscopy and Spectral Analysis* 31 (10): 2629–33.

Kannan, R., A. Solaimalai, M. Jayakumar, and U. Surendran. 2022. "Advance Molecular Tools to Detect Plant Pathogens." In *Biopesticides*, Vol. 2 Advances in Bio-Inoculants, edited by A. Rakshit, V.S. Meena, P.C. Abhilash, et al., 401–16. Woodhead Publishing. https://doi.org/10.1016/C2019-0-04053-8.

Kar, S. 2020. "Terahertz Technology – Emerging Trends and Application Viewpoints." *Terahertz Biomedical and Healthcare Technologies*, 89–111. Amsterdam: Elsevier.

Kar, S., S. Jayanthi, E. Freysz, and A.K. Sood. 2014. "Time Resolved Terahertz Spectroscopy of Low Frequency Electronic Resonances and Optical Pump-Induced Terahertz Photoconductivity in Reduced Graphene Oxide Membrane." *Carbon* 80: 762–70.

Karaliunas, M., K.E. Nasser, A. Urbanowicz, et al. 2018. "Non-Destructive Inspection of Food and Technical Oils by Terahertz Spectroscopy." *Scientific Reports* 8 (1): 1–11.

Karaman, I. 2017. "Preprocessing and Pretreatment of Metabolomics Data for Statistical Analysis." *Metabolomics: From Fundamentals to Clinical Applications* 965: 145–61.

Karpowicz, N., D. Dawes, M.J. Perry, and X.C. Zhang. 2006. "Fire Damage on Carbon Fiber Materials Characterized by THz Waves." In *Terahertz for Military and Security Applications IV*, 6212:62120G.

Karpowicz, N., H. Zhong, J. Xu, K. Lin, J. Hwang, and X.C. Zhang. 2005. "Comparison between Pulsed Terahertz Time-Domain Imaging and Continuous Wave Terahertz Imaging." *Semiconductor Science and Technology* 20 (7): S293.

Kasap, S., and P. Capper. 2017. *Springer Handbook of Electronic and Photonic Materials*. New York, NY: Springer.

Kashima, M., S. Tsuchikawa, and T. Inagaki. 2020. "Simultaneous Detection of Density, Moisture Content and Fiber Direction of Wood by THz Time-Domain Spectroscopy." *Journal of Wood Science* 66 (1): 1–8.

Katz, M.B. 2013. *Questions of Uniqueness and Resolution in Reconstruction from Projections*. Vol. 26. New York, NY: Springer Science & Business Media.

Kaufman, L. 1993. "Maximum Likelihood, Least Squares, and Penalized Least Squares for PET." *IEEE Transactions on Medical Imaging* 12 (2): 200–14.

Khaled, A.Y., C.A. Parrish, and A. Adedeji. 2021. "Emerging Nondestructive Approaches for Meat Quality and Safety Evaluation – A Review." *Comprehensive Reviews in Food Science and Food Safety* 20 (4): 3438–63.

Khan, K.M., H. Krishna, S.K. Majumder, and P.K. Gupta. 2015. "Detection of Urea Adulteration in Milk Using Near-Infrared Raman Spectroscopy." *Food Analytical Methods* 8 (1): 93–102.

Khodadadi, M., N. Nozhat, and S.M.M. Moshiri. 2022. "Theoretical Analysis of a Graphene Quantum Well Hybrid Plasmonic Waveguide to Design an Inter/Intra-Chip Nano-Antenna." *Carbon* 189: 443–58.

Khushbu, S., M. Yashini, A. Rawson, and C.K. Sunil. 2021. "Recent Advances in Terahertz Time-Domain Spectroscopy and Imaging Techniques for Automation in Agriculture and Food Sector." *Food Analytical Methods* 15: 498–526.

Kim, G., J. Kim, S. Jeon, J. Kim, K. Park, and C. Oh. 2012. "Enhanced Continuous-Wave Terahertz Imaging with a Horn Antenna for Food Inspection." *Journal of Infrared, Millimeter, and Terahertz Waves* 33 (6): 657–64.

Kim, J., O. Kwon, F.D.J. Brunner, M. Jazbinsek, S. Lee, and P. Guunter. 2015. "Phonon Modes of Organic Electro-Optic Molecular Crystals for Terahertz Photonics." *The Journal of Physical Chemistry C* 119 (18): 10031–39.

Kim, J., D. Yoon, J. Yun, et al. 2018. "Three-Dimensional Terahertz Tomography with Transistor-Based Signal Source and Detector Circuits Operating Near 300 GHz." *IEEE Transactions on Terahertz Science and Technology* 8 (5): 482–91.

Kitaeva, G.K. 2008. "Terahertz Generation by Means of Optical Lasers." *Laser Physics Letters* 5 (8): 559–76.

Kittlaus, E.A., D. Eliyahu, S. Ganji, et al. 2021. "A Low-Noise Photonic Heterodyne Synthesizer and Its Application to Millimeter-Wave Radar." *Nature Communications* 12 (1): 1–10.

Klarskov, P., H. Kim, V.L. Colvin, and D.M. Mittleman. 2017. "Nanoscale Laser Terahertz Emission Microscopy." *ACS Photonics* 4 (11): 2676–80.

Kleine-Ostmann, T., P. Knobloch, M. Koch, et al. 2001. "Continuous-Wave THz Imaging." *Electronics Letters* 37 (24): 1461–63.

Kohen, S., B.S. Williams, and Q. Hu. 2005. "Electromagnetic Modeling of Terahertz Quantum Cascade Laser Waveguides and Resonators." *Journal of Applied Physics* 97 (5): 53106.

Koul, S.K., and P. Kaurav. 2022. "Terahertz Spectrum in Biomedical Engineering." In *Sub-Terahertz Sensing Technology for Biomedical Applications*, 1–29. Singapore: Springer.

Kovacic, P., and R. Somanathan. 2014. "Toxicity of Imine – Iminium Dyes and Pigments: Electron Transfer, Radicals, Oxidative Stress and Other Physiological Effects." *Journal of Applied Toxicology* 34 (8): 825–34.

Kowalski, M. 2019. "Hidden Object Detection and Recognition in Passive Terahertz and Mid-Wavelength Infrared." *Journal of Infrared, Millimeter, and Terahertz Waves* 40 (11): 1074–91.

Kowalski, M., N. Palka, M. Piszczek, and M. Szustakowski. 2013. "Hidden Object Detection System Based on Fusion of THz and VIS Images." *Acta Physica Polonica A* 124 (3): 490–93.

Krasnok, A.E., I.S. Maksymov, A.I. Denisyuk, et al. 2013. "Optical Nanoantennas." *Physics-Uspekhi* 56 (6): 539.

Krugener, K., J. Ornik, L.M. Schneider, et al. 2020. "Terahertz Inspection of Buildings and Architectural Art." *Applied Sciences* 10 (15): 5166.

Kuba, A., and G. Hermann. 2008. "Some Mathematical Concepts for Tomographic Reconstruction." *Advanved Tomographic Methods in Materials Science and Engineering*. Oxford: Oxford University Press, 19–36.

Kulya, M.S., N.S. Balbekin, I.V. Gredyuhina, M.V. Uspenskaya, A.P. Nechiporenko, and N.V. Petrov. 2017. "Computational Terahertz Imaging with Dispersive Objects." *Journal of Modern Optics* 64 (13): 1283–88.

Kumar, S., Q. Hu, and J.L. Reno. 2009. "186 K Operation of Terahertz Quantum-Cascade Lasers Based on a Diagonal Design." *Applied Physics Letters* 94 (13): 131105.

Kundu, P.K., and Pragti. 2022. "THz Image Processing and Its Applications." In *Generation, Detection and Processing of Terahertz Signals*, edited by Aritra Acharyya, Arindam Biswas, and Palash Das, 123–37. Singapore: Springer Singapore. https://doi.org/10.1007/978-981-16-4947-9_9.

Kurita, I., T. Suzuki, N. Kondo, et al. 2014. "Specific Detection of Escherichia Coli by Using Metallic Mesh Sensor in THz Region." In *2014 XXXIth URSI General Assembly and Scientific Symposium (URSI GASS)*, 1–4.

Kyoung, J.S., M.A. Seo, H.R. Park, et al. 2009. "Terahertz Transmission through Nanogaps Using Both Pulsed and CW Sources." In *2009 34th International Conference on Infrared, Millimeter, and Terahertz Waves*, 1–2.

Lasch, P. 2012. "Spectral Pre-Processing for Biomedical Vibrational Spectroscopy and Microspectroscopic Imaging." *Chemometrics and Intelligent Laboratory Systems* 117: 100–14.

Lavadiya, S., and V. Sorathiya. 2021. "Terahertz Antenna: Fundamentals, Types, Fabrication, and Future Scope." In *Advances in Terahertz Technology and Its Applications*, edited by Sudipta Das, N Anveshkumar, Joydeep Dutta, and Arindam Biswas, 113–35. Singapore: Springer Singapore. https://doi.org/10.1007/978-981-16-5731-3_7.

Law, J.W., N.A. Mutalib, K. Chan, and L. Lee. 2015. "Rapid Methods for the Detection of Foodborne Bacterial Pathogens: Principles, Applications, Advantages and Limitations." *Frontiers in Microbiology* 5: 770.

Lawler, N.B., D. Ho, C.W. Evans, V.P. Wallace, and K.S. Iyer. 2020. "Convergence of Terahertz Radiation and Nanotechnology." *Journal of Materials Chemistry C* 8 (32): 10942–55.

Lazcka, O., F.J.D. Campo, and F.X. Munoz. 2007. "Pathogen Detection: A Perspective of Traditional Methods and Biosensors." *Biosensors and Bioelectronics* 22 (7): 1205–17.

Lee, D., J. Kang, J. Lee, et al. 2015. "Highly Sensitive and Selective Sugar Detection by Terahertz Nano-Antennas." *Scientific Reports* 5 (1): 1–7.

Lee, D., G. Kim, J. Son, and M. Seo. 2016. "Highly Sensitive Terahertz Spectroscopy of Residual Pesticide Using Nano-Antenna." In *Terahertz, RF, Millimeter, and Submillimeter-Wave Technology and Applications IX*, 9747:97470S.

Lee, K., J.H. Timothy, and U. Yun. 2014. "Application of Raman Spectroscopy for Qualitative and Quantitative Analysis of Aflatoxins in Ground Maize Samples." *Journal of Cereal Science* 59 (1): 70–78.

Lee, W.S.L., F. Ariel, W. Withayachumnankul, and J.A. Able. 2020. "Assessing Frost Damage in Barley Using Terahertz Imaging." *Optics Express* 28 (21): 30644–55.

Lee, Y. 2009. *Principles of Terahertz Science and Technology*, 170. New York, NY: Springer Science & Business Media.

Lee, Y., S. Choi, S. Han, D. Woo, and H.S. Chun. 2012. "Detection of Foreign Bodies in Foods Using Continuous Wave Terahertz Imaging." *Journal of Food Protection* 75 (1): 179–83.

Lei, T., Q. Li, and D. Sun. 2022. "A Dual AE-GAN Guided THz Spectral Dehulling Model for Mapping Energy and Moisture Distribution on Sunflower Seed Kernels." *Food Chemistry* 380: 131971.

Lent, A. 1976. "Maximum Entropy and Multiplicative ART." In *Proc. Conf. Image Analysis and Evaluation, SPSE, Toronto*.

Lepeshov, S., A. Gorodetsky, A. Krasnok, et al. 2018. "Boosting Terahertz Photoconductive Antenna Performance with Optimised Plasmonic Nanostructures." *Scientific Reports* 8 (1): 1–7.

Lepeshov, S., A. Gorodetsky, A. Krasnok, E. Rafailov, and P. Belov. 2017. "Enhancement of Terahertz Photoconductive Antenna Operation by Optical Nanoantennas." *Laser and Photonics Reviews* 11 (1): 1600199.

Leulescu, M., A. Rotaru, A. Moancta, et al. 2021. "Azorubine: Physical, Thermal and Bioactive Properties of the Widely Employed Food, Pharmaceutical and Cosmetic Red Azo Dye Material." *Journal of Thermal Analysis and Calorimetry* 143 (6): 3945–67.

Leuschner, R.G.K., and P.J. Lillford. 2001. "Investigation of Bacterial Spore Structure by High Resolution Solid-State Nuclear Magnetic Resonance Spectroscopy and Transmission Electron Microscopy." *International Journal of Food Microbiology* 63 (1–2): 35–50.

Lew, T.T.S., R. Sarojam, I. Jang, et al. 2020. "Species-Independent Analytical Tools for Next-Generation Agriculture." *Nature Plants* 6 (12): 1408–17.

Lewis, R. 2017. "Materials for Terahertz Engineering." In *Springer Handbook of Electronic and Photonic Materials*, edited by S. Kasap and P. Capper. Cham: Springer Handbooks. https://doi.org/10.1007/978-3-319-48933-9_55.

Lewis, R.A. 2014. "A Review of Terahertz Sources." *Journal of Physics D: Applied Physics* 47 (37): 374001.

Li, B., J. Bai, and S. Zhang. 2021b. "Low Concentration Noroxin Detection Using Terahertz Spectroscopy Combined with Metamaterial." *Spectrochimica Acta Part A: Molecular and Biomolecular Spectroscopy* 247: 119101.

Li, B., W. Cao, S. Mathanker, W. Zhang, and N. Wang. 2010. "Preliminary Study on Quality Evaluation of Pecans with Terahertz Time-Domain Spectroscopy." In *Infrared, Millimeter Wave, and Terahertz Technologies*, 7854:78543V.

Li, B., K. Hu, and Y. Shen. 2020a. "A Scientometric Analysis of Global Terahertz Research by Web of Science Data." *IEEE Access* 8: 56092–112.

Li, B., C. Li, C. Dong, P. Li, J. Ma, and D. Ye. 2021a. "Mechanism of Lead Pollution Detection in Soil Using Terahertz Spectrum." *International Journal of Environmental Science and Technology* 19: 7243–50

Li, B., Y. Long, and H. Yang. 2018. "Measurements and Analysis of Water Content in Winter Wheat Leaf Based on Terahertz Spectroscopy." *International Journal of Agricultural and Biological Engineering* 11 (3): 178–82.

Li, B., and X. Shen. 2020. "Preliminary Study on Discrimination of Transgenic Cotton Seeds Using Terahertz Time-Domain Spectroscopy." *Food Science & Nutrition* 8 (10): 5426–33.

Li, B., M. Wang, W. Cao, and Z. Zhang. 2011. "Research on Heavy Metal Ions Detection in Soil with Terahertz Time-Domain Spectroscopy." In *International Symposium on Photoelectronic Detection and Imaging 2011: Terahertz Wave Technologies and Applications*, 8195:81951V.

Li, B., D. Zhang, and Y. Shen. 2020b. "Study on Terahertz Spectrum Analysis and Recognition Modeling of Common Agricultural Diseases." *Spectrochimica Acta Part A: Molecular and Biomolecular Spectroscopy* 243: 118820.

Li, B., X. Zhao, Y. Zhang, S. Zhang, and B. Luo. 2020c. "Prediction and Monitoring of Leaf Water Content in Soybean Plants Using Terahertz Time-Domain Spectroscopy." *Computers and Electronics in Agriculture* 170: 105239.

Li, C., B. Li, and D. Ye. 2020d. "Analysis and Identification of Rice Adulteration Using Terahertz Spectroscopy and Pattern Recognition Algorithms." *IEEE Access* 8: 26839–50.

Li, C., M. Li, and L. Jiang. 2015a. "Research on L-Ascorbic Acid and Thiamine Based on Wide-Band Terahertz Spectroscopy Technique." *Spectroscopy and Spectral Analysis* 35 (3): 595–98.

Li, F., Z. Liu, T. Sun, Y. Ma, and X. Ding. 2015b. "Confocal Three-Dimensional Micro X-Ray Scatter Imaging for Non-Destructive Detecting Foreign Bodies with Low Density and Low-Z Materials in Food Products." *Food Control* 54: 120–25.

Li, F., J. Zhang, and Y. Wang. 2022. "Vibrational Spectroscopy Combined with Chemometrics in Authentication of Functional Foods." *Critical Reviews in Analytical Chemistry* 27: 1–22.

Li, H., H. Ye, and Y. Yang. 2017. "Characterizing Polymorphism and Crystal Transformation of Polylactide by Terahertz Time-Domain Spectroscopy." *Polymer Testing* 57: 52–57.

Li, H., Y. Zhe, D. Yu, J. Zhou, and H. Yang. 2012. "Study on Terahertz Radiation Test of Blackbody." In *6th International Symposium on Advanced Optical Manufacturing and Testing Technologies: Optical Test and Measurement Technology and Equipment*, 8417:841730.

Li, J., and J. Li. 2020. "Terahertz (THz) Generator and Detection." *Electrical Science & Engineering* 2: 11–25.

Li, L., F. Xue, D. Liang, and X. Chen. 2021c. "A Hard Example Mining Approach for Concealed Multi-Object Detection of Active Terahertz Image." *Applied Sciences* 11 (23): 11241.

Li, Q., J. Kolbel, M.P. Davis, et al. 2022. "In-Situ Observation of the Structure of Crystallizing Magnesium Sulfate Heptahydrate Solutions with Terahertz Transmission Spectroscopy." *Crystal Growth & Design*. 22 (6): 3961–72. https://doi.org/10.1021/acs.cgd.2c00352.

Li, S., Y. Tian, P. Jiang, Y. Lin, X. Liu, and H. Yang. 2021d. "Recent Advances in the Application of Metabolomics for Food Safety Control and Food Quality Analyses." *Critical Reviews in Food Science and Nutrition* 61 (9): 1448–69.

Li, T., J. He, L. Zhang, et al. 2021e. "Fast Quantitative Analysis of Hidden Dangerous Substances in Mail Based on Specific Interval PLS." *Journal of Infrared, Millimeter, and Terahertz Waves* 42 (5): 572–87.

Li, Z., N. Rothbart, X. Deng, et al. 2020e. "Qualitative and Quantitative Analysis of Terahertz Gas-Phase Spectroscopy Using Independent Component Analysis." *Chemometrics and Intelligent Laboratory Systems* 206: 104129.

Lian, F.Y., H.Y. Ge, X.J. Ju, Y. Zhang, and M.X. Fu. 2019. "Quantitative Analysis of Trans Fatty Acids in Cooked Soybean Oil Using Terahertz Spectrum." *Journal of Applied Spectroscopy* 86 (5): 917–24.

Lian, F., D. Xu, M. Fu, H. Ge, Y. Jiang, and Y. Zhang. 2017. "Identification of Transgenic Ingredients in Maize Using Terahertz Spectra." *IEEE Transactions on Terahertz Science and Technology* 7 (4): 378–84.

Liang, W., J. Zuo, Q. Zhou, and C. Zhang. 2022. "Quantitative Determination of Glycerol Concentration in Aqueous Glycerol Solutions by Metamaterial-Based Terahertz Spectroscopy." *Spectrochimica Acta Part A: Molecular and Biomolecular Spectroscopy* 270: 120812.

Liebermeister, L., S. Nellen, R.B. Kohlhaas, et al. 2021. "Optoelectronic Frequency-Modulated Continuous-Wave Terahertz Spectroscopy with 4 THz Bandwidth." *Nature Communications* 12 (1): 1–10.

Liebermeister, L., S. Nellen, R. Kohlhaas, S. Breuer, M. Schell, and B. Globisch. 2019. "Ultra-Fast, High-Bandwidth Coherent Cw THz Spectrometer for Non-Destructive Testing." *Journal of Infrared, Millimeter, and Terahertz Waves* 40 (3): 288–96.

Lin, H., Y. Dong, D. Markl, et al. 2017. "Measurement of the Intertablet Coating Uniformity of a Pharmaceutical Pan Coating Process with Combined Terahertz and Optical Coherence Tomography In-Line Sensing." *Journal of Pharmaceutical Sciences* 106 (4): 1075–84.

Lin, H., R.K. May, M.J. Evans, et al. 2015. "Impact of Processing Conditions on Inter-Tablet Coating Thickness Variations Measured by Terahertz in-Line Sensing." *Journal of Pharmaceutical Sciences* 104 (8): 2513–22.

Liu, H., G. Plopper, S. Earley, Y. Chen, B. Ferguson, and X.C. Zhang. 2007. "Sensing Minute Changes in Biological Cell Monolayers with THz Differential Time-Domain Spectroscopy." *Biosensors and Bioelectronics* 22 (6): 1075–80.

Liu, J. 2017a. "Terahertz Spectroscopy and Chemometric Tools for Rapid Identification of Adulterated Dairy Product." *Optical and Quantum Electronics* 49 (1): 1.

Liu, J. 2017b. "Terahertz Spectroscopy and Chemometrics Classification of Transgenic Corn Oil from Corn Edible Oil." *Microwave and Optical Technology Letters* 59 (3): 654–58.

Liu, J., and L. Fan. 2020. "Qualitative and Quantitative Determination of Potassium Aluminum Sulfate Dodecahydrate in Potato Starch Based on Terahertz Spectroscopy." *Microwave and Optical Technology Letters* 62 (2): 525–30.

Liu, J., L. Fan, Y. Liu, L. Mao, and J. Kan. 2019a. "Application of Terahertz Spectroscopy and Chemometrics for Discrimination of Transgenic Camellia Oil." *Spectrochimica Acta Part A: Molecular and Biomolecular Spectroscopy* 206: 165–69.

Liu, J., L. Mao, J. Ku, et al. 2017a. "Using Terahertz Spectroscopy to Identify Transgenic Cottonseed Oil According to Physicochemical Quality Parameters." *Optik* 142: 483–88.

Liu, M., M. Wang, J. Wang, and D. Li. 2013. "Comparison of Random Forest, Support Vector Machine and Back Propagation Neural Network for Electronic Tongue Data Classification: Application to the Recognition of Orange Beverage and Chinese Vinegar." *Sensors and Actuators B: Chemical* 177: 970–80.

Liu, T., T. Zhang, D. Wang, and Z. Huang. 2017b. "Compact Beam Transport System for Free-Electron Lasers Driven by a Laser Plasma Accelerator." *Physical Review Accelerators and Beams* 20 (2): 20701.

Liu, W., C. Liu, F. Chen, J. Yang, and L. Zheng. 2016a. "Discrimination of Transgenic Soybean Seeds by Terahertz Spectroscopy." *Scientific Reports* 6 (1): 1–7.

Liu, W., C. Liu, X. Hu, J. Yang, and L. Zheng. 2016b. "Application of Terahertz Spectroscopy Imaging for Discrimination of Transgenic Rice Seeds with Chemometrics." *Food Chemistry* 210: 415–21.

Liu, W., C. Liu, J. Yu, et al. 2018a. "Discrimination of Geographical Origin of Extra Virgin Olive Oils Using Terahertz Spectroscopy Combined with Chemometrics." *Food Chemistry* 251: 86–92.

Liu, W., Y. Zhang, M. Li, D. Han, and W. Liu. 2020. "Determination of Invert Syrup Adulterated in Acacia Honey by Terahertz Spectroscopy with Different Spectral Features." *Journal of the Science of Food and Agriculture* 100 (5): 1913–21.

Liu, W., Y. Zhang, S. Yang, and D. Han. 2018b. "Terahertz Time-Domain Attenuated Total Reflection Spectroscopy Applied to the Rapid Discrimination of the Botanical Origin of Honeys." *Spectrochimica Acta Part A: Molecular and Biomolecular Spectroscopy* 196: 123–30.

Liu, W., P. Zhao, Y. Shi, C. Liu, and L. Zheng. 2021. "Rapid Determination of Peroxide Value of Peanut Oils during Storage Based on Terahertz Spectroscopy." *Food Analytical Methods* 14 (6): 1269–77.

Liu, W., P. Zhao, C. Wu, C. Liu, J. Yang, and L. Zheng. 2019b. "Rapid Determination of Aflatoxin B1 Concentration in Soybean Oil Using Terahertz Spectroscopy with Chemometric Methods." *Food Chemistry* 293: 213–19.

Liu, Y., Y. Cao, T. Wang, Q. Dong, J. Li, and C. Niu. 2019c. "Detection of 12 Common Food-Borne Bacterial Pathogens by TaqMan Real-Time PCR Using a Single Set of Reaction Conditions." *Frontiers in Microbiology* 10: 222.

Long, Y., B. Li, and H. Liu. 2018. "Analysis of Fluoroquinolones Antibiotic Residue in Feed Matrices Using Terahertz Spectroscopy." *Applied Optics* 57 (3): 544–50.

Lu, M., J. Shen, N. Li, et al. 2006. "Detection and Identification of Illicit Drugs Using Terahertz Imaging." *Journal of Applied Physics* 100 (10): 103104.

Lu, S.H., B.Q. Li, H.L. Zhai, X. Zhang, and Z.Y. Zhang. 2018. "An Effective Approach to Quantitative Analysis of Ternary Amino Acids in Foxtail Millet Substrate Based on Terahertz Spectroscopy." *Food Chemistry* 246: 220–27.

Lu, W., H. Luo, L. He, et al. 2022. "Detection of Heavy Metals in Vegetable Soil Based on THz Spectroscopy." *Computers and Electronics in Agriculture* 197: 106923.

Lu, W., and F. Yin. 2004. "Adaptive Algebraic Reconstruction Technique: Adaptive Algebraic Reconstruction Technique." *Medical Physics* 31 (12): 3222–30.

Lu, X., H. Sun, T. Chang, J. Zhang, and H. Cui. 2020. "Terahertz Detection of Porosity and Porous Microstructure in Pharmaceutical Tablets: A Review." *International Journal of Pharmaceutics* 591: 120006.

Lucyszyn, S., F. Hu, and W.J. Otter. 2013. "Technology Demonstrators for Low-Cost Terahertz Engineering." In *2013 Asia-Pacific Microwave Conference Proceedings (APMC)*, 518–20.

Luo, H., J. Zhu, W. Xu, and M. Cui. 2019. "Identification of Soybean Varieties by Terahertz Spectroscopy and Integrated Learning Method." *Optik* 184: 177–84.

Luo, L., I. Chatzakis, J. Wang, et al. 2014. "Broadband Terahertz Generation from Metamaterials." *Nature Communications* 5 (1): 1–6.

Ma, C.C., Y.B. Gao, H.Y. Guo, J.L. Wang, J.B. Wu, and J.S. Xu. 2008. "Physiological Adaptations of Four Dominant Caragana Species in the Desert Region of the Inner Mongolia Plateau." *Journal of Arid Environments* 72 (3): 247–54.

Ma, Q., Y. Teng, C. Li, and L. Jiang. 2022. "Simultaneous Quantitative Determination of Low-Concentration Ternary Pesticide Mixtures in Wheat Flour Based on Terahertz Spectroscopy and BPNN." *Food Chemistry* 377: 132030.

Maeng, I., S.H. Baek, H.Y. Kim, G. Ok, S. Choi, and H.S. Chun. 2014. "Feasibility of Using Terahertz Spectroscopy to Detect Seven Different Pesticides in Wheat Flour." *Journal of Food Protection* 77 (12): 2081–87.

Maestrini, A., B. Thomas, H. Wang, et al. 2010. "Schottky Diode-Based Terahertz Frequency Multipliers and Mixers." *Comptes Rendus Physique* 11 (7–8): 480–95.

Mahesh, S., D.S. Jayas, J. Paliwal, and N.D.G. White. 2015. "Hyperspectral Imaging to Classify and Monitor Quality of Agricultural Materials." *Journal of Stored Products Research* 61: 17–26.

Malhotra, I., K.R. Jha, and G. Singh. 2018. "Terahertz Antenna Technology for Imaging Applications: A Technical Review." *International Journal of Microwave and Wireless Technologies* 10 (3): 271–90.

Malhotra, I., and G. Singh. 2021. "Terahertz Technology for Biomedical Application." In *Terahertz Antenna Technology for Imaging and Sensing Applications*, 235–64. Cham: Springer. https://doi.org/10.1007/978-3-030-68960-5_10.

Mamrashev, A., F. Minakov, L. Maximov, N. Nikolaev, and P. Chapovsky. 2019. "Correction of Optical Delay Line Errors in Terahertz Time-Domain Spectroscopy." *Electronics* 8 (12): 1408.

Mansourzadeh, S., D. Damyanov, T. Vogel, et al. 2021. "High-Power Lensless THz Imaging of Hidden Objects." *IEEE Access* 9: 6268–76.

Mantsch, H.H., and D. Naumann. 2010. "Terahertz Spectroscopy: The Renaissance of Far Infrared Spectroscopy." *Journal of Molecular Structure* 964 (1–3): 1–4.

Markl, D., J. Sauerwein, D.J. Goodwin, S.V.D. Ban, and J.A. Zeitler. 2017. "Non-Destructive Determination of Disintegration Time and Dissolution in Immediate Release Tablets by Terahertz Transmission Measurements." *Pharmaceutical Research* 34 (5): 1012–22.

Markl, D., A. Strobel, R. Schlossnikl, et al. 2018. "Characterisation of Pore Structures of Pharmaceutical Tablets: A Review." *International Journal of Pharmaceutics* 538 (1–2): 188–214.

Massaouti, M., C. Daskalaki, A. Gorodetsky, A.D. Koulouklidis, and S. Tzortzakis. 2013. "Detection of Harmful Residues in Honey Using Terahertz Time-Domain Spectroscopy." *Applied Spectroscopy* 67 (11): 1264–69.

Mathanker, S.K., P.R. Weckler, and N. Wang. 2013. "Terahertz (THz) Applications in Food and Agriculture: A Review." *Transactions of the ASABE* 56 (3): 1213–26.

Matsui, T., and S. Kidera. 2020. "Virtual Source Extended Range Points Migration Method for Auto-Focusing 3-D Terahertz Imaging." *IEEE Geoscience and Remote Sensing Letters* 18 (6): 989–93.

Maurer, L., and H. Leuenberger. 2009. "Terahertz Pulsed Imaging and Near Infrared Imaging to Monitor the Coating Process of Pharmaceutical Tablets." *International Journal of Pharmaceutics* 370 (1–2): 8–16.

May, R., and P.F. Taday. 2013. "Crystallization of Sucrose Monitored by Terahertz Pulsed Spectroscopy." In *2013 38th International Conference on Infrared, Millimeter, and Terahertz Waves (IRMMW-THz)*, 1–1. 10.1109/IRMMW-THz.2013.6665545.

May, R.K., M.J. Evans, S. Zhong, et al. 2011. "Terahertz In-Line Sensor for Direct Coating Thickness Measurement of Individual Tablets during Film Coating in Real-Time." *Journal of Pharmaceutical Sciences* 100 (4): 1535–44.

May, R.K., K.E. Su, L. Han, et al. 2013. "Hardness and Density Distributions of Pharmaceutical Tablets Measured by Terahertz Pulsed Imaging." *Journal of Pharmaceutical Sciences* 102 (7): 2179–86.

McIntosh, A.I., B. Yang, S.M. Goldup, M. Watkinson, and R.S. Donnan. 2013. "Crystallization of Amorphous Lactose at High Humidity Studied by Terahertz Time Domain Spectroscopy." *Chemical Physics Letters* 558: 104–8.

Medina, S., R. Perestrelo, P. Silva, J.A.M. Pereira, and J.S. Camara. 2019. "Current Trends and Recent Advances on Food Authenticity Technologies and Chemometric Approaches." *Trends in Food Science & Technology* 85: 163–76.

Mei, J., F. Zhao, R. Xu, and Y. Huang. 2021. "A Review on the Application of Spectroscopy to the Condiments Detection: From Safety to Authenticity." *Critical Reviews in Food Science and Nutrition* 62: 6374–89.

MenloSystems. 2022. "Terahertz Time Domain Solutions." Menlo Systems Inc. 2022. Accessed 14th May 2022, https://www.menlosystems.com/products/thz-time-domain-solutions/.

Mikerov, M., R. Shrestha, P.V. Dommelen, D.M. Mittleman, and M. Koch. 2020. "Analysis of Ancient Ceramics Using Terahertz Imaging and Photogrammetry." *Optics Express* 28 (15): 22255–63.

Mikulics, M., M. Marso, M. Lepsa, D. Grutzmacher, and P. Kordos. 2008. "Output Power Improvement in MSM Photomixers by Modified Finger Contacts Configuration." *IEEE Photonics Technology Letters* 21 (3): 146–48.

Mishra, G.K., A.B.F. Tehrani, and R.K. Mishra. 2018. "Food Safety Analysis Using Electrochemical Biosensors." *Foods* 7 (9): 141.

Mitchell, G.A. 1990. "Methods of Starch Analysis." *Starch-Stärke* 42 (4): 131–34.

Mitrofanov, O., L. Viti, E. Dardanis, et al. 2017. "Near-Field Terahertz Probes with Room-Temperature Nanodetectors for Subwavelength Resolution Imaging." *Scientific Reports* 7 (1): 1–10.

Mittleman, D. 2013. *Sensing with Terahertz Radiation*. Vol. 85. New York, NY: Springer.

Mittleman, D.M. 2017. "Perspective: Terahertz Science and Technology." *Journal of Applied Physics* 122 (23): 230901.

Mittleman, D.M., S. Hunsche, L. Boivin, and M.C. Nuss. 1997. "T-Ray Tomography." *Optics Letters* 22 (12): 904–06.

Mittleman, D.M., R.H. Jacobsen, and M.C. Nuss. 1996. "T-Ray Imaging." *IEEE Journal of Selected Topics in Quantum Electronics* 2 (3): 679–92.

Mohammadpour, Z., S.H. Abdollahi, A. Omidvar, A. Mohajeri, and A. Safavi. 2020. "Aqueous Solutions of Carbohydrates Are New Choices of Green Solvents for Highly Efficient Exfoliation of Two-Dimensional Nanomaterials." *Journal of Molecular Liquids* 309: 113087.

Mohan, S., E. Kato, J.K. Drennen III, and C.A. Anderson. 2019. "Refractive Index Measurement of Pharmaceutical Solids: A Review of Measurement Methods and Pharmaceutical Applications." *Journal of Pharmaceutical Sciences* 108 (11): 3478–95.

Moller, U., J.R. Folkenberg, and P.U. Jepsen. 2010. "Dielectric Properties of Water in Butter and Water–AOT–Heptane Systems Measured Using Terahertz Time-Domain Spectroscopy." *Applied Spectroscopy* 64 (9): 1028–36.

Molloy, J., and M. Naftaly. 2014. "Wool Textile Identification by Terahertz Spectroscopy." *The Journal of the Textile Institute* 105 (8): 794–98.

Morales-Hernandez, J.A., A.K. Singh, S.J. Villanueva-Rodriguez, and E. Castro-Camus. 2019. "Hydration Shells of Carbohydrate Polymers Studied by Calorimetry and Terahertz Spectroscopy." *Food Chemistry* 291: 94–100.

Morita, Y., A. Dobroiu, C. Otani, and K. Kawase. 2005. "A Real-Time Inspection System Using a Terahertz Technique to Detect Microleak Defects in the Seal of Flexible Plastic Packages." *Journal of Food Protection* 68 (4): 833–37.

Mousavi, P., F. Haran, D. Jez, F. Santosa, and J.S. Dodge. 2009. "Simultaneous Composition and Thickness Measurement of Paper Using Terahertz Time-Domain Spectroscopy." *Applied Optics* 48 (33): 6541–46.

Mueckstein, R., C. Graham, C.C. Renaud, A.J. Seeds, J.A. Harrington, and O. Mitrofanov. 2011. "Imaging and Analysis of THz Surface Plasmon Polariton Waves with the Integrated Sub-Wavelength Aperture Probe." *Journal of Infrared, Millimeter, and Terahertz Waves* 32 (8): 1031–42.

Mumcuoglu, E.U., R. Leahy, S.R. Cherry, and Z. Zhou. 1994. "Fast Gradient-Based Methods for Bayesian Reconstruction of Transmission and Emission PET Images." *IEEE Transactions on Medical Imaging* 13 (4): 687–701.

Murate, K., H. Kanai, and K. Kawase. 2021. "Application of Machine Learning to Terahertz Spectroscopic Imaging of Reagents Hidden by Thick Shielding Materials." *IEEE Transactions on Terahertz Science and Technology* 11 (6): 620–25.

Murate, K., and K. Kawase. 2018. "Perspective: Terahertz Wave Parametric Generator and Its Applications." *Journal of Applied Physics* 124 (16): 160901.

Naftaly, M., G. Savvides, F. Alshareef, et al. 2022. "Non-Destructive Porosity Measurements of 3D Printed Polymer by Terahertz Time-Domain Spectroscopy." *Applied Sciences* 12 (2): 927.

Nagarajan, S., C.F. Neese, and F.C.D. Lucia. 2017. "Cavity-Based Medium Resolution Spectroscopy (CBMRS) in the THz: A Bridge between High-and Low-Resolution Techniques for Sensor and Spectroscopy Applications." *IEEE Transactions on Terahertz Science and Technology* 7 (3): 233–43.

Naito, H., Y. Ogawa, K. Shiraga, et al. 2011. "Inspection of Milk Components by Terahertz Attenuated Total Reflectance (THz-ATR) Spectrometer Equipped Temperature Controller." In *2011 IEEE/SICE International Symposium on System Integration (SII)*, 192–96.

Nakajima, S., S. Horiuchi, A. Ikehata, and Y. Ogawa. 2021. "Determination of Starch Crystallinity with the Fourier-Transform Terahertz Spectrometer." *Carbohydrate Polymers* 262: 117928.

Nakajima, S., K. Shiraga, T. Suzuki, N. Kondo, and Y. Ogawa. 2019. "Quantification of Starch Content in Germinating Mung Bean Seedlings by Terahertz Spectroscopy." *Food Chemistry* 294: 203–08.

Nakanishi, A., and H. Takahashi. 2018. "Terahertz Optical Material Based on Wood-Plastic Composites." *Optical Materials Express* 8 (12): 3653–58.

Nallappan, K., Y. Cao, G. Xu, H. Guerboukha, C. Nerguizian, and M. Skorobogatiy. 2020. "Dispersion-Limited versus Power-Limited Terahertz Communication Links Using Solid Core Subwavelength Dielectric Fibers." *Photonics Research* 8 (11): 1757–75.

Nallappan, K., J. Dash, S. Ray, and B. Pesala. 2013. "Identification of Adulterants in Turmeric Powder Using Terahertz Spectroscopy." In *2013 38th International Conference on Infrared, Millimeter, and Terahertz Waves (IRMMW-THz)*, 1–2.

Nam, I., M. Kim, T.H. Lee, S.W. Lee, and H. Suk. 2015. "Highly-Efficient 20 TW Ti: Sapphire Laser System Using Optimized Diverging Beams for Laser Wakefield Acceleration Experiments." *Current Applied Physics* 15 (4): 468–72.

Nan, Z., L.S. Jie, T.J. Min, W. Yuefan, and K. Lin. 2022. "Investigation of Spoilage in Salmon by Electrochemical Impedance Spectroscopy and Time-Domain Terahertz Spectroscopy." *ChemPhysMater* 2: 148–54. https://doi.org/10.1016/j.chphma.2021.12.003.

Nekvapil, F., I. Brezestean, D. Barchewitz, B. Glamuzina, V. Chis, and S.C. Pinzaru. 2018. "Citrus Fruits Freshness Assessment Using Raman Spectroscopy." *Food Chemistry* 242: 560–67.

Nelson, R., T. Wiesner-Hanks, R. Wisser, and P. Balint-Kurti. 2018. "Navigating Complexity to Breed Disease-Resistant Crops." *Nature Reviews Genetics* 19 (1): 21–33.

Nguyen, D.T., A. Pissard, J.A.F. Pierna, et al. 2022. "A Method for Non-Destructive Determination of Cocoa Bean Fermentation Levels Based on Terahertz Hyperspectral Imaging." *International Journal of Food Microbiology* 365: 109537.

Nguyen, K.L., T. Friscic, G.M. Day, L.F. Gladden, and W. Jones. 2007. "Terahertz Time-Domain Spectroscopy and the Quantitative Monitoring of Mechanochemical Cocrystal Formation." *Nature Materials* 6 (3): 206–9.

Nie, P., F. Qu, L. Lin, et al. 2017. "Detection of Water Content in Rapeseed Leaves Using Terahertz Spectroscopy." *Sensors* 17 (12): 2830.

Niehues, G., M. Heyden, D.A. Schmidt, and M. Havenith. 2011. "Exploring Hydrophobicity by THz Absorption Spectroscopy of Solvated Amino Acids." *Faraday Discussions* 150: 193–207.

Nikitkina, A.I., P.Y. Bikmulina, E.R. Gafarova, et al. 2021. "Terahertz Radiation and the Skin: A Review." *Journal of Biomedical Optics* 26 (4): 43005.

Niu, L., K. Wang, Z. Yang, and J. Liu. 2019. "Terahertz Tomography." *Terahertz Science and Technology* 12 (3): 77–92.

Niwa, M., and Y. Hiraishi. 2014. "Quantitative Analysis of Visible Surface Defect Risk in Tablets During Film Coating Using Terahertz Pulsed Imaging." *International Journal of Pharmaceutics* 461 (1–2): 342–50.

Niwa, M., Y. Hiraishi, N. Iwasaki, and K. Terada. 2013. "Quantitative Analysis of the Layer Separation Risk in Bilayer Tablets Using Terahertz Pulsed Imaging." *International Journal of Pharmaceutics* 452 (1–2): 249–56.

Novikova, A., D. Markl, J.A. Zeitler, T. Rades, and C.S. Leopold. 2018. "A Non-Destructive Method for Quality Control of the Pellet Distribution within a MUPS Tablet by Terahertz Pulsed Imaging." *European Journal of Pharmaceutical Sciences* 111: 549–55.

Oda, N. 2010. "Uncooled Bolometer-Type Terahertz Focal Plane Array and Camera for Real-Time Imaging." *Comptes Rendus Physique* 11 (7–8): 496–509.

Ogawa, Y., L. Cheng, S. Hayashi, and K. Fukunaga. 2009. "Attenuated Total Reflection Spectra of Aqueous Glycine in the Terahertz Region." *IEICE Electronics Express* 6 (2): 117–21.

O'Hara, J., and D. Grischkowsky. 2002. "Synthetic Phased-Array Terahertz Imaging." *Optics Letters* 27 (12): 1070–72.

O'Hara, J., and D. Grischkowsky. 2004. "Quasi-Optic Synthetic Phased-Array Terahertz Imaging." *JOSA B* 21 (6): 1178–91.

Ok, G., S. Choi, K.H. Park, and H.S. Chun. 2013. "Foreign Object Detection by Sub-Terahertz Quasi-Bessel Beam Imaging." *Sensors* 13 (1): 71–85.

Ok, G., H.J. Kim, H.S. Chun, and S. Choi. 2014. "Foreign-Body Detection in Dry Food Using Continuous Sub-Terahertz Wave Imaging." *Food Control* 42: 284–89.

Ok, G., K. Park, M. Lim, H. Jang, and S. Choi. 2018. "140-GHz Subwavelength Transmission Imaging for Foreign Body Inspection in Food Products." *Journal of Food Engineering* 221: 124–31.

Ok, G., H.J. Shin, M. Lim, and S. Choi. 2019. "Large-Scan-Area Sub-Terahertz Imaging System for Nondestructive Food Quality Inspection." *Food Control* 96: 383–89.

Okamoto, K., K. Tsuruda, S. Diebold, S. Hisatake, M. Fujita, and T. Nagatsuma. 2017. "Terahertz Sensor Using Photonic Crystal Cavity and Resonant Tunneling Diodes." *Journal of Infrared, Millimeter, and Terahertz Waves* 38 (9): 1085–97.

Ollmann, Z., J.A. Fulop, J. Hebling, and G. Almasi. 2014. "Design of a High-Energy Terahertz Pulse Source Based on ZnTe Contact Grating." *Optics Communications* 315: 159–63.

Orio, M., and D.A. Pantazis. 2009. "Density Functional Theory." *Photosynthesis Research* 102 (2): 443–53.

Ornik, J., D. Knoth, M. Koch, and C.M. Keck. 2020. "Terahertz-Spectroscopy for Non-Destructive Determination of Crystallinity of L-Tartaric Acid in SmartFilms® and Tablets Made from Paper." *International Journal of Pharmaceutics* 581: 119253.

Osman, O.B., and M.H. Arbab. 2019. "Mitigating the Effects of Granular Scattering Using Cepstrum Analysis in Terahertz Time-Domain Spectral Imaging." *PLOS ONE* 14 (5): e0216952.

Ostrowski, R., A. Cywinski, and M. Strzelec. 2021. "Electronic Warfare in the Optical Band: Main Features, Examples and Selected Measurement Data." *Defence Technology* 17 (5): 1636–49.

Ozanyan, K.B. 2015. "Tomography Defined as Sensor Fusion." In *2015 IEEE SENSORS*, 1–4.

Ozturk, T., O. Morikawa, I. Unal, and I. Uluer. 2017. "Comparison of Free Space Measurement Using a Vector Network Analyzer and Low-Cost-Type THz-TDS Measurement Methods between 75 and 325 GHz." *Journal of Infrared, Millimeter, and Terahertz Waves* 38 (10): 1241–51.

Pagano, M., L. Baldacci, A. Ottomaniello, et al. 2019. "THz Water Transmittance and Leaf Surface Area: An Effective Nondestructive Method for Determining Leaf Water Content." *Sensors* 19 (22): 4838.

Pahlow, S., S. Meisel, D. Cialla-May, K. Weber, P. Rosch, and J. Popp. 2015. "Isolation and Identification of Bacteria by Means of Raman Spectroscopy." *Advanced Drug Delivery Reviews* 89: 105–20.

Paidhungat, M., B. Setlow, A. Driks, and P. Setlow. 2000. "Characterization of Spores of Bacillus Subtilis Which Lack Dipicolinic Acid." *Journal of Bacteriology* 182 (19): 5505–12.

Palermo, R., R.P. Cogdill, S.M. Short, J.K. Drennen III, and P.F. Taday. 2008. "Density Mapping and Chemical Component Calibration Development of Four-Component Compacts via Terahertz Pulsed Imaging." *Journal of Pharmaceutical and Biomedical Analysis* 46 (1): 36–44.

Palka, N. 2013. "Identification of Concealed Materials, Including Explosives, by Terahertz Reflection Spectroscopy." *Optical Engineering* 53 (3): 31202.

Pan, R., S. Zhao, and J. Shen. 2010. "Terahertz Spectra Applications in Identification of Illicit Drugs Using Support Vector Machines." *Procedia Engineering* 7: 15–21.

Pandiselvam, R., V.P. Mayookha, A. Kothakota, S.V. Ramesh, R. Thirumdas, and P. Juvvi. 2020. "Biospeckle Laser Technique – A Novel Non-Destructive Approach for Food Quality and Safety Detection." *Trends in Food Science & Technology* 97: 1–13.

Paoloni, C., D. Gamzina, R. Letizia, Y. Zheng, and N.C. Luhmann Jr. 2021. "Millimeter Wave Traveling Wave Tubes for the 21st Century." *Journal of Electromagnetic Waves and Applications* 35 (5): 567–603.

Parasoglou, P., E.P.J. Parrott, J.A. Zeitler, et al. 2009. "Quantitative Moisture Content Detection in Food Wafers." In *2009 34th International Conference on Infrared, Millimeter, and Terahertz Waves*, 1–2. 10.1109/ICIMW.2009.5324623.

Parasoglou, P., E.P.J. Parrott, J.A. Zeitler, et al. 2010. "Quantitative Water Content Measurements in Food Wafers Using Terahertz Radiation." *Terahertz Science and Technology* 3 (4): 176–82.

Park, S.J., J.T. Hong, S.J. Choi, et al. 2014. "Detection of Microorganisms Using Terahertz Metamaterials." *Scientific Reports* 4: 4988. https://doi.org/10.1038/srep04988.

Park, Y.W., and L.N. Bell. 2004. *Determination of Moisture and Ash Contents of Foods. Food Science and Technology*. Vol. 138. New York, NY: Marcel Dekker AG.

Pasquazi, A., M. Peccianti, L. Razzari, et al. 2018. "Micro-Combs: A Novel Generation of Optical Sources." *Physics Reports* 729: 1–81.

Patil, M.R., S.B. Ganorkar, A.S. Patil, and A.A. Shirkhedkar. 2022. "Terahertz Spectroscopy: Encoding the Discovery, Instrumentation, and Applications toward Pharmaceutical Prospectives." *Critical Reviews in Analytical Chemistry* 52 (2): 343–55.

Patras, A., N.P. Brunton, G. Downey, A. Rawson, K. Warriner, and G. Gernigon. 2011. "Application of Principal Component and Hierarchical Cluster Analysis to Classify Fruits and Vegetables Commonly Consumed in Ireland Based on in Vitro Antioxidant Activity." *Journal of Food Composition and Analysis* 24 (2): 250–56.

Pawar, A.Y., D.D. Sonawane, K.B. Erande, and D.V. Derle. 2013. "Terahertz Technology and Its Applications." *Drug Invention Today* 5 (2): 157–63.

Pearce, J., H. Choi, D.M. Mittleman, J. White, and D. Zimdars. 2005. "T-Ray Reflection Computed Tomography." In *Conference on Lasers and Electro-Optics*, CFD5.

Peiponen, K., P. Bawuah, M. Chakraborty, M. Juuti, J.A. Zeitler, and J. Ketolainen. 2015. "Estimation of Young's Modulus of Pharmaceutical Tablet Obtained by Terahertz Time-Delay Measurement." *International Journal of Pharmaceutics* 489 (1–2): 100–05.

Peiponen, K., A. Zeitler, and M. Kuwata-Gonokami. 2012. *Terahertz Spectroscopy and Imaging*. Vol. 171. Berlin: Springer.

Peng, K., P. Parkinson, J.L. Boland, et al. 2016. "Broadband Phase-Sensitive Single InP Nanowire Photoconductive Terahertz Detectors." *Nano Letters* 16 (8): 4925–31.

Peng, Y., J. Huang, J. Luo, et al. 2021. "Three-Step One-Way Model in Terahertz Biomedical Detection." *PhotoniX* 2 (1): 1–18.

Peng, Y., C. Shi, Y. Zhu, M. Gu, and S. Zhuang. 2020. "Terahertz Spectroscopy in Biomedical Field: A Review on Signal-to-Noise Ratio Improvement." *PhotoniX* 1 (1): 1–18.

Penkov, N.V., M.V. Goltyaev, M.E. Astashev, et al. 2021. "The Application of Terahertz Time-Domain Spectroscopy to Identification of Potato Late Blight and Fusariosis." *Pathogens* 10 (10): 1336.

Peters, O., S.F. Busch, B.M. Fischer, and M. Koch. 2012. "Determination of the Carbon Nanotube Concentration and Homogeneity in Resin Films by THz Spectroscopy and Imaging." *Journal of Infrared, Millimeter, and Terahertz Waves* 33 (12): 1221–26.

Peters, O., M. Schwerdtfeger, S. Wietzke, et al. 2013. "Terahertz Spectroscopy for Rubber Production Testing." *Polymer Testing* 32 (5): 932–36.

Petrov, N.V., M.S. Kulya, A.N. Tsypkin, V.G. Bespalov, and A. Gorodetsky. 2016. "Application of Terahertz Pulse Time-Domain Holography for Phase Imaging." *IEEE Transactions on Terahertz Science and Technology* 6 (3): 464–72.

Pfeiffer, M., and T. Pfeil. 2018. "Deep Learning with Spiking Neurons: Opportunities and Challenges." *Frontiers in Neuroscience* 12: 774.

Picque, N., and T.W. Hansch. 2019. "Frequency Comb Spectroscopy." *Nature Photonics* 13 (3): 146–57.

Praveena, A., V.A. Ponnapalli, and G. Umamaheswari. 2022. "Terahertz Antenna Technology for Detection of Explosives and Weapons: A Concise Review." *Smart Antennas* 1: 331–42.

Preu, S., G.H. Dohler, S. Malzer, L.J. Wang, and A.C. Gossard. 2011. "Tunable, Continuous-Wave Terahertz Photomixer Sources and Applications." *Journal of Applied Physics* 109 (6): 4.

Probst, T., A. Rehn, and M. Koch. 2015. "Compact and Low-Cost THz QTDS System." *Optics Express* 23 (17): 21972–82.

Puc, U., A. Abina, A. Jeglic, et al. 2018. "Spectroscopic Analysis of Melatonin in the Terahertz Frequency Range." *Sensors* 18 (12): 4098.

Qi, X., L. Zhu, C. Wang, H. Zhang, L. Wang, and H. Qian. 2017. "Development of Standard Fingerprints of Naked Oats Using Chromatography Combined with Principal Component Analysis and Cluster Analysis." *Journal of Cereal Science* 74: 224–30.

Qin, B., Z. Li, T. Chen, and Y. Chen. 2017a. "Identification of Genetically Modified Cotton Seeds by Terahertz Spectroscopy with MPGA-SVM." *Optik* 142: 576–82.

Qin, B., Z. Li, F. Hu, et al. 2018. "Highly Sensitive Detection of Carbendazim by Using Terahertz Time-Domain Spectroscopy Combined with Metamaterial." *IEEE Transactions on Terahertz Science and Technology* 8 (2): 149–54.

Qin, H., J. Sun, S. Liang, et al. 2017b. "Room-Temperature, Low-Impedance and High-Sensitivity Terahertz Direct Detector Based on Bilayer Graphene Field-Effect Transistor." *Carbon* 116: 760–65.

Qin, J., L. Xie, and Y. Ying. 2015. "Determination of Tetracycline Hydrochloride by Terahertz Spectroscopy with PLSR Model." *Food Chemistry* 170: 415–22.

Qin, J., L. Xie, and Y. Ying. 2016. "A High-Sensitivity Terahertz Spectroscopy Technology for Tetracycline Hydrochloride Detection Using Metamaterials." *Food Chemistry* 211: 300–05.

Qin, J., L. Xie, and Y. Ying. 2017c. "Rapid Analysis of Tetracycline Hydrochloride Solution by Attenuated Total Reflection Terahertz Time-Domain Spectroscopy." *Food Chemistry* 224: 262–69.

Qin, J., Y. Ying, and L. Xie. 2013. "The Detection of Agricultural Products and Food Using Terahertz Spectroscopy: A Review." *Applied Spectroscopy Reviews* 48 (6): 439–57.

Qiu, Q., and Z. Huang. 2021. "Photodetectors of 2D Materials from Ultraviolet to Terahertz Waves." *Advanced Materials* 33 (15): 2008126.

Qu, F., L. Lin, C. Cai, T. Dong, Y. He, and P. Nie. 2018a. "Molecular Characterization and Theoretical Calculation of Plant Growth Regulators Based on Terahertz Time-Domain Spectroscopy." *Applied Sciences* 8 (3): 420.

Qu, F., L. Lin, Y. He, et al. 2018b. "Spectral Characterization and Molecular Dynamics Simulation of Pesticides Based on Terahertz Time-Domain Spectra Analyses and Density Functional Theory (DFT) Calculations." *Molecules* 23 (7): 1607.

Qu, F., L. Lin, Y. He, et al. 2018c. "Terahertz Multivariate Spectral Analysis and Molecular Dynamics Simulations of Three Pyrethroid Pesticides." *Journal of Infrared, Millimeter, and Terahertz Waves* 39 (11): 1148–61.

Qu, F., Y. Pan, L. Lin, et al. 2018d. "Experimental and Theoretical Study on Terahertz Absorption Characteristics and Spectral De-Noising of Three Plant Growth Regulators." *Journal of Infrared, Millimeter, and Terahertz Waves* 39 (10): 1015–27.

Quodbach, J., and P. Kleinebudde. 2016. "A Critical Review on Tablet Disintegration." *Pharmaceutical Development and Technology* 21 (6): 763–74.

Ramirez-Garcia, E., E. Garduno-Nolasco, L.M. Rodriguez-Mendez, et al. 2019. "DC Current-Crowding Estimation for SiGe: C Heterojunction Bipolar Transistors." *Solid-State Electronics* 153: 1–7.

Recur, B., P. Desbarats, J. Domenger, et al. 2012a. "Terahertz Radiation for Tomographic Inspection." *Optical Engineering* 51 (9): 91609.

Recur, B., J. Guillet, I. Manek-Honninger, et al. 2012b. "Propagation Beam Consideration for 3D THz Computed Tomography." *Optics Express* 20 (6): 5817–29.

Recur, B., A. Younus, S. Salort, et al. 2011. "Investigation on Reconstruction Methods Applied to 3D Terahertz Computed Tomography." *Optics Express* 19 (6): 5105–17.

Redo-Sanchez, A., G. Salvatella, R. Galceran, et al. 2011. "Assessment of Terahertz Spectroscopy to Detect Antibiotic Residues in Food and Feed Matrices." *Analyst* 136 (8): 1733–38.

Ren, A., A. Zahid, D. Fan, et al. 2019a. "State-of-the-Art in Terahertz Sensing for Food and Water Security – A Comprehensive Review." *Trends in Food Science & Technology* 85: 241–51.

Ren, A., A. Zahid, X. Yang, A. Alomainy, M.A. Imran, and Q.H. Abbasi. 2019b. "Terahertz (THz) Application in Food Contamination Detection." In *Nano-Electromagnetic Communication at Terahertz and Optical Frequencies: Principles and Applications*, edited by A. Alomainy, K. Yang, M.A. Imran, Q.H. Abbasi, and X. Yao, 77–100. London: SciTech Publishing, Institution of Engineering and Technology.

Ren, G., Z. Zhu, J. Zhang, H. Zhao, Y. Li, and J. Han. 2020. "Broadband Terahertz Spectroscopy of Paper and Banknotes." *Optics Communications* 475: 126267.

Rothbart, N., H. Richter, M. Wienold, L. Schrottke, H.T. Grahn, and H. Hubers. 2013. "Fast 2-D and 3-D Terahertz Imaging with a Quantum-Cascade Laser and a Scanning Mirror." *IEEE Transactions on Terahertz Science and Technology* 3 (5): 617–24.

Ruffin, A.B., J. Decker, L. Sanchez-Palencia, et al. 2001. "Time Reversal and Object Reconstruction with Single-Cycle Pulses." *Optics Letters* 26 (10): 681–83.

Ruffin, A.B., J.V. Rudd, J. Decker, et al. 2002. "Time Reversal Terahertz Imaging." *IEEE Journal of Quantum Electronics* 38 (8): 1110–19.

Ruggiero, M.T. 2020. "Invited Review: Modern Methods for Accurately Simulating the Terahertz Spectra of Solids." *Journal of Infrared, Millimeter, and Terahertz Waves* 41 (5): 491–528.

Runge, M., D. Engel, M. Schneider, K. Reimann, M. Woerner, and T. Elsaesser. 2020. "Spatial Distribution of Electric-Field Enhancement across the Gap of Terahertz Bow-Tie Antennas." *Optics Express* 28 (17): 24389–98.

Rutz, F., T. Hasek, M. Koch, H. Richter, and U. Ewert. 2006. "Terahertz Birefringence of Liquid Crystal Polymers." *Applied Physics Letters* 89 (22): 221911.

Rutz, F., S. Wietzke, M. Koch, H. Richter, and U. Ewert. 2007. "Non-Destructive Terahertz Testing of Textured Liquid Crystal Polymers." In *Optical Terahertz Science and Technology*, MD1.

Ryu, C., S. Park, D. Kim, K. Jhang, and H. Kim. 2016. "Nondestructive Evaluation of Hidden Multi-Delamination in a Glass-Fiber-Reinforced Plastic Composite Using Terahertz Spectroscopy." *Composite Structures* 156: 338–47.

Saeedkia, D., and S. Safavi-Naeini. 2008. "Terahertz Photonics: Optoelectronic Techniques for Generation and Detection of Terahertz Waves." *Journal of Lightwave Technology* 26 (15): 2409–23.

Safian, R., G. Ghazi, and N. Mohammadian. 2019. "Review of Photomixing Continuous-Wave Terahertz Systems and Current Application Trends in Terahertz Domain." *Optical Engineering* 58 (11): 110901.

Sakamoto, T., A. Portieri, P.F. Taday, et al. 2009. "Detection of Tulobuterol Crystal in Transdermal Patches Using Terahertz Pulsed Spectroscopy and Imaging." *Die Pharmazie – An International Journal of Pharmaceutical Sciences* 64 (6): 361–65.

Samanta, D., M.P. Karthikeyan, D. Agarwal, A. Biswas, A. Acharyya, and A. Banerjee. 2022. "Trends in Terahertz Biomedical Applications." In *Generation, Detection and Processing of Terahertz Signals*, edited by Aritra Acharyya, Arindam Biswas, and Palash Das, 285–99. Singapore: Springer Singapore. https://doi.org/10.1007/978-981-16-4947-9_19.

Samoska, L.A., T.C. Gaier, A. Peralta, et al. 2000. "MMIC Power Amplifiers as Local Oscillator Drivers for FIRST." In *UV, Optical, and IR Space Telescopes and Instruments*, 4013:275–84.

Sampaolo, A., C. Yu, T. Wei, et al. 2021. "H_2S Quartz-Enhanced Photoacoustic Spectroscopy Sensor Employing a Liquid-Nitrogen-Cooled THz Quantum Cascade Laser Operating in Pulsed Mode." *Photoacoustics* 21: 100219.

Sanchez, P.D.C., N. Hashim, R. Shamsudin, and M.Z.M. Nor. 2020. "Applications of Imaging and Spectroscopy Techniques for Non-Destructive Quality Evaluation of Potatoes and Sweet Potatoes: A Review." *Trends in Food Science & Technology* 96: 208–21.

Santesteban, L.G., I. Palacios, C. Miranda, J.C. Iriarte, J.B. Royo, and R. Gonzalo. 2015. "Terahertz Time Domain Spectroscopy Allows Contactless Monitoring of Grapevine Water Status." *Frontiers in Plant Science* 6: 404.

Santitewagun, S., R. Thakkar, J.A. Zeitler, and M. Maniruzzaman. 2022. "Detecting Crystallinity Using Terahertz Spectroscopy in 3D Printed Amorphous Solid Dispersions." *Molecular Pharmaceutics*. 19 (7), 2380–89. 10.1021/acs.molpharmaceut.2c00163.

Schmidt, L.P., S. Biber, G. Rehm, and K. Huber. 2002. "THz Measurement Technologies and Applications." In *14th International Conference on Microwaves, Radar and Wireless Communications. MIKON-2002. Conference Proceedings (IEEE Cat. No. 02EX562)*, 2:581–87.

Seifert, T., S. Jaiswal, M. Sajadi, et al. 2017. "Ultrabroadband Single-Cycle Terahertz Pulses with Peak Fields of 300 KV Cm^{-1} from a Metallic Spintronic Emitter." *Applied Physics Letters* 110 (25): 252402.

Sethy, P.K., P.R. Mishra, and S. Behera. 2015. "An Introduction to Terahertz Technology, Its History, Properties and Application." In *International Conference on Computing and Communication*. Trivandrum, India.

Shalini, M., and M.G. Madhan. 2022. "Photoconductive Bowtie Dipole Antenna Incorporating Photonic Crystal Substrate for Terahertz Radiation." *Optics Communications* 517: 128327.

Shao, Y., J. Liu, Z. Zhu, Y. Wang, Y. Zhu, and Y. Peng. 2021. "Quantitative Detection on Metabolites of Haematococcus Pluvialis by Terahertz Spectroscopy." *Computers and Electronics in Agriculture* 186: 106223.

Shao, Y., Y. Wang, D. Zhu, et al. 2022. "Measuring Heavy Metal Ions in Water Using Nature Existed Microalgae as Medium Based on Terahertz Technology." *Journal of Hazardous Materials* 435: 129028.

Shchepetilnikov, A.V., P.A. Gusikhin, V.M. Muravev, et al. 2020. "New Ultra-Fast Sub-Terahertz Linear Scanner for Postal Security Screening." *Journal of Infrared, Millimeter, and Terahertz Waves* 41 (6): 655–64.

Shen, Y. 2011. "Terahertz Pulsed Spectroscopy and Imaging for Pharmaceutical Applications: A Review." *International Journal of Pharmaceutics* 417 (1–2): 48–60.

Shen, Y., B. Li, G. Li, et al. 2022b. "Rapid Identification of Producing Area of Wheat Using Terahertz Spectroscopy Combined with Chemometrics." *Spectrochimica Acta Part A: Molecular and Biomolecular Spectroscopy* 269: 120694.

Shen, Y., X. Qiao, Z. Song, S. Zhong, and D. Wei. 2022a. "Terahertz Spectroscopy of Citrate Salts: Effects of Crystalline State and Crystallization Water." *Spectrochimica Acta Part A: Molecular and Biomolecular Spectroscopy* 277: 121288.

Shen, Y., C. Zhao, B. Li, G. Li, Y. Yin, and B. Pang. 2021. "Determination of Wheat Moisture Using Terahertz Spectroscopy Combined with the Tabu Search Algorithm." *Analytical Methods* 13 (36): 4120–30.

Shepp, L.A., and B.F. Logan. 1974. "The Fourier Reconstruction of a Head Section." *IEEE Transactions on Nuclear Science* 21 (3): 21–43.

Shepp, L.A., and Y. Vardi. 1982. "Maximum Likelihood Reconstruction for Emission Tomography." *IEEE Transactions on Medical Imaging* 1 (2): 113–22.

Shi, Q., K. Tian, H. Zhu, et al. 2020. "Flexible and Giant Terahertz Modulation Based on Ultra-Strain-Sensitive Conductive Polymer Composites." *ACS Applied Materials & Interfaces* 12 (8): 9790–96.

Shin, H.J., S. Choi, and G. Ok. 2018a. "Qualitative Identification of Food Materials by Complex Refractive Index Mapping in the Terahertz Range." *Food Chemistry* 245: 282–88.

Shin, H.J., S. Kim, K. Park, M. Lim, S. Choi, and G. Ok. 2017. "Free-Standing Guided-Mode Resonance Humidity Sensor in Terahertz." *Sensors and Actuators A: Physical* 268: 27–31.

Shin, H.J., S.J. Oh, M. Lim, S. Choi, and G. Ok. 2018b. "Dielectric Traces of Food Materials in the Terahertz Region." *Infrared Physics & Technology* 92: 128–33.

Shiraga, K., Y. Ogawa, N. Kondo, A. Irisawa, and M. Imamura. 2013. "Evaluation of the Hydration State of Saccharides Using Terahertz Time-Domain Attenuated Total Reflection Spectroscopy." *Food Chemistry* 140 (1–2): 315–20.

Shumyatsky, P., and R.R. Alfano. 2011. "Terahertz Sources." *Journal of Biomedical Optics* 16 (3): 33001.

Sibik, J., and J.A. Zeitler. 2016. "Direct Measurement of Molecular Mobility and Crystallisation of Amorphous Pharmaceuticals Using Terahertz Spectroscopy." *Advanced Drug Delivery Reviews* 100: 147–57.

Siegel, P.H. 2002. "Terahertz Technology." *IEEE Transactions on Microwave Theory and Techniques* 50 (3): 910–28.

Singh, A., M. Andrello, N. Thawdar, and J.M. Jornet. 2020b. "Design and Operation of a Graphene-Based Plasmonic Nano-Antenna Array for Communication in the Terahertz Band." *IEEE Journal on Selected Areas in Communications* 38 (9): 2104–17.

Singh, A., J. Li, A. Pashkin, et al. 2021. "High-Field THz Pulses from a GaAs Photoconductive Emitter for Non-Linear THz Studies." *Optics Express* 29 (13): 19920–27.

Singh, A.K., A.V. Perez-Lopez, J. Simpson, and E. Castro-Camus. 2020a. "Three-Dimensional Water Mapping of Succulent *Agave victoriae-reginae* Leaves by Terahertz Imaging." *Scientific Reports* 10 (1): 1–9.

Singh, C.B., and D.S. Jayas. 2011. "Spectroscopic Techniques for Fungi and Mycotoxins Detection." In *Determining Mycotoxins and Mycotoxigenic Fungi in Food and Feed*, 401–14. Cambridge: Elsevier.

Singh, P.K., R.P. Singh, P. Singh, and R.L. Singh. 2019. "Food Hazards: Physical, Chemical, and Biological." In *Food Safety and Human Health*, edited by R.L. Singh and S. Mondal, 15–65. Academic Press. https://doi.org/10.1016/C2017-0-04079-X.

Singhal, S., and M. Jena. 2013. "A Study on WEKA Tool for Data Preprocessing, Classification and Clustering." *International Journal of Innovative Technology and Exploring Engineering (IJItee)* 2 (6): 250–53.

Sinyukov, A.M., Z. Liu, Y.L. Hor, et al. 2008. "Rapid-Phase Modulation of Terahertz Radiation for High-Speed Terahertz Imaging and Spectroscopy." *Optics Letters* 33 (14): 1593–95.

Sirtori, C. 2021. "Terahertz Race Heats Up." *Nature Photonics* 15 (1): 1–2.

Sitnikov, D.S., S.A. Romashevskiy, A.A. Pronkin, and I.V. Ilina. 2019. "Open-Path Gas Detection Using Terahertz Time-Domain Spectroscopy." *Journal of Physics: Conference Series* 1147: 12061.

Smith, R.M. 2012. "*Terahertz Frequency Analysis of Gaseous and Solid Samples Using Terahertz Time-Domain Spectroscopy*." Iowa city, IA: The University of Iowa, Proquest Dissertations Publishing.

Smith, R.M., and M.A. Arnold. 2011. "Terahertz Time-Domain Spectroscopy of Solid Samples: Principles, Applications, and Challenges." *Applied Spectroscopy Reviews* 46 (8): 636–79.

Smye, S.W., J.M. Chamberlain, A.J. Fitzgerald, and E. Berry. 2001. "The Interaction between Terahertz Radiation and Biological Tissue." *Physics in Medicine & Biology* 46 (9): R101.

Sominskii, G., V. Sezonov, T. Tumareva, and E. Taradaev. 2018. "Perspective Field Emitters for Electron-Beam Microwave Devices of Short-Wave Millimeter and Submillimeter Range." In *2018 43rd International Conference on Infrared, Millimeter, and Terahertz Waves (IRMMW-THz)*, 1–2.

Son, J.H. 2013. "Terahertz Bio-Sensing Techniques." In *Handbook of Terahertz Technology for Imaging, Sensing and Communications*, 217–30. Cambridge: Elsevier.

Song, Z., S. Yan, Z. Zang, et al. 2018. "Temporal and Spatial Variability of Water Status in Plant Leaves by Terahertz Imaging." *IEEE Transactions on Terahertz Science and Technology* 8 (5): 520–27.

Srivastava, V. 2015. "Vacuum Microelectronic Devices for THz Communication Systems." In *2015 Annual IEEE India Conference (INDICON)*, 1–5.

Stanley, R. 1983. "Deans: The Radon Transform and Some of Its Applications."

Stantchev, R.I., K. Li, and E. Pickwell-MacPherson. 2021. "Rapid Imaging of Pulsed Terahertz Radiation with Spatial Light Modulators and Neural Networks." *ACS Photonics* 8 (11): 3150–55.

Stantchev, R.I., B. Sun, S.M. Hornett, et al. 2016. "Noninvasive, Near-Field Terahertz Imaging of Hidden Objects Using a Single-Pixel Detector." *Science Advances* 2 (6): e1600190.

Stefanuto, P., A. Smolinska, and J. Focant. 2021. "Advanced Chemometric and Data Handling Tools for GC × GC-TOF-MS: Application of Chemometrics and Related Advanced Data Handling in Chemical Separations." *TrAC Trends in Analytical Chemistry* 139: 116251.

Stoik, C.D., M.J. Bohn, and J.L. Blackshire. 2008. "Nondestructive Evaluation of Aircraft Composites Using Transmissive Terahertz Time Domain Spectroscopy." *Optics Express* 16 (21): 17039–51.

Strachan, C.J., P.F. Taday, D.A. Newnham, et al. 2005. "Using Terahertz Pulsed Spectroscopy to Quantify Pharmaceutical Polymorphism and Crystallinity." *Journal of Pharmaceutical Sciences* 94 (4): 837–46.

Stranzinger, S., E. Faulhammer, J. Li, et al. 2019. "Measuring Bulk Density Variations in a Moving Powder Bed via Terahertz In-Line Sensing." *Powder Technology* 344: 152–60.

Stubling, E., A. Rehn, T. Siebrecht, et al. 2019. "Application of a Robotic THz Imaging System for Sub-Surface Analysis of Ancient Human Remains." *Scientific Reports* 9 (1): 1–8.

Su, R., H. Zheng, S. Dong, et al. 2019. "Facile Detection of Melamine by a FAM–Aptamer–G-Quadruplex Construct." *Analytical and Bioanalytical Chemistry* 411 (12): 2521–30.

Su, W., H. He, and D. Sun. 2017. "Non-Destructive and Rapid Evaluation of Staple Foods Quality by Using Spectroscopic Techniques: A Review." *Critical Reviews in Food Science and Nutrition* 57 (5): 1039–51.

Suhandy, D., M. Yulia, Y. Ogawa, and N. Kondo. 2011. "Prediction of Vitamin C Using FTIR-ATR Terahertz Spectroscopy Combined with Interval Partial Least Squares (IPLS) Regression." In *2011 IEEE/SICE International Symposium on System Integration (SII)*, 202–06.

Sulovska, K., and M. Lehocky. 2014. "Characterization of Plasma Treated Surfaces for Food Safety by Terahertz Spectroscopy." In *Millimetre Wave and Terahertz Sensors and Technology VII*, 9252:32–38.

Suman, M., P.D. Sangma, D.R. Meghawal, and O.P. Sahu. 2017. "Effect of Plant Growth Regulators on Fruit Crops." *Journal of Pharmacognosy and Phytochemistry* 6 (2): 331–37.

Sun, H., J. Liu, C. Zhou, et al. 2021a. "Enhanced Transmission from Visible to Terahertz in ZnTe Crystals with Scalable Subwavelength Structures." *ACS Applied Materials & Interfaces* 13 (14): 16997–7005.

Sun, L., L. Zhao, and R. Peng. 2021b. "Research Progress in the Effects of Terahertz Waves on Biomacromolecules." *Military Medical Research* 8 (1): 1–8.

Sun, P. and Y. Zou. 2016. "Complex Dielectric Properties of Anhydrous Polycrystalline Glucose in the Terahertz Region." *Optical and Quantum Electronics* 48, 1–10.

Sun, Q., Y. He, K. Liu, S. Fan, E.P.J. Parrott, and E. Pickwell-MacPherson. 2017. "Recent Advances in Terahertz Technology for Biomedical Applications." *Quantitative Imaging in Medicine and Surgery* 7 (3): 345.

Sun, X., D. Cui, Y. Shen, W. Li, and J. Wang. 2022. "Non-Destructive Detection for Foreign Bodies of Tea Stalks in Finished Tea Products Using Terahertz Spectroscopy and Imaging." *Infrared Physics & Technology* 121: 104018.

Sun, X., and J. Liu. 2020a. "Analysis of Response of THz Spectroscopy to Insect Foreign Bodies Hidden in Tea Products." In *Infrared, Millimeter-Wave, and Terahertz Technologies VII*, 11559:115590Z.

Sun, X., and J. Liu. 2020b. "Measurement of Plumpness for Intact Sunflower Seed Using Terahertz Transmittance Imaging." *Journal of Infrared, Millimeter, and Terahertz Waves* 41 (3): 307–21.

Sun, X., K. Zhu, J. Liu, et al. 2019. "Terahertz Spectroscopy Determination of Benzoic Acid Additive in Wheat Flour by Machine Learning." *Journal of Infrared, Millimeter, and Terahertz Waves* 40 (4): 466–75.

Sun, Y., J. Huang, L. Shan, S. Fan, Z. Zhu, and X. Liu. 2021c. "Quantitative Analysis of Bisphenol Analogue Mixtures by Terahertz Spectroscopy Using Machine Learning Method." *Food Chemistry* 352: 129313.

Suryanarayana, T.M.V., and P.B. Mistry. 2016. *Principal Component Regression for Crop Yield Estimation.* Singapore: Springer.

Sypek, M., and J. Starobrat. 2021. "Non-Destructive Testing THz Systems: Fast Postal Scanner Case Study." In *Terahertz (THz), Mid Infrared (MIR) and Near Infrared (NIR) Technologies for Protection of Critical Infrastructures against Explosives and CBRN*, 89–100. Dordrecht: Springer. https://doi.org/10.1007/978-94-024-2082-1_7.

Tahhan, S.R., and H.K. Aljobouri. 2020. "Sensing of Illegal Drugs by Using Photonic Crystal Fiber in Terahertz Regime." *Journal of Optical Communications*. https://doi.org/10.1515/joc-2019-0291.

Tahir, M.A., N.E. Dina, H. Cheng, V.K. Valev, and L. Zhang. 2021. "Surface-Enhanced Raman Spectroscopy for Bioanalysis and Diagnosis." *Nanoscale* 13 (27): 11593–634.

Taiber, J., M. Kahlmeyer, A. Winkel, et al. 2021. "Ageing Condition Determination of Bonded Joints by Terahertz Spectroscopy." In *Industrial Applications of Adhesives*, 127–38. Singapore: Springer.

Tajima, T., M. Nakamura, K. Shiraga, Y. Ogawa, K. Ajito, and H. Koizumi. 2016. "Double-Beam CW THz System with Photonic Phase Modulator for Sub-THz Glucose Hydration Sensing." In *2016 IEEE MTT-S International Microwave Symposium (IMS)*, 1–4.

Takai, M., K. Shibata, M. Uemoto, and S. Watanabe. 2016. "Spatial Polarization Variation in Terahertz Electromagnetic Wave Focused by Off-Axis Parabolic Mirror." *Applied Physics Express* 9 (5): 52206.

Tanaka, K., K. Harada, and K.M.T. Yamada. 2011. "THz and Submillimeter-Wave Spectroscopy of Molecular Complexes." In *Handbook of High-Resolution Spectroscopy*, edited by M. Quack and F. Merkt. Hoboken, NJ: John Wiley & Sons. https://doi.org/10.1002/9780470749593.hrs029.

Tang, E.K., P.N. Suganthan, X. Yao, and A.K. Qin. 2005. "Linear Dimensionality Reduction Using Relevance Weighted LDA." *Pattern Recognition* 38 (4): 485–93.

Tanoto, H., J.H. Teng, Q.Y. Wu, et al. 2013. "Nano-antenna in a photoconductive photomixer for highly efficient continuous wave terahertz emission." *Scientific Reports* 3(1): 1–6.

Tao, C. 2016. "Terahertz Spectra Identification of Biomolecules Based on PCA and Fuzzy Recognition." *Chinese Journal of Quantum Electronics* 33 (4): 392.

Taraskin, S.N., S.I. Simdyankin, S.R. Elliott, J.R. Neilson, and T. Lo. 2006. "Universal Features of Terahertz Absorption in Disordered Materials." *Physical Review Letters* 97 (5): 55504.

TeraSense. 2022. "Terahertz Inspection of Agricultural Products." TeraSense – Terahertz Imaging, Terasense Group Inc. 2022, https://terasense.com/applications/terahertz-agriculture/.

Toft, P. 1996. "The Radon Transform." *Theory and Implementation (Ph. D. Dissertation) (Copenhagen: Technical University of Denmark)*.

Toh, R., K. Suizu, and Y. Tojima. 2017. "Terahertz Pulse Reflection Imaging Using the Time-Domain Correlating Synthesis Method." *IEEE Transactions on Terahertz Science and Technology* 7 (4): 385–92.

Tong, Y., S. Wang, K. Han, et al. 2021. "Development of a Novel Metal Grating and Its Applications of Terahertz Spectroscopic Detection of $CuSO_4$ in Fruit." *Food Analytical Methods* 14 (8): 1590–99.

Tonouchi, M. 2005. "New Frontier of Terahertz Technology." In *2005 18th International Conference on Applied Electromagnetics and Communications*, 1–4.

Tonouchi, M. 2007. "Cutting-Edge Terahertz Technology." *Nature Photonics* 1 (2): 97–105.

Tonouchi, M. 2019. "Laser Terahertz Emission Microscope for Real World Application (Conference Presentation)." In *Terahertz Emitters, Receivers, and Applications X*, 11124:111240T.

Tonouchi, M. 2020. "Simplified Formulas for the Generation of Terahertz Waves from Semiconductor Surfaces Excited with a Femtosecond Laser." *Journal of Applied Physics* 127 (24): 245703.

Toptica. 2022. "Toptica Terahertz Systems." Toptica Photonics AG, Germany. 2022. Accessed 14th May 2022, https://www.toptica.com/contact-us.

Tros, M., L. Zheng, J. Hunger, et al. 2017. "Picosecond Orientational Dynamics of Water in Living Cells." *Nature Communications* 8 (1): 1–7.

Tu, W., S. Zhong, M. Luo, and Q. Zhang. 2021. "Non-Destructive Evaluation of Hidden Defects beneath the Multilayer Organic Protective Coatings Based on Terahertz Technology." *Frontiers in Physics* 9: 304.

Tzydynzhapov, G., P. Gusikhin, V. Muravev, A. Dremin, Y. Nefyodov, and I. Kukushkin. 2020. "New Real-Time Sub-Terahertz Security Body Scanner." *Journal of Infrared, Millimeter, and Terahertz Waves* 41: 632–41.

Ueno, Y., K. Ajito, N. Kukutsu, and E. Tamechika. 2011. "Quantitative Analysis of Amino Acids in Dietary Supplements Using Terahertz Time-Domain Spectroscopy." *Analytical Sciences* 27 (4): 351.

Ung, B.S.Y., B.M. Fischer, B.H.W. Ng, and D. Abbott. 2009. "Comparative Investigation of Detection of Melamine in Food Powders." In *2009 34th International Conference on Infrared, Millimeter, and Terahertz Waves*, 1–2.

Urbanczyk, M., J. Gora, and N.L.R. Sewald. 2017. "Antifreeze Glycopeptides: From Structure and Activity Studies to Current Approaches in Chemical Synthesis." *Amino Acids* 49 (2): 209–22.

Vadivambal, R., and D.S. Jayas. 2016. *Bio-Imaging: Principles, Techniques, and Applications*. Boca Raton, FL: CRC Press, Taylor and Francis Group, LLC. New York.

Vaks, V. 2012. "High-Precise Spectrometry of the Terahertz Frequency Range: The Methods, Approaches and Applications." *Journal of Infrared, Millimeter, and Terahertz Waves* 33 (1): 43–53.

Vaks, V., E. Domracheva, E. Sobakinskaya, and M. Chernyaeva. 2014b. "High-Precision Terahertz Spectroscopy for Noninvasive Medicine Diagnostics." *Photonics and Lasers in Medicine* 3 (4): 373–80.

Vaks, V.L., E.G. Domracheva, S.I. Pripolzin, et al. 2014a. "Methods and Instruments of High-Resolution Transient THz Spectroscopy for Diagnostics of Socially Important Diseases." *Physics of Wave Phenomena* 22 (3): 177–84.

Valk, N.C.J.D., W.A.M.V.D. Marel, and P.C.M. Planken. 2005. "Terahertz Polarization Imaging." *Optics Letters* 30 (20): 2802–04.

Vandewal, M., E. Cristofani, A. Brook, et al. 2013. "Structural Health Monitoring Using a Scanning THz System." In *2013 38th International Conference on Infrared, Millimeter, and Terahertz Waves (IRMMW-THz)*, 1–2.

Vithu, P., and J.A. Moses. 2016. "Machine Vision System for Food Grain Quality Evaluation: A Review." *Trends in Food Science & Technology* 56: 13–20.

Vitiello, M.S., L. Consolino, M. Inguscio, and P.D. Natale. 2021. "Toward New Frontiers for Terahertz Quantum Cascade Laser Frequency Combs." *Nanophotonics* 10 (1): 187–94.

Vitiello, M.S., G. Scamarcio, and V. Spagnolo. 2006. "Electronic and Thermal Properties of THz Quantum Cascade." In *2006 Joint 31st International Conference on Infrared Millimeter Waves and 14th International Conference on Teraherz Electronics*, 558.

Vynckier, A.K., H. Lin, J.A. Zeitler, et al. 2015. "Calendering as a Direct Shaping Tool for the Continuous Production of Fixed-Dose Combination Products via Co-Extrusion." *European Journal of Pharmaceutics and Biopharmaceutics* 96: 125–31.

Wagner, C., C. Peveling-Oberhag, C. Heinen, et al. 2011. "*In Vivo* Chlorophyll Monitoring of Biological Samples with THz-Time-Domain-Spectroscopy." In *2011 International Conference on Infrared, Millimeter, and Terahertz Waves*, 1–2.

Wahaia, F., G. Valusis, L.M. Bernardo, et al. 2011. "Detection of Colon Cancer by Terahertz Techniques." *Journal of Molecular Structure* 1006 (1–3): 77–82.

Wan, M., J.J. Healy, and J.T. Sheridan. 2020. "Terahertz Phase Imaging and Biomedical Applications." *Optics and Laser Technology* 122: 105859.

Wang, C., J. Gong, Q. Xing, et al. 2010a. "Application of Terahertz Time-Domain Spectroscopy in Intracellular Metabolite Detection." *Journal of Biophotonics* 3 (10–11): 641–45.

Wang, C., J. Qin, W. Xu, M. Chen, L. Xie, and Y. Ying. 2018a. "Terahertz Imaging Applications in Agriculture and Food Engineering: A Review." *Transactions of the ASABE* 61 (2): 411–24.

Wang, C., R. Zhou, Y. Huang, L. Xie, and Y. Ying. 2019a. "Terahertz Spectroscopic Imaging with Discriminant Analysis for Detecting Foreign Materials among Sausages." *Food Control* 97: 100–04.

Wang, H., P. Chen, J. Dai, et al. 2022a. "Recent Advances of Chemometric Calibration Methods in Modern Spectroscopy: Algorithms, Strategy, and Related Issues." *TrAC Trends in Analytical Chemistry* 153: 116648.

Wang, J., J. Gou, Z. Wu, and Y. Jiang. 2016. "Design and Imaging Application of Room-Temperature Terahertz Detector with Micro-Bolometer Focal Plane Array." *Journal of Electronic Science and Technology* 14 (2): 98–102.

Wang, J., T. Xu, L. Zhang, et al. 2022b. "Nondestructive Damage Evaluation of Composites Based on Terahertz and X-Ray Image Fusion." *NDT & E International*, 127: 102616.

Wang, J., B. Zhong, and J.L. Zhou. 2017a. "Quality-Relevant Fault Monitoring Based on Locality-Preserving Partial Least-Squares Statistical Models." *Industrial & Engineering Chemistry Research* 56 (24): 7009–20.

Wang, K., H. Pu, and D. Sun. 2018b. "Emerging Spectroscopic and Spectral Imaging Techniques for the Rapid Detection of Microorganisms: An Overview." *Comprehensive Reviews in Food Science and Food Safety* 17 (2): 256–73.

Wang, K., D. Sun, and H. Pu. 2017b. "Emerging Non-Destructive Terahertz Spectroscopic Imaging Technique: Principle and Applications in the Agri-Food Industry." *Trends in Food Science & Technology* 67: 93–105.

Wang, Q., S. Hameed, L. Xie, and Y. Ying. 2020a. "Non-Destructive Quality Control Detection of Endogenous Contaminations in Walnuts Using Terahertz Spectroscopic Imaging." *Journal of Food Measurement and Characterization* 14: 2453–60.

Wang, Q., X. Li, T. Chang, et al. 2019b. "Nondestructive Imaging of Hidden Defects in Aircraft Sandwich Composites Using Terahertz Time-Domain Spectroscopy." *Infrared Physics and Technology* 97: 326–40.

Wang, Q., Y. Shan, X. Shi, et al. 2022d. "High-Sensitivity Detection of Trace Imidacloprid and Tetracycline Hydrochloride by Multi-Frequency Resonance Metamaterials." *Journal of Food Measurement and Characterization*, 16: 2041–48.

Wang, Q., L. Xie, and Y. Ying. 2022c. "Overview of Imaging Methods Based on Terahertz Time-Domain Spectroscopy." *Applied Spectroscopy Reviews* 57 (3): 249–64.

Wang, R., L. Xie, S. Hameed, C. Wang, and Y. Ying. 2018c. "Mechanisms and Applications of Carbon Nanotubes in Terahertz Devices: A Review." *Carbon* 132: 42–58.

Wang, S., and X.C. Zhang. 2004. "Pulsed Terahertz Tomography." *Journal of Physics D: Applied Physics* 37 (4): R1.

Wang, X., Y. Cui, D. Hu, W. Sun, J. Ye, and Y. Zhang. 2009. "Terahertz Quasi-Near-Field Real-Time Imaging." *Optics Communications* 282 (24): 4683–87.

Wang, X., Y. Cui, W. Sun, J. Ye, and Y. Zhang. 2010b. "Terahertz Real-Time Imaging With Balanced Electro-Optic Detection." *Optics Communications* 283 (23): 4626–32.

Wang, Y., Y. Wang, D. Xu, et al. 2020b. "Interference Elimination Based on the Inversion Method for Continuous-Wave Terahertz Reflection Imaging." *Optics Express* 28 (15): 21926–39.

Wang, Y., Z. Zhao, J. Qin, H. Liu, A. Liu, and M. Xu. 2020c. "Rapid in Situ Analysis of L-Histidine and α-Lactose in Dietary Supplements by Fingerprint Peaks Using Terahertz Frequency-Domain Spectroscopy." *Talanta* 208: 120469.

Wei, X., S. Li, S. Zhu, et al. 2021. "Quantitative Analysis of Soybean Protein Content by Terahertz Spectroscopy and Chemometrics." *Chemometrics and Intelligent Laboratory Systems* 208: 104199.

Wei, X., W. Zheng, S. Zhu, S. Zhou, W. Wu, and Z. Xie. 2020. "Application of Terahertz Spectrum and Interval Partial Least Squares Method in the Identification of Genetically Modified Soybeans." *Spectrochimica Acta Part A: Molecular and Biomolecular Spectroscopy* 238: 118453.

Weijs, P.J.M., W.G.P.M. Looijaard, I.M. Dekker, et al. 2020. "Imaging." In *Post-Intensive Care Syndrome*, edited by Jean-Charles Preiser, Margaret Herridge, and Elie Azoulay, 109–24. Cham: Springer International Publishing. https://doi.org/10.1007/978-3-030-24250-3_8.

Wessel, J., K. Schmalz, B.P. Cahill, G. Gastrock, and C. Meliani. 2013. "Contactless Characterization of Yeast Cell Cultivation at 7 GHz and 240 GHz." In *2013 IEEE Topical Conference on Biomedical Wireless Technologies, Networks, and Sensing Systems*, 70–72.

Wiegand, C., M. Herrmann, S. Bachtler, et al. 2010. "A Pulsed THz Imaging System with a Line Focus and a Balanced 1-D Detection Scheme with Two Industrial CCD Line – Scan Cameras." *Optics Express* 18 (6): 5595–601.

Wietzke, S., C. Jansen, C. Jordens, et al. 2009a. "Industrial Applications of THz Systems." In *International Symposium on Photoelectronic Detection and Imaging 2009: Terahertz and High Energy Radiation Detection Technologies and Applications*, 7385:738506.

Wietzke, S., C. Jansen, T. Jung, et al. 2009b. "Terahertz Time-Domain Spectroscopy as a Tool to Monitor the Glass Transition in Polymers." *Optics Express* 17 (21): 19006–14.

Wietzke, S., C. Jansen, N. Krumbholz, et al. 2010. "Terahertz Spectroscopy: A Powerful Tool for the Characterization of Plastic Materials." In *2010 10th IEEE International Conference on Solid Dielectrics*, 1–4.

Wietzke, S., C. Jansen, M. Reuter, et al. 2011. "Terahertz Spectroscopy on Polymers: A Review of Morphological Studies." *Journal of Molecular Structure* 1006 (1–3): 41–51.

Wietzke, S., C. Jansen, F. Rutz, D.M. Mittleman, and M. Koch. 2007a. "Determination of Additive Content in Polymeric Compounds with Terahertz Time-Domain Spectroscopy." *Polymer Testing* 26 (5): 614–18.

Wietzke, S., C. Jordens, N. Krumbholz, B. Baudrit, M. Bastian, and M. Koch. 2007b. "Terahertz Imaging: A New Non-Destructive Technique for the Quality Control of Plastic Weld Joints." *Journal of the European Optical Society-Rapid Publications*. Vol. 2, Europe. http://dx.doi.org/10.2971/jeos.2007.07013.

Wilke, I., and S. Sengupta. 2017. "Nonlinear Optical Techniques for Terahertz Pulse Generation and Detection – Optical Rectification and Electrooptic Sampling." In *Terahertz Spectroscopy*, 59–90. New York, NY: CRC Press.

Williams, G.P. 2002. "Far-IR/THz Radiation from the Jefferson Laboratory, Energy Recovered Linac, Free Electron Laser." *Review of Scientific Instruments* 73 (3): 1461–63.

Wilmink, G.J., and J.E. Grundt. 2011. "Invited Review Article: Current State of Research on Biological Effects of Terahertz Radiation." *Journal of Infrared, Millimeter, and Terahertz Waves* 32 (10): 1074–122.

Wu, Q., T.D. Hewitt, and X.C. Zhang. 1996. "Two-Dimensional Electro-Optic Imaging of THz Beams." *Applied Physics Letters* 69 (8): 1026–28.

Wu, Y., and Y. Zhang. 2013. "Analytical Chemistry, Toxicology, Epidemiology and Health Impact Assessment of Melamine in Infant Formula: Recent Progress and Developments." *Food and Chemical Toxicology* 56: 325–35.

Xia, Y., W. Liu, Y. Shi, S. Younas, C. Liu, and L. Zheng. 2022. "Rapid Determination of Capsaicin Concentration in Soybean Oil by Terahertz Spectroscopy." *Journal of Food Science* 87: 567–75.

Xi-Ai, C., Z. Guang-Xin, H. Ping-Jie, H. Di-Bo, K. Xu-Sheng, and Z. Ze-Kui. 2011. "Classification of the Green Tea Varieties Based on Support Vector Machines Using Terahertz Spectroscopy." In *2011 IEEE International Instrumentation and Measurement Technology Conference*, 1–5.

Xiao-li, Z., and L. Jiu-sheng. 2011. "Diagnostic Techniques of Talc Powder in Flour Based on the THz Spectroscopy." *Journal of Physics-Conference Series* 276: 12234.

Xie, L., W. Gao, J. Shu, Y. Ying, and J. Kono. 2015. "Extraordinary Sensitivity Enhancement by Metasurfaces in Terahertz Detection of Antibiotics." *Scientific Reports* 5 (1): 1–4.

Xie, L., C. Wang, M. Chen, et al. 2019. "Temperature-Dependent Terahertz Vibrational Spectra of Tetracycline and Its Degradation Products." *Spectrochimica Acta Part A: Molecular and Biomolecular Spectroscopy* 222: 117179.

Xie, L., Y. Yao, and Y. Ying. 2014. "The Application of Terahertz Spectroscopy to Protein Detection: A Review." *Applied Spectroscopy Reviews* 49 (6): 448–61.

Xu, J., H. Wang, Y. Duan, Y. He, S. Chen, and Z. Zhang. 2020. "Terahertz Imaging and Vibro-Thermography for Impact Response in Carbon Fiber Reinforced Plastics." *Infrared Physics & Technology* 109: 103413.

Xu, W., L. Xie, Z. Ye, et al. 2015. "Discrimination of Transgenic Rice Containing the Cry1Ab Protein Using Terahertz Spectroscopy and Chemometrics." *Scientific Reports* 5 (1): 1–9.

Xu, Y., H. Hao, D.S. Citrin, X. Wang, L. Zhang, and X. Chen. 2021. "Three-Dimensional Nondestructive Characterization of Delamination in GFRP by Terahertz Time-of-Flight Tomography with Sparse Bayesian Learning-Based Spectrum-Graph Integration Strategy." *Composites Part B: Engineering* 225: 109285.

Xu-dong, S., and L. Jun-bin. 2021. "THz Spectroscopy Detection of Insect Foreign Body Hidden in Tea Products." *Spectroscopy and Spectral Analysis* 41 (9): 2723.

Yada, H., M. Nagai, and K. Tanaka. 2009. "The Intermolecular Stretching Vibration Mode in Water Isotopes Investigated with Broadband Terahertz Time-Domain Spectroscopy." *Chemical Physics Letters* 473 (4–6): 279–83.

Yakovlev, E.V., K.I. Zaytsev, I.N. Dolganova, and S.O. Yurchenko. 2015. "Non-Destructive Evaluation of Polymer Composite Materials at the Manufacturing Stage Using Terahertz Pulsed Spectroscopy." *IEEE Transactions on Terahertz Science and Technology* 5 (5): 810–16.

Yamahara, K., A. Mannan, I. Kawayama, H. Nakanishi, and M. Tonouchi. 2020. "Ultrafast Spatiotemporal Photocarrier Dynamics Near GaN Surfaces Studied by Terahertz Emission Spectroscopy." *Scientific Reports* 10 (1): 1–10.

Yan, H., W. Fan, X. Chen, L. Liu, H. Wang, and X. Jiang. 2021. "Terahertz Signatures and Quantitative Analysis of Glucose Anhydrate and Monohydrate Mixture." *Spectrochimica Acta Part A: Molecular and Biomolecular Spectroscopy* 258: 119825.

Yan, L., C. Liu, H. Qu, et al. 2018. "Discrimination and Measurements of Three Flavonols with Similar Structure Using Terahertz Spectroscopy and Chemometrics." *Journal of Infrared, Millimeter, and Terahertz Waves* 39 (5): 492–504.

Yan, S., D. Wei, M. Tang, et al. 2016. "Determination of Critical Micelle Concentrations of Surfactants by Terahertz Time-Domain Spectroscopy." *IEEE Transactions on Terahertz Science and Technology* 6 (4): 532–40.

Yan, Z., and W. Shi. 2022. "Detection of Aging in the Common Explosive RDX Using Terahertz Time-Domain Spectroscopy." *Journal of the Optical Society of America B* 39 (3): A9–12.

Yan, Z., Y. Ying, H. Zhang, and H. Yu. 2006. "Research Progress of Terahertz Wave Technology in Food Inspection." In *Terahertz Physics, Devices, and Systems*, 6373:63730R.

Yan, Z., L. Zhu, K. Meng, W. Huang, and Q. Shi. 2022. "THz Medical Imaging: From in Vitro to in Vivo." *Trends in Biotechnology* 40: 816–30.

Yang, J., H. Qin, and K. Zhang. 2018. "Emerging Terahertz Photodetectors Based on Two-Dimensional Materials." *Optics Communications* 406: 36–43.

Yang, R., Y. Li, B. Qin, D. Zhao, Y. Gan, and J. Zheng. 2022. "Pesticide Detection Combining the Wasserstein Generative Adversarial Network and the Residual Neural Network Based on Terahertz Spectroscopy." *RSC Advances* 12 (3): 1769–76.

Yang, S., C. Li, Y. Mei, et al. 2021a. "Determination of the Geographical Origin of Coffee Beans Using Terahertz Spectroscopy Combined with Machine Learning Methods." *Frontiers in Nutrition* 8: 313.

Yang, S., C. Li, Y. Mei, et al. 2021b. "Discrimination of Corn Variety Using Terahertz Spectroscopy Combined with Chemometrics Methods." *Spectrochimica Acta Part A: Molecular and Biomolecular Spectroscopy* 252: 119475.

Yang, S.H., R. Watts, X. Li, et al. 2015. "Tunable Terahertz Wave Generation through a Bimodal Laser Diode and Plasmonic Photomixer." *Optics Express* 23 (24): 31206–15.

Yang, X., J. Shi, K. Yang, et al. 2017. "Label-Free and Reagentless Bacterial Detection and Assessment by Continous-Wave Terahertz Imaging." In *2017 42nd International Conference on Infrared, Millimeter, and Terahertz Waves (IRMMW-THz)*, 1–1. https://10.1109/IRMMW-THz.2017.8067149.

Yang, X., D. Wei, S. Yan, et al. 2016a. "Rapid and Label-Free Detection and Assessment of Bacteria by Terahertz Time-Domain Spectroscopy." *Journal of Biophotonics* 9 (10): 1050–58.

Yang, X., K. Yang, Y. Luo, and W. Fu. 2016b. "Terahertz Spectroscopy for Bacterial Detection: Opportunities and Challenges." *Applied Microbiology and Biotechnology* 100 (12): 5289–99.

Yassin, S., D.J. Goodwin, A. Anderson, et al. 2015. "The Disintegration Process in Microcrystalline Cellulose Based Tablets, Part 1: Influence of Temperature, Porosity and Superdisintegrants." *Journal of Pharmaceutical Sciences* 104 (10): 3440–50.

Yasui, T., and T. Araki. 2005. "Sensitive Measurement of Water Content in Dry Material Based on Low-Frequency Terahertz Time-Domain Spectroscopy." In *ICO20: Optical Devices and Instruments*, 6024:60240A.

Yi, C., S. Tuo, L. Zhang, and H. Xiao. 2022. "Improved Kernel Entropy Composition Analysis Method for Transgenic Cotton Seeds Recognition Based on Terahertz Spectroscopy." *Chemometrics and Intelligent Laboratory Systems* 225: 104575.

Yin, J., S. Hameed, L. Xie, and Y. Ying. 2021. "Non-Destructive Detection of Foreign Contaminants in Toast Bread with Near Infrared Spectroscopy and Computer Vision Techniques." *Journal of Food Measurement and Characterization* 15 (1): 189–98.

Yin, M., S. Tang, and M. Tong. 2016. "Identification of Edible Oils Using Terahertz Spectroscopy Combined with Genetic Algorithm and Partial Least Squares Discriminant Analysis." *Analytical Methods* 8 (13): 2794–98.

Yin, X., S. Hadjiloucas, and Y. Zhang. 2017. *Pattern Classification of Medical Images: Computer Aided Diagnosis*. Cham: Springer.

Yoneyama, H., M. Yamashita, S. Kasai, H. Ito, and T. Ouchi. 2007. "Application of Terahertz Spectrum in the Detection of Harmful Food Additive." In *2007 Joint 32nd International Conference on Infrared and Millimeter Waves and the 15th International Conference on Terahertz Electronics*, 281–82.

You, B., C. Chen, C. Yu, P. Wang, and J. Lu. 2018. "Frequency-Dependent Skin Penetration Depth of Terahertz Radiation Determined by Water Sorption-Desorption." *Optics Express* 26 (18): 22709–21.

Yu, B., A. Alimova, A. Katz, and R.R. Alfano. 2005. "THz Absorption Spectrum of Bacillus subtilis Spores." *Proceedings SPIE, Terahertz and Gigahertz Electronics and Photonics IV*, 5727: 20–23. https://doi.org/10.1117/12.590951

Yu, J. 2021. "Generation and Detection of Terahertz Signal." In *Broadband Terahertz Communication Technologies*, 25–45. Singapore: Springer Singapore. https://doi.org/10.1007/978-981-16-3160-3_2.

Yulia, M., D. Suhandy, Y. Ogawa, and N. Kondo. 2014. "Investigation on the Influence of Temperature in L-Ascorbic Acid Determination Using FTIR-ATR Terahertz Spectroscopy: Calibration Model with Temperature Compensation." *Engineering in Agriculture, Environment and Food* 7 (4): 148–54.

Zahid, A., K. Dashtipour, H.T. Abbas, et al. 2022. "Machine Learning Enabled Identification and Real-Time Prediction of Living Plant's Stress Using Terahertz Waves." *Defence Technology.* 18 (8): 1330–39. https://doi.org/10.1016/j.dt.2022.01.003.

Zang, Z., Z. Li, X. Lu, et al. 2021. "Terahertz Spectroscopy for Quantification of Free Water and Bound Water in Leaf." *Computers and Electronics in Agriculture* 191: 106515.

Zang, Z., J. Wang, H. Cui, and S. Yan. 2019. "Terahertz Spectral Imaging Based Quantitative Determination of Spatial Distribution of Plant Leaf Constituents." *Plant Methods* 15 (1): 1–11.

Zappia, S., L. Crocco, and I. Catapano. 2021. "THz Imaging for Food Inspections: A Technology Review and Future Trends." *Terahertz Technology.*

Zatta, R., R. Jain, J. Grzyb, and U.R. Pfeiffer. 2021. "Resolution Limits of Hyper-Hemispherical Silicon Lens-Integrated THz Cameras Employing Geometrical Multiframe Super-Resolution Imaging." *IEEE Transactions on Terahertz Science and Technology* 11 (3): 277–86.

Zeitler, J.A. 2016. "Pharmaceutical Terahertz Spectroscopy and Imaging." In *Analytical Techniques in the Pharmaceutical Sciences*, edited by A Mullertz, Y Perrie, and T Rades, 171–222. New York, NY: Springer. https://doi.org/10.1007/978-1-4939-4029-5_5.

Zeitler, J.A., and L.F. Gladden. 2009. "In-Vitro Tomography and Non-Destructive Imaging at Depth of Pharmaceutical Solid Dosage Forms." *European Journal of Pharmaceutics and Biopharmaceutics* 71 (1): 2–22.

Zeitler, J.A., Y. Shen, C. Baker, P.F. Taday, M. Pepper, and T. Rades. 2007a. "Analysis of Coating Structures and Interfaces in Solid Oral Dosage Forms by Three Dimensional Terahertz Pulsed Imaging." *Journal of Pharmaceutical Sciences* 96 (2): 330–40.

Zeitler, J.A., P.F. Taday, K.C. Gordon, M. Pepper, and T. Rades. 2007b. "Solid-State Transition Mechanism in Carbamazepine Polymorphs by Time-Resolved Terahertz Spectroscopy." *ChemPhysChem* 8 (13): 1924–27.

Zeitler, J.A., P.F. Taday, M. Pepper, and T. Rades. 2007c. "Relaxation and Crystallization of Amorphous Carbamazepine Studied by Terahertz Pulsed Spectroscopy." *Journal of Pharmaceutical Sciences* 96 (10): 2703–09.

Zeng, Y., B. Qiang, and Q.J. Wang. 2020. "Photonic Engineering Technology for the Development of Terahertz Quantum Cascade Lasers." *Advanced Optical Materials* 8 (3): 1900573.

Zhai, M., A. Locquet, and D.S. Citrin. 2020. "Pulsed THz Imaging for Thickness Characterization of Plastic Sheets." *NDT & E International* 116: 102338.

Zhan, H., J. Xi, K. Zhao, R. Bao, and L. Xiao. 2016. "A Spectral-Mathematical Strategy for the Identification of Edible and Swill-Cooked Dirty Oils Using Terahertz Spectroscopy." *Food Control* 67: 114–18.

Zhang, H., and Z. Li. 2018. "Terahertz Spectroscopy Applied to Quantitative Determination of Harmful Additives in Medicinal Herbs." *Optik* 156: 834–40.

Zhang, H., Z. Li, T. Chen, and B. Qin. 2017b. "Quantitative Determination of Auramine O by Terahertz Spectroscopy with 2DCOS-PLSR Model." *Spectrochimica Acta Part A: Molecular and Biomolecular Spectroscopy* 184: 335–41.

Zhang, H., S. Sfarra, K. Saluja, et al. 2017a. "Non-Destructive Investigation of Paintings on Canvas by Continuous Wave Terahertz Imaging and Flash Thermography." *Journal of Nondestructive Evaluation* 36 (2): 1–12.

Zhang, H., L. Zhang, and X. Yu. 2021a. "Terahertz Band: Lighting up Next-Generation Wireless Communications." *China Communications* 18 (5): 153–74.

Zhang, J., Y. Yang, X. Feng, H. Xu, J. Chen, and Y. He. 2020a. "Identification of Bacterial Blight Resistant Rice Seeds Using Terahertz Imaging and Hyperspectral Imaging Combined with Convolutional Neural Network." *Frontiers in Plant Science* 11: 821. https://doi.org/10.3389/fpls.2020.00821.

Zhang, L., M. Zhang, and A.S. Mujumdar. 2021b. "Terahertz Spectroscopy: A Powerful Technique for Food Drying Research." *Food Reviews International* 39: 1–18.

Zhang, M., Z. Guo, X. Mi, Z. Li, and Y. Liu. 2022a. "Ultrafast Imaging of Molecular Dynamics Using Ultrafast Low-Frequency Lasers, X-Ray Free Electron Lasers, and Electron Pulses." *The Journal of Physical Chemistry Letters* 13 (7): 1668–80.

Zhang, T., H. Huang, Z. Zhang, H. Gao, L. Gao, and Z. Zheng. 2021c. "Sensitive Characterizations of Polyvinyl Chloride Using Terahertz Time-Domain Spectroscopy." *Infrared Physics & Technology* 118: 103878.

Zhang, X., S. Lu, Y. Liao, and Z. Zhang. 2017d. "Simultaneous Determination of Amino Acid Mixtures in Cereal by Using Terahertz Time Domain Spectroscopy and Chemometrics." *Chemometrics and Intelligent Laboratory Systems* 164: 8–15.

Zhang, X., A. Shkurinov, and Y. Zhang. 2017c. "Extreme Terahertz Science." *Nature Photonics* 11 (1): 16–18.

Zhang, X., Q. Xu, L. Xia, et al. 2020b. "Terahertz Surface Plasmonic Waves: A Review." *Advanced Photonics* 2 (1): 14001.

Zhang, X.C., and J. Xu. 2010a. "THz Wave Near-Field Imaging." In *Introduction to THz Wave Photonics*, 149–74. Boston, MA: Springer US. https://doi.org/10.1007/978-1-4419-0978-7_7.

Zhang, X.C., and J. Xu. 2010b. "THz Wave 3D Imaging and Tomography." In *Introduction to THz Wave Photonics*, 127–48. Boston, MA: Springer US. https://doi.org/10.1007/978-1-4419-0978-7_6.

Zhang, Y., K. Li, and H. Zhao. 2021d. "Intense Terahertz Radiation: Generation and Application." *Frontiers of Optoelectronics* 14 (1): 4–36.

Zhang, Z., and T. Buma. 2009. "Improved THz Imaging with a Virtual-Source Based Synthetic Aperture Focusing Technique and Coherence Weighting." In *Conference on Lasers and Electro-Optics*, JWA16.

Zhang, Z., T. Zhang, F. Fan, Y. Ji, and S. Chang. 2022b. "Terahertz Polarization Sensing of Bovine Serum Albumin Proteolysis on Curved Flexible Metasurface." *Sensors and Actuators A: Physical* 338: 113499.

Zhao, R., B. Zou, G. Zhang, D. Xu, and Y. Yang. 2020. "High-Sensitivity Identification of Aflatoxin B1 and B2 Using Terahertz Time-Domain Spectroscopy and Metamaterial-Based Terahertz Biosensor." *Journal of Physics D: Applied Physics* 53 (19): 195401.

Zhao, Y., M. Gouda, G. Yu, et al. 2021. "Analyzing Cadmium-Phytochelatin2 Complexes in Plant Using Terahertz and Circular Dichroism Information." *Ecotoxicology and Environmental Safety* 225: 112800.

Zheng, Z., W. Fan, Y. Liang, and H. Yan. 2012. "Application of Terahertz Spectroscopy and Molecular Modeling in Isomers Investigation: Glucose and Fructose." *Optics Communications* 285 (7): 1868–71.

Zhong, H., J. Xu, X. Xie, et al. 2005. "Nondestructive Defect Identification with Terahertz Time-of-Flight Tomography." *IEEE Sensors Journal* 5 (2): 203–08.

Zhong, J., T. Mori, Y. Fujii, et al. 2020a. "Molecular Vibration and Boson Peak Analysis of Glucose Polymers and Ester via Terahertz Spectroscopy." *Carbohydrate Polymers* 232: 115789.

Zhong, L., L. Gao, L. Li, and H. Zang. 2020b. "Trends-Process Analytical Technology in Solid Oral Dosage Manufacturing." *European Journal of Pharmaceutics and Biopharmaceutics* 153: 187–99.

Zhong, S., and W. Nsengiyumva. 2022. "Terahertz Testing Technique for Fiber-Reinforced Composite Materials." In *Nondestructive Testing and Evaluation of Fiber-Reinforced Composite Structures*, 273–314. Singapore: Springer.

Zhong, S., Y. Shen, L. Ho, et al. 2011. "Non-Destructive Quantification of Pharmaceutical Tablet Coatings Using Terahertz Pulsed Imaging and Optical Coherence Tomography." *Optics and Lasers in Engineering* 49 (3): 361–65.

Zhong, Y., L. Du, Q. Liu, L. Zhu, and B. Zhang. 2020c. "Metasurface-Enhanced ATR Sensor for Aqueous Solution in the Terahertz Range." *Optics Communications* 465: 125508.

Zhou, L., C. Zhang, Z. Qiu, and Y. He. 2020. "Information Fusion of Emerging Non-Destructive Analytical Techniques for Food Quality Authentication: A Survey." *TrAC Trends in Analytical Chemistry* 127: 115901.

Zhou, R., C. Wang, Q. Wang, L. Xie, and Y. Ying. 2022. "Rapid Analysis of Fruit Acids by Laser-Engraved Free-Standing Terahertz Metamaterials." *Food Analytical Methods* 15 (4): 961–69.

Zielinski, A.A.F., C.W.I. Haminiuk, C.A. Nunes, E. Schnitzler, S.M.V. Ruth, and D. Granato. 2014. "Chemical Composition, Sensory Properties, Provenance, and Bioactivity of Fruit Juices as Assessed by Chemometrics: A Critical Review and Guideline." *Comprehensive Reviews in Food Science and Food Safety* 13 (3): 300–16.

Index

For Product Safety Concerns and Information please contact our EU
representative GPSR@taylorandfrancis.com
Taylor & Francis Verlag GmbH, Kaufingerstraße 24, 80331 München, Germany

www.ingramcontent.com/pod-product-compliance
Lightning Source LLC
Chambersburg PA
CBHW060344220326
41598CB00023B/2799